Nanostructure Science and Technology

Series Editor

David J. Lockwood, FRSC
National Research Council of Canada
Ottawa, Canada

Nanostructure science and technology now forms a common thread that runs through all physical and materials sciences and is emerging in industrial applications as nanotechnology. The breadth of the subject material is demonstrated by the fact that it covers and intertwines many of the traditional areas of physics, chemistry, biology, and medicine. Within each main topic in this field there can be many subfields. For example, the electrical properties of nanostructured materials is a topic that can cover electron transport in semiconductor quantum dots, self-assembled molecular nanostructures, carbon nanotubes, chemically tailored hybrid magnetic-semiconductor nanostructures, colloidal quantum dots, nanostructured superconductors, nanocrystalline electronic junctions, etc. Obviously, no one book can cope with such a diversity of subject matter. The nanostructured material system is, however, of increasing significance in our technology-dominated economy and this suggests the need for a series of books to cover recent developments.

The scope of the series is designed to cover as much of the subject matter as possible – from physics and chemistry to biology and medicine, and from basic science to applications. At present, the most significant subject areas are concentrated in basic science and mainly within physics and chemistry, but as time goes by more importance will inevitably be given to subjects in applied science and will also include biology and medicine. The series will naturally accommodate this flow of developments in the sciences and technology of nanostructures and maintain its topicality by virtue of its broad emphasis. It is important that emerging areas in the biological and medical sciences, for example, not be ignored as, despite their diversity, developments in this field are often interlinked. The series will maintain the required cohesiveness from a judicious mix of edited volumes and monographs that while covering subfields in depth will also contain more general and interdisciplinary texts.

Thus the series is planned to cover in a coherent fashion the developments in basic research from the distinct viewpoints of physics, chemistry, biology, and materials science and also the engineering technologies emerging from this research. Each volume will also reflect this flow from science to technology. As time goes by, the earlier series volumes will then serve as reference texts to subsequent volumes.

Neetu Talreja · Divya Chauhan ·
Mohammad Ashfaq
Editors

Waste-Derived Carbon Nanostructures

Synthesis and Applications

Springer

Editors
Neetu Talreja
Department of Science
Alliance University
Bengaluru, Karnataka, India

Divya Chauhan
Ernst and Young Global LLP
New Delhi, Delhi, India

Mohammad Ashfaq
University Centre for Research
and Development (UCRD)
Chandigarh University
Mohali, Punjab, India

ISSN 1571-5744 ISSN 2197-7976 (electronic)
Nanostructure Science and Technology
ISBN 978-3-031-75246-9 ISBN 978-3-031-75247-6 (eBook)
https://doi.org/10.1007/978-3-031-75247-6

This Springer imprint is published by the registered company Springer Nature Switzerland AG
The registered company address is: Gewerbestrasse 11, 6330 Cham, Switzerland

Preface

The incessant exploration of technological advancement in material science and engineering has led to the development of newer classes of nanomaterials (NMs) that are significantly increasing, especially carbon nanostructured materials (CNs) due to their exceptional properties and endless applications. CNs have gained significant interest from researchers due to their mechanical, electrical, and thermal characteristics. Indeed, CNs have shown potential in end applications mainly environmental remediation, energy storage, catalysis, sensors, agriculture, and biomedical applications. Researchers continue searching for a cost-effective and sustainable way to produce CNs. In this aspect, the development of waste-derived carbon nanostructures (WD-CNs) from different waste materials, including municipal solid garbage, plastics, industrial wastes, and agricultural residues, might be long-term cost-effective solutions, which also reduces the burden of solid waste management.

WD-CNs are made from trash, which offers newer possibilities for recycling garbage into goods. WD-CNs effectively eliminate the need for waste product cleansing, by establishing a closed-loop system. The garbage is converted into useful resources, and CNs created from waste that support the tenets of the circular economy. Additionally, by adjusting the surface functional groups, metal/non-metal ions, and polymers significantly improved their applicability towards end application. This book highlights the synthesis and characterization of the WD-CNs, especially carbon nanotubes (CNTs), graphene, and carbon dots (CDots) from various sources including agricultural waste, plastic waste, and organic waste. The WD-CNs-based composite materials are effectively applied in various applications including environmental, energy, biomedical, and microbial fuel cells.

The area of WD-CNs is still in its early development, but the potential it holds is enormous. By turning waste into a treasury, we can reduce our dependence on non-renewable resources, and reduce environmental pollution, simultaneously establishing a new hope for society as well as the development of the country. In addition, employing trash as a starting point for synthesizing CNs can produce a lower carbon footprint than conventional techniques that rely on fossil fuels or energy-intensive procedures. The production and commercialization of CNs obtained from waste can foster creativity, open up new business prospects, and advance the advancement of green technologies.

This book is intended to serve as a comprehensive resource for researchers, academicians, engineers, students, and policymakers, who are interested in this exciting and rapidly evolving field **"waste to treasury"**. We believe that the knowledge shared within this book will inspire further invention and collaboration, thrusting the development of sustainable WD-CNs, which can meet the challenges of the future.

Bengaluru, India Dr. Neetu Talreja
New Delhi, India Dr. Divya Chauhan
Mohali, India Dr. Mohammad Ashfaq

Contents

Chapter 1
Waste-Derived Carbon Nanomaterials (WD-CNMs): Synthesis and Characterization

Kulwinder Singh, Sawini, and Deepak Kumar

Abstract The rapid urbanization and industrialization of the past few decades have resulted in real-time ecological challenges, particularly concerning water contamination and environmental pollution. In addition to the rapid decrease of the normal environment, serious issues that impact all ecological systems include energy scarcity, confined availability, and too much use of unrenewable resources. The production of waste materials (WMs) is increasing exponentially day by day. Scientists are now working on the production of nanomaterials (NMs) for various applications from WMs such as plastic, biomass, and industrial waste to minimize the burden of environmental pollution. This chapter is mainly focused on carbon-based nanomaterials (CNMs) including carbon nanotubes (CNTs), and graphene derived from the different types of WMs. Different synthesis processes and their characterization techniques for the determination of various properties such as structural, microstructural, compositional, thermal properties, etc. have also been discussed in detail. A thorough literature study on WM-derived (WD) graphene (WD-graphene) and WD-CNTs from synthesis processes to their characterization of various properties is incorporated in this chapter.

Keywords Waste materials · Carbon nanomaterials · CNTs · Graphene

K. Singh (✉)
University Centre for Research and Development, Chandigarh University, Mohali 140 413, Punjab, India
e-mail: ksingh.nano@gmail.com

Sawini
Department of Physics, Chandigarh University, Mohali, Punjab 140 413, India

D. Kumar
Department of Chemistry, School of Chemical Engineering and Physical Sciences, Lovely Professional University, Phagwara 144 411, Punjab, India

N. Talreja et al. (eds.), *Waste-Derived Carbon Nanostructures*, Nanostructure Science and Technology, https://doi.org/10.1007/978-3-031-75247-6_1

1.1 Introduction

The exponential and unconventional expansion in universal development brought about by the global population explosion leads to the generation of various solid wastes which is expanding exponentially daily. Recently, several reports claim that approximately 7–8 billion tons of solid waste materials (WMs) are generated yearly, globally. This presents a significant problem to the scientific community because it has various negative effects on the environment and human health [1]. Approximately 11.2 billion tons of WMs are generated annually, according to the UN Environment Program. WMs are one of the major causes of environmental contamination that directly affects the health of human beings as well as aquatic animals, especially in developing countries, where over ~ 90% of WMs are burned/dumped in open sites [2]. Moreover, enormous amounts of WMs (plastics, electric, and electronic) imports from developed nations worsen the WMs-related problems in different countries including Southeast Asian nations (Vietnam, Malaysia, the Philippines, India, and Thailand) with inadequate waste management capabilities [3–5]. Furthermore, ecological problems mainly water contamination significantly increased with increasing industrialization and urbanization [6], which directly affects society in various aspects like lack of energy, frequent use of non-renewable resources, and limited accessibility [7]. Researchers are continuously focused on the development of novel methods/technologies for eco-friendly, clean, and renewable sources to develop next-generation materials [8]. Consequently, it's imperative to construct advanced, eco-friendly next-generation materials from renewable resources that fulfill our everyday requirements [9].

Agricultural WMs have accumulated significant consideration for the production of carbon sources due to their profuse allocation, affordability, and easy renewable sources [10]. However, implementing padlocked-loop management of engineered materials (EMs) on a wide level presents several severe technological, financial, environmental, and societal challenges. For instance, the use of toxic chemicals for extraction and purification in plastics and metals recycling processes creates new health hazards for humans as well as the environment [11]. The domains of materials sciences and nanotechnology are becoming more and more inattentive with sustainability tactics, frameworks, and metrics to diminish the adverse effects on the environment and health hazards connected to the synthesis, use, and discarding of innovative nanomaterials (NMs) [12, 13].

1.2 Waste-Derived Carbon Nanomaterials (WD-CNMs)

WD-CNMs are a thrilling area of research due to their potential for recycling WMs into valuable NMs. CNMs can be synthesized from waste materials. By converting WMs into valuable CNMs, it's possible to address both environmental issues and resource scarcity, making this field highly impactful and sustainable. Before going into the details, it is necessary to give a brief overview of CNMs and their properties.

1.2.1 CNMs

Carbon has so many different types and strengths of bonds with itself and many other elements, it is the greatest useful element in the periodic table [14]. Diamond and graphite are the two crystalline allotropic forms of carbon that occur naturally [15]. CNMs like graphene, carbon nanotubes (CNTs), and fullerenes are a few low-dimensional allotropes of carbon that can be produced synthetically. Although different NMs have been found, CNMs in particular are prime important to nanotechnology [16]. Nanotechnology has given a lot of devotion to CNMs since the discovery of several CNMs (fullerenes in 1985, CNTs in 1991, and graphene in 2004) [17]. Carbon is the primary component in all CNMs and is categorized based on dimensions like zero-dimensional (0D) like fullerenes, one-dimensional (1D), i.e., CNTs, two-dimensional (2D) such as graphene; and three-dimensional (3D), i.e., graphite [18]. Among all of them, graphene and CNTs stand out among them and hold the most promise for use in nanotechnology [19]. Different allotropic forms of CNMs are shown in Fig. 1.1.

The sp^2-hybridization of carbon atoms is arranged in a hexagonal pattern to produce graphene and CNTs structures. In addition to the recently discovered thick,

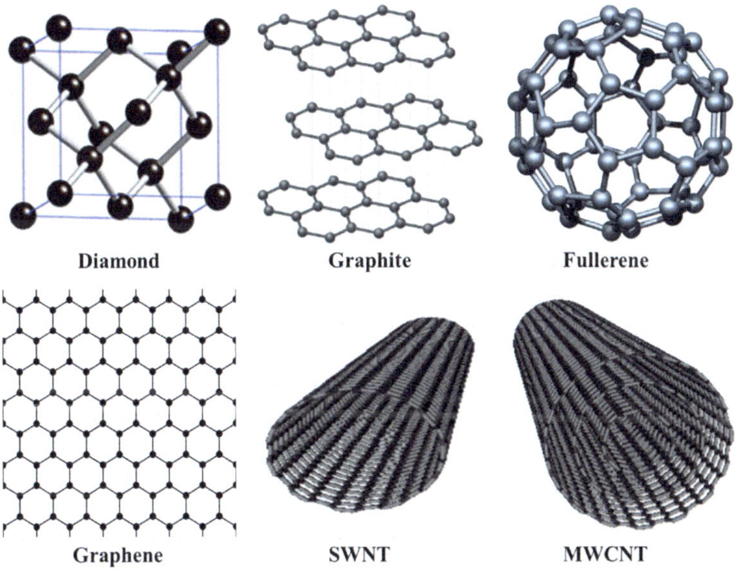

Fig. 1.1 Structures of different allotropic forms of carbon including CNMs

flawless 2D single-atom structures: Graphene, a monolayer of carbon atoms organized hexagonally, has drawn a lot of consideration due to their high thermal conductivity, elasticity, stiffness, and electron mobility [20–23]. Because of these exceptional characteristics, graphene is an appropriate CNMs for diverse applications such as energy storage, sensors, and nano-electronic devices [24–28]. A multitude of applications of CNMs especially sensing devices (electrochemical, fluorescence, and colorimetric), energy storage (battery, and supercapacitors), catalysis, and biomedical (drug delivery, chemotherapy, cell imaging, and biosensors), and environmental remediation (Adsorption/removal and detection of contamination) have been possible due to development of CNMs especially graphene [29–33]. The fabrication of next-generation CNMs, such as smart materials with excellent performance and versatility, has drawn attention. The utilization of precursors in the growth of CNMs such as graphene and CNTs is crucial in determining a technology's feasibility. In order to create CNTs, a variety of carbon precursors have been employed such as methane, acetylene, benzene, xylene, toluene, etc. [34–38]. These carbon precursors are connected to fossil fuels, and there might be a crisis with these precursors. Thus, it is essential to think about the production of CNMs from natural resources/WMs to create more competitive CNMs. Employing these sources predominantly serves to supply "green" substitutes and inexpensive raw materials for the large-scale fabrication of CNMs. According to the green chemistry principle, all industrial processes require renewable feedstock in contrast to natural resources. Additionally, the procedure needs to be planned to maximize atom-by-atom integration of the feedstock's constituents into the finished produced products. A lot of emphasis has recently been paid to the safe and environmentally benign process of synthesizing functional CNMs using natural precursors, which opens up an extensive scope of prospective usage in diverse applications. The natural precursors and WMs-based on hydrocarbons are widely accessible on Earth to prepare significant amounts of CNMs such as graphene and CNTs. Agricultural WMs, including seeds, fibers, oils, and bagasse, have produced various kinds of carbon using pyrolysis. These organic precursors offer a sustainable process for producing CNMs like graphene and CNTs for various applications [39]. The economic and environmental benefits for using of WMs to produce products like CNMs. The advancement in the field of nanotechnology, recycling WMs to fabricate newer CNMs for various purposes is a good example of converting "waste to wealth" [40]. Researchers have made remarkable attempts to produce CNMs from WMs with excellent characteristics like higher chemical stability, and electrical and thermal conductivity, in response to the increasing demand for these materials [41].

1.2.1.1 Graphene

Graphene serves as the building block for fullerenes and CNTs. It is an atomic-thick layer of graphite and a 2D-crystal structure is made up of two kinds of sp^2-hybridized carbon atoms which bond together to form a honeycomb lattice, with delocalized electrons present due to the p orbitals of the carbon atoms [42–45]. Single-layer graphene was first isolated in 2004 using micromechanical cleavage. CNMs have been extensively used in various applications due to their unique properties. Graphene is extensively used in numerous cutting-edge devices and has opened the door to many interesting areas of physics [46, 47]. However, once the number of layers of graphene surpasses more than ten, it is no longer considered 2D graphene but rather 3D thin graphite. The number of graphene layers is often determined using Raman spectroscopy [48]. Usually, three major peaks in Raman spectra were observed, indicating single-layer graphene. The location of the D band is at ~ 1350 cm^{-1}, with a higher peak intensity indicating the existence of more faults and disorders; the G band is around at 1580 cm^{-1} attributed to bond stretching of sp^2 carbon. On the other hand, a peak at ~ 2700 cm^{-1} is attributed to 2D peaks. The intensity ratio of the G and D peaks estimate the number of layers of any 2D materials. It is indicated that single-layer graphene is formed if the I_G/I_{2D} ratio is approximately equal to 0.5 [49]. The brief of graphene's properties is given as follows:

Different Properties of Graphene

(a) Optical
Single-layered graphene has a transparency of ~ 97.7% and is practically transparent due to its zero/nearly zero band gap value and 2D structure. It is important to mention that transparency decreases with increasing the number of layers. Transmittance drops 2.3% with each additional layer [50]. Furthermore, fast-moving carriers offer higher photoreaction without producing a strong electrical field. Therefore, graphene is expected to be used in optoelectronic devices mainly transparent electrodes and sensors due to its favorable optical properties [44, 45].

(b) Mechanical
The compressed structure of graphene and incredibly strong covalent bonds exhibit higher mechanical capabilities that are equally astounding. Graphene has been demonstrated to be the strongest material that has Young's modulus up to 1 Tpa (five times higher than steel (0.2 Tpa)). According to reports, monolayer graphene has the highest intrinsic strength of 130 Gpa [22]. However, graphene's critical stress intensity factor (~4 Mpa) revealed that it has an ordinary fracture toughness [51]. Extensive research is being done on graphene-based fillers and composites because of their strong characteristics and light density.

(c) Electronic
Graphene exhibits notable advantages in electronic aspects due to its layered structure. The unique characteristics of graphene easily tune band gap values. The charge carriers can exhibit remarkably high concentration and fast mobility (> 15,000 cm^2V^{-1} s^{-1}) due to the consequence of an ambipolar electric field [20]. Significantly, the mobility can be increased to 100,000 cm^2V^{-1} s^{-1} and is almost temperature-independent. Additionally, the quantum Hall effect (QHE) of graphene was observed even at room temperature [52]. Therefore, graphene has appeared

as the next-generation material of choice for applications involving nanoscale electronics and transistors that require large current sustainability (>10^8 A/cm^2), ballistic transport, linear current–voltage characteristics, and massive in-plane electrical conductivity.

(d) Thermal

Furthermore, due to its strong covalent bonding, graphene has excellent thermal characteristics. Even at high temperatures, graphene exhibits notable thermal stability. Phonon transport, which involves diffusive conduction at higher temperatures and ballistic conduction at lower temperatures, varies and determines its thermal conductivity [53]. Once the temperature rises above 100 K, thermal conductivity decreases with increasing temperature. The Cu is recognized for having good thermal conductivity (400 W/mK). Graphene is synthesized by using mechanical exfoliation that has as high as 600 W/mK [54]. Graphene can be used with different/tunable thermal conductivity in thermoelectric applications (where poor thermal conductivity is preferable) or heat-sinking applications.

(e) Physico-Chemical

Graphene exhibits high specific surface area (~2630 m^2/g), strong inertness, and noticeable non-permeability, graphene is set to become one of the most promising materials for biomedicine (e.g., encapsulating drugs or catalysts), coating (improving corrosion resistance), and energy storage (supporting electron/hole transfer).

Synthesis of WD-Graphene

An overview of graphene manufacturing from WMs is given in the next section. Figure 1.2 shows the graphical illustration of the different WMs used in the production of WD-graphene.

(a) Synthesis of WD-Graphene from Waste Plastics (WPs)

WPs have been used for the synthesis of graphene during the breakdown of APs, aromatic compounds, and hydrocarbons. The different processes were used to turn a variety of waste plastics (WPs) into graphene. In order to achieve highly yielding graphene flakes (GFs), Gong et al. used WPs (polypropylene (PP)) to catalyzed by modified Montmorillonite (OMMT). Initially, the mixture of PP (~89 wt%), talcum (~11 wt%), and OMMT was kept in a crucible boat and heated at 700 °C for 15 min to produce carbonized char (CC). The produced CC was dipped in HF and HNO_3 solution. HNO_3 oxidized the amorphous carbon while HF dissolved impurities (such as MMT and talc additives) and then collected the produced graphene by centrifugation process [55]. Another study focuses on the unique synthesis process to produce single-layered graphene from polystyrene (PS)-based WPs [56]. Another study demonstrated that the chemical vapor deposition (CVD) grown graphene [57] by utilizing WPs, which contained approximately 86% PE and 14% PS. Manukyan et al. reported an energy-efficient combustion process [58] that uses SiC and waste

Fig. 1.2 A graphical illustration of the different WMs used in the production of WD-graphene

polytetrafluoroethylene (PTFE) to create graphene sheets. In a different study, Takami et al. used the CVD process to produce graphene from polymethylmethacrylate (PMMA) film [59]. Table 1.1 summarizes the different WMs used for the production of CNMs such as graphene.

(b) Synthesis of WD-Graphene from Biomass Waste

Biomass mainly composed of cellulose, and lignin, is a nontoxic, feasible, and sustainable carbon source. Biomass has been extensively used as a feedstock for the synthesis of CNMs including graphene [41]. The corn straw, sugarcane bagasse, wheat stems, and rice husk (RH) are the main sources of agricultural WMs. The RH, rice bran (RB), rice straw (RS), and rice husk ash (RHA) are the main by-products of the rice industry [82]. Moreover, Paddy rice mainly contains ~ 69% starchy endosperm, ~ 20% RH, and ~ 11% grain/bran layer. Furthermore, ~ 18–20% of RH is converted into RHA after burning. Approximately 685 million tons of RH and ~ 137 million tons RS were produced annually, worldwide [83]. The RH is mainly composed of ~ 15–20% inorganic components (silica), and ~ 65–75% organic compounds (like cellulose, hemicellulose, and lignin). To fabricate highly valuable NMs including CNMs using agricultural WMs, are first processed into RH. Next, RH is converted into the different types of CNMs [84]. Agricultural WMs have the potential to be employed as carbon sources for the production of CNMs mainly CNTs, graphene, and graphene derivatives for numerous applications including sustainable energy resources [85]. Sun et al. produced porous graphene nanosheets [64] using coconut shells (a type of biomass waste). N-doped graphene was synthesized catalyst-free by Primo et al. using chitosan which was removed from the waste skin of crustaceans [65]. To synthesize graphene quantum dots, Suryawanshi et al. extracted turbo-stratic carbon from dead neem leaves [66]. A novel process for synthesizing

Table 1.1 List of different WMs used for the production of WD-CNMs

Type of WMs	WMs	Output structure (s)	Ref
WPs	PP	Graphene flakes	[55]
	PS (grass blades, feces of dogs, cockroach legs, cookies and chocolate)	Monolayer graphene	[56]
	PE (~86%)-PS (~14%)	Graphene	[57]
	PTFE (SiC)	Graphene sheets	[58]
	PMMA	Thin graphene films	[59]
	PET bottles	Graphene	[60]
	PP waste	Graphene	[55]
	Plastics	Graphene	[61–63]
Biomass waste	Coconut shell	Graphene nanosheets	[64]
	Crustacean skin wastes	Monolayer N doped-graphene	[65]
	Dead neem leaves	Graphene oxide (GO) sheets	[66]
	Hemp	Interconnected graphene carbon nanosheets	[67]
	RH	Graphene sheets	[68–70]
	RH	MWCNTs	[71]
	RH	Few-layered graphene	[72]
	RS	Graphene	[73]
	RH	Multi-layered GO	[74]
	RH	Graphene	[75]
	RB	GO	[76]
	RS	GO	[77]
	RH	GO nanoflakes	[78]
	Internal part of oak seeds	K-doped GO	[79]
Other WMs	Wood, leaf, bagasse, fruit, newspaper, chicken bone, cow dung, industrial soot	rGO sheets	[80]
	Waste paper	GO sheets	[81]

crystalline and corrugated graphene was suggested by Muramatsu et al. [68]. The graphene was prepared by using RH with KOH and calcination at 850 °C to produce monolayer/bilayer graphene nanosheets with clean and active edges as well as excellent crystallinity. The data suggested that by using this process authors developed low-temperature and catalyst-free graphene. Muramatsu et al. [68] used carbon black as an additional barrier against oxidation in addition to KOH. Subsequently, though, Singh et al. [72] removed carbon black by modifying the synthesis process and substituting RH for carbon black. Similarly, Uda et al. [73] synthesize graphene from RSA at 700 °C using KOH with an impregnation ratio of 1:2.5. Authors have characterized the produced graphene using different characterizations for different properties. Othman et al. [75] synthesized porous graphene by using RHAs at different stabilization temperatures. Another study focused on the four-step procedure for the synthesis of graphene oxide (GO) was reported by Seitzhanova et al. [74] and Rhee et al. [69]. The synthesized graphene layers were rich in inorganic materials and had a high carbon content. Azizovna et al. [86] also used RH to synthesize GO by activating it with KOH and then dehydrating it with alkaline solutions. The synthesized GO was characterized by using electron microscopic and Raman spectroscopy techniques. Next, a desalination membrane that could filter out small particles, common salts, and organic molecules was fabricated using the produced GO. A microwave furnace approach was used by Kumar et al. [87] to synthesize graphene from RH. According to Hashmi et al. [76], GO can be produced efficiently from agricultural WMs by using sugarcane bagasse (SCB), RB, and orange peel (OP) under optimized conditions with relatively lower temperatures. The carbonaceous precursor, powdered RB, and ferrocene were heated to 300 °C for 15 min. The produced solid products were studied using various characterization techniques. The synthesis of GO is facile, quick, one-step, and eco-friendly, which demonstrates that the solid carbonaceous agricultural WMs might be utilized as feedstock to create high-quality, value-added CNMs including GO. Tohamy et al. [77] used agricultural WMs (RS, SCB, lignin, and mature beech pinewood sawdust (MW)) to produce GO by utilizing ferrocene as a catalyst. The produced GO from different sources was characterized using various techniques for micro-structural, morphological, and elemental properties. Naik et al. [78] used agricultural WMs, such as waste newspaper, RH, and SCB, to produce GO nanoflakes by using a pyrolytic synthesis process without using an inert atmosphere. The precursors were subjected to pyrolysis in the atmospheric atmosphere, which resulted in the breakdown of glycosidic linkages and cellulosic structures into monomeric glucose, thereby GO forms nanoflakes-like structures.

(c) Synthesis of WD-graphene from other WMs

Akhavan et al. synthesized GO and rGO from highly pyrolytic graphite (HPG) with carbonaceous wastes from industry and nature [80]. Adolfsson et al. recently synthesized GO-quantum dots from scrap paper [81].

1.2.1.2 CNTs

CNMs like CNTs gain significant attention from researchers due to their applicability in numerous applications [88]. CNTs were created as early as 1958 [89]. However, Radush-Kevich and Lukyanovich made the first observation of CNTs in 1952 [90], and Oberlin et al. followed up in 1976 [91]. Prior to the discovery of graphene, CNTs were the substance under close examination. Iijima [92] reported the discovery of CNTs, while Ebbesen and Ajayan [93] reported their plentiful separation. The sp^2 hybridized carbon atoms are assembled in a hexagonal network to produce cylindrical CNTs [94]. CNTs are divided into two groups: (1) multi-walled

CNTs (MWCNTs), made up of multiple layers of graphene sheets, and (2) single-walled CNTs (SWCNTs), made up of a single layer of graphene sheets rolled up to produce multilayers of the graphene sheet [95–97]. Whereas MWCNTs have a diameter of around ten nanometers (nm) SWCNTs have a diameter of a few nm [98]. Strong mechanical, electrical, optical, thermal, chemical stability, and electrochemical characteristics are possessed by CNTs [99, 100]. CNTs extensively used in numerous applications mainly environmental remediation, energy, antibiotic materials, agriculture, sensors, supercapacitors, catalysis, and biomedical due to their unique characteristics [97]. The brief of properties of CNTs is given below:

Different Properties of CNTs

(a) Mechanical

CNTs have outstanding mechanical properties like stiffness, strength, and toughness due to their strong bonding [101]. CNTs have an exceptional tensile strength (~100 GPa) and Young's modulus (more than 1 TPa) [95, 102]. These outstanding abilities of the CNTs play critical roles in structural applications (i.e., CNTs-reinforced polymer-based composite) [103].

(b) Electrical

CNTs can exhibit a metallic/semiconducting nature upon the modification of their structures [104]. The electrical conductivity of CNTs can theoretically be orders of magnitude greater than Cu, particularly for metallic CNTs [105, 106]. In addition to their excellent electronic and quantum effects, CNTs have excellent magnetic and lattice properties. Furthermore, CNTs are capable of changing from a metallic to an insulating condition in response to strain [105].

(c) Thermal

The distinct structure of CNTs offers higher thermal conductivity when temperature rises over ambient conditions due to quantum confinement effects. The thermal characteristics of CNTs might be tuned upon changing their temperatures [101]. By utilizing these characteristics, CNTs outperform traditional materials in sophisticated electronics that require an internal operating temperature of more than 100 °C.

(d) Optical

The straight band gap in 1D band structure/CNTs provides exceptional support for applications employed under wavelength ranges (300–3000 nm) [95]. Additionally, due to the enormous exciton binding energy of SWCNTs, these can be useful in communication devices [107].

(e) Other Properties

CNTs can be used as effective catalysts due to their large surface area and high catalytic activity [95]. When it comes to supercapacitor applications, CNTs-based electrodes are superior due to their enormous specific surface area, and exceptional

Fig. 1.3 A graphical illustration of the different sources of WMs used for the synthesis of CNTs

electrical and mechanical characteristics. Furthermore, adding CNTs into polymers, coating materials, and films can improve mechanical performance or boost corrosion resistance [97].

Synthesis of WD-CNTs Using WMs

CNTs can be produced by utilizing WMs that act as catalysts or carbon precursors [108]. The synthesis processes are covered in detail concerning WMs used in the synthesis of CNTs. The graphical representation of the different sources of WMs used for the synthesis of CNTs (Fig. 1.3).

(a) Synthesis of WD-CNTs from WPs

CNTs can be synthesized from WPs under optimized conditions [109]. WPs, consisting primarily of PP and polyethylene (PE), are known as polyolefins (carbon content: approximately 85.7 wt%). Two prototypes of waste polyolefins have been used extensively in green synthesis methods for CNMs [110]. Mishra et al. synthesized CNTs from PP-based WMs by using the CVD process [111]. Gong et al. also synthesized CNTs from PE-based WMs (HDPE (high-density PE), LDPE (low-density PE), and LLDPE (linear low-density PE)) by using CuBr/NiO as catalysts [112]. The data suggested that the yield of CNTs is to be ~ 56.5 wt% from LLDPE, ~ 36.8 wt% from LDPE, and ~ 30.9 wt% from HDPE. Zhou et al. synthesized CNTs from PE-based WMs by using a simple combustion process. For growing CNTs, the arrangement of the setup is unique in that it uses stainless steel wire mesh, which serves as both a substrate and a catalyst. Through this wire mesh, combustion effluent can be suitably filtered to prevent the development of MWCNTs. Moreover, soot and the ensuing high fuel/oxygen ratio can be eliminated using the high-temperature ceramic filter. The results showed that CNTs yield exceeded ~ 10% weight [113]. Zhang et al. synthesized CNTs from PP-based WMs. The synthesized CNTs have large diameter (~160 nm) and exceptional yield of up to 80% [114]. Bajad et al. produced MWCNTs from PP-based WMs by using Ni/Mo/MgO (mole ratio: 4/0.2/1) as a catalyst. The produced MWCNTs show a yield (~45.8%) [115]. The synthesis of MWCNTs from PP-based WMs under atmospheric conditions was established by Tang et al. [116]. Similarly, CNTs synthesized from PVA film with a high yield (~54%) by Nath et al. [117]. Berkmans et al. synthesized nano-channeled CNTs from PET-based WMs without using any catalyst with a yield of ~ 39% [118]. Ganesh et al. synthesized CNTs from WPs by using a catalyst (Ni/Mo/MgO) [119]. Similarly, Pol et al. [120] have described the pyrolytic method for the synthesis of CNTs from WPs. Wu et al. reported the synthesis of CNTs and hydrogen gas through gasification of PP-based WMs [121]. Jiang et al. employed organically modified clay and nickel dioxides, trioxides, and hydroxides to synthesize MWCNTs qualitatively at 630–830 °C [122]. Gong et al. synthesized the CNTs using a one-pot direct synthesis process [123]. Zhuo et al. described using a high-temperature pyrolytic up-cyclization method for the production of CNTs from PE/PET-based WMs [124]. Similarly, Liu et al. used a binary synthetic pyrolytic method to convert waste PP into CNTs [125].

(b) Synthesis of WD-CNTs from Waste Vegetable Oils (WVO) and Biomass
To create extremely pure (~90%) CNTs, WVO of different sources (coconut, neem, turpentine, eucalyptus, and palm), which have a significant carbon concentration [126]. Of them, palm oil ($C_{55}H_{100}O_6$) is the most notable for giving CNTs a notable quality. Suriani et al. [127] synthesized Vertically aligned CNTs (VACNTs) by using palm oil-based WVO using a floating catalytic CVD process. In another study of the same group, Suriani et al. used WVO (chicken fat) for the production of CNTs. Authors have used chicken fat due to its frequent production from the WMs of chickens [126]. A two-step procedure for the synthesis of CNTs bundles was described by Fathy [128]. Bundles of CNTs have been synthesized by carbonizing hydrothermally treated RS by using ferrocene–nickel nitrate catalyst (RS-H/Fe–Ni) through the CVD process. Wang et al. [71] synthesized graphene CNTs (g-CNTs) from RH by using the microwave plasma irradiation (MPI) process under inert atmospheric conditions and H_2 atmosphere. A novel method for the synthesis of CNTs (different sizes and shapes) from hydrothermally treated RS was described by Lotfy et al. [129]. This process involved the pyrolytic decomposition of RS in a vertical bed reactor for 60 min at 830 °C under atmospheric conditions. The process yields high-quality CNTs.

(c) Synthesis of CNTs from Industrial Waste
In addition to employing WVO, Suriani et al. synthesized quasi-aligned-CNTs from waste engine oil (WEO) that have higher carbon content compared with other lubricants [130]. Oscar et al. synthesized the MWCNTs from red mud (WMs produced during the Bayer process of alumina manufacturing) as catalysts [131].

Quan et al. [132] discovered the synthesis process of CNTs from printed circuit board (PCB)-based WMs. Research on turning solid waste tires (SWTs) into MWCNTs, onion fullerenes, and nanospheres was presented by Richardo et al. [133]. Mittal et al. demonstrated the production of water-soluble CNTs from soot collected from bitumen-based WMs [134]. The primary waste product in the production of fullerenes, fullerene waste soot (FWS), can also yield high-quality SWCNTs by employing the arc discharge technique. Hu et al. synthesized a high yield of SWCNTs (yield of 6 g/h) [135]. Using a CVD process, Yousefi et al. proposed a rapid approach for the synthesis of millimeter-wave nanotubes from waste camphor oil [136]. Wu et al. synthesized the CNTs using a catalyst (Ni-Mg–Al) to convert 1 ton of waste glycerol into ~ 2.8 kg CNTs [137]. Table 1.2 summarized different types of WMs used for the synthesis of CNTs.

1.3 Characterization of WD-CNMs

1.3.1 Morphological Properties

Morphological properties can be investigated using scanning electron microscopy (SEM) and transmission electron microscopy (TEM). The SEM provides detailed images of the surface texture, revealing the size, shape, and distribution of CNMs. SEM images show surface features at high magnification, useful for assessing the quality and uniformity of the synthesized NMs including CNTs, and graphene. On the other hand, TEM offers high-resolution images of the internal structure, allowing the observation of lattice fringes and defects. It transmits electrons through a thin sample and captures the interactions to form an image that provides insights into the crystallinity, layer stacking (in graphene), and defects in the CNMs. TEM is essential for characterizing CNTs, graphene, and nanofibers at the atomic level. Sharma et al. [145] have investigated the morphological properties of the graphene film. SEM results (as shown in Fig. 1.4a) confirmed the domain formation of graphene having a size range of ~ 2 to 3 µm. Also, the high-resolution TEM results (Fig. 1.4b) of graphene show that the suspended film edge can fold back which provides the information on layer number. Whereas, Acomb et al. [146] studied the impact of steam flow rate on the morphological properties of LDPE WD-CNTs (Fig. 1.4c and d). Steam flow leads to the formation of less dense carbon structures. HRTEM results (Fig. 1.4e–g) confirmed the formation of MWCNTs in the obtained product derived from PE, PP, and PS waste. The steam flow leads to the generation of irregularities and deformities in the structure of CNTs.

Table 1.2 Different types of WMs are used for the synthesis of CNTs

Type of WMs	WMs	Output structure (s)	Ref
WPs	PP	CNTs	[110]
	PP	MWNTs	[111]
	HDPE	MWNTs	[113]
	PP	CNTs	[114]
	PP	CNTs	[115]
	PP	MWNTs	[116]
	PVA	CNTs	[117, 138, 139]
	PET	CNTs and solid carbon spheres	[118]
	PP	CNTs	[119, 121]
	HDPE	MWCNTs	[120]
	PE, PET bottles	CNTs	[124]
	PP	MWCNTs	[125]
Waste vegetable and animal oils as well as agricultural waste	Palm oil	CNTs	[127]
	Chicken fat	CNTs	[126, 140]
	Biochar from Wheat straw	MWCNTs	[141]
	SCB	Graphitic flakes and MWCNTs	[142]
	Wood sawdust	CNTs	[143]
	RS	CNTs	[128]
	RS	CNTs on Fe–Ni/Al_2O_3	[129]
	RH	g-CNTs	[71]
	RH	Spherical and tubular structures of CNTs	[144]
Industrial waste	Waste engine oil	Quasi aligned CNTs	[130]
	PCB	Porous carbons wrapped inside CNTs	[132]
	Solid waste tires	MWCNTs (~25 nm (diameter), ~ 100 nm (length)); onion fullerenes; nanospheres	[133]
	Soot of combusted bitumen	Water-soluble CNTs	[134]
	Fullerene waste soot	SWNTs	[135]
	Camphor oil	MWNTs	[136]
	Glycerol	CNTs	[137]

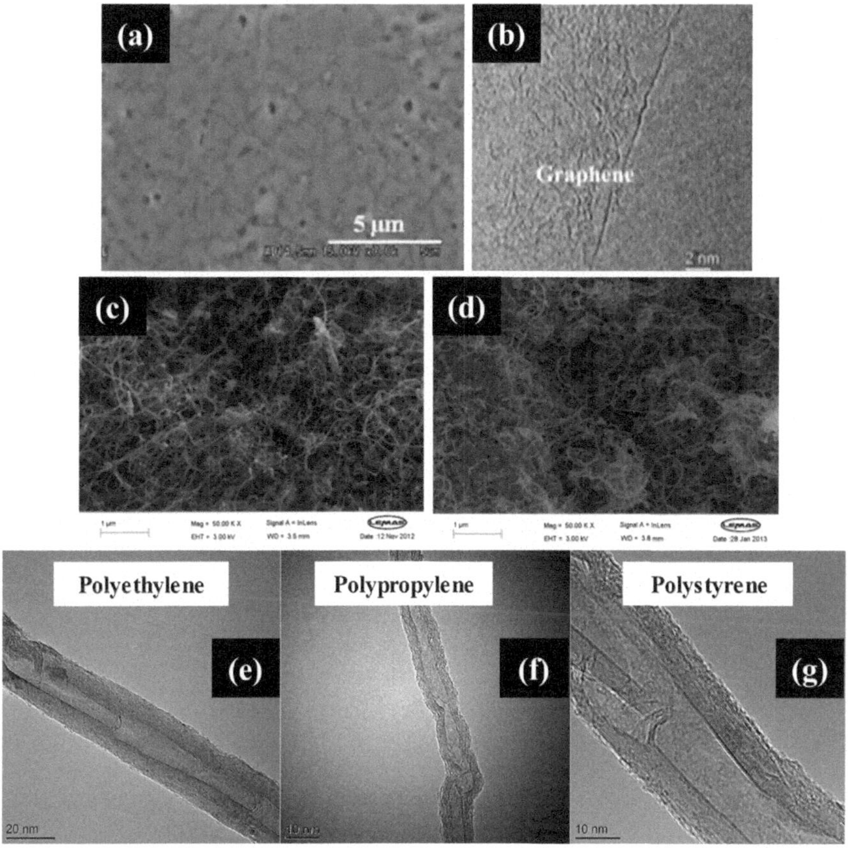

Fig. 1.4 **a** SEM and **b** TEM images of WD-graphene from solid camphor, respectively, reproduced with permission from ref. [145]; SEM image **c**, **d** CNTs derived from LDPE at different steam flow rates, reproduced with permission from ref. [146], **e–g** TEM image of CNTs derived from PE, PP, and PS-based WMs, respectively, reproduced with permission from ref. [146]

1.3.2 Structural Properties

1.3.2.1 X-ray Diffraction (XRD)

XRD determines the crystalline structure and phase composition of the CNMs, indicating the degree of graphitization and the presence of amorphous phases. It reveals information about the lattice spacing and crystallinity. This technique is used to identify the phases present and assess the quality of graphitic structures. Garg et al. [147] determined the structural properties of the graphene nanosheets derived from WPs. XRD results shown in Fig. 1.5a confirmed the formation of graphene nanosheets in the final product.

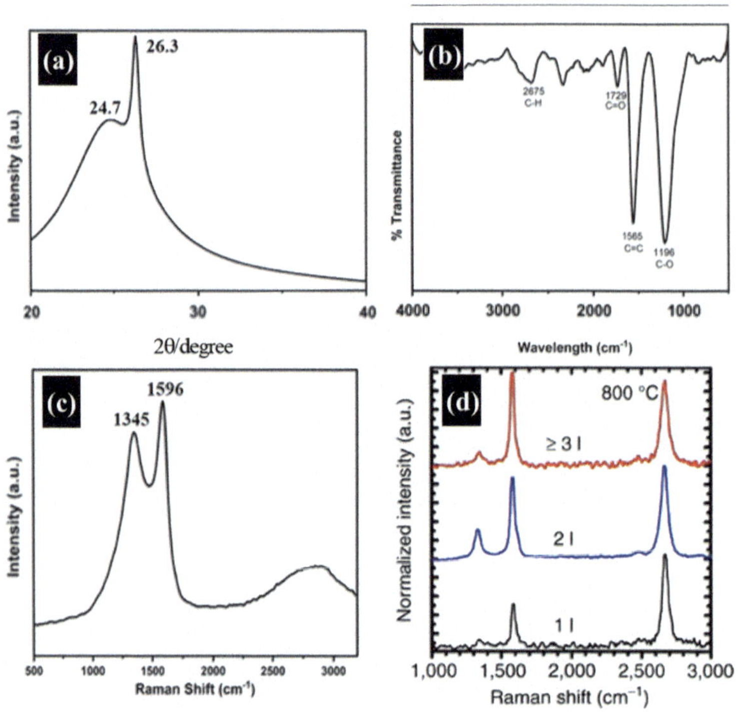

Fig. 1.5 **a** XRD pattern, **b** FTIR spectrum, **c** Raman spectra corresponding to plastic waste-derived graphene nanosheets, respectively, **a–c** are reproduced with permission from ref. [147]; **d** Raman spectra of different layered regions in the graphene film, reproduced with permission from ref. [148]

On the other hand, Mishra et al. [111] have synthesized and studied the structural properties of the CNTs derived from PP waste at different temperatures as shown in Fig. 1.6a–c. Results revealed that the high purity of CNTs was obtained for synthesizing temperature of 800 °C.

1.3.2.2 Raman Spectroscopy

Raman spectroscopy provides information on the vibrational modes of the carbon atoms, revealing the degree of disorder and the presence of defects in the carbon lattice. Raman spectra show the characteristic peaks (D, G, and 2D bands) that provide insights into the structural quality of graphene, CNTs, and other CNMs. The intensity and position of these peaks indicate the level of defects, doping, and graphitization. Garg et al. [147] determined the vibrational properties of the graphene nanosheets derived from plastic waste. Raman's results shown in Fig. 1.5c confirmed the synthesis of graphene nanosheets as a final product. The structure of the obtained

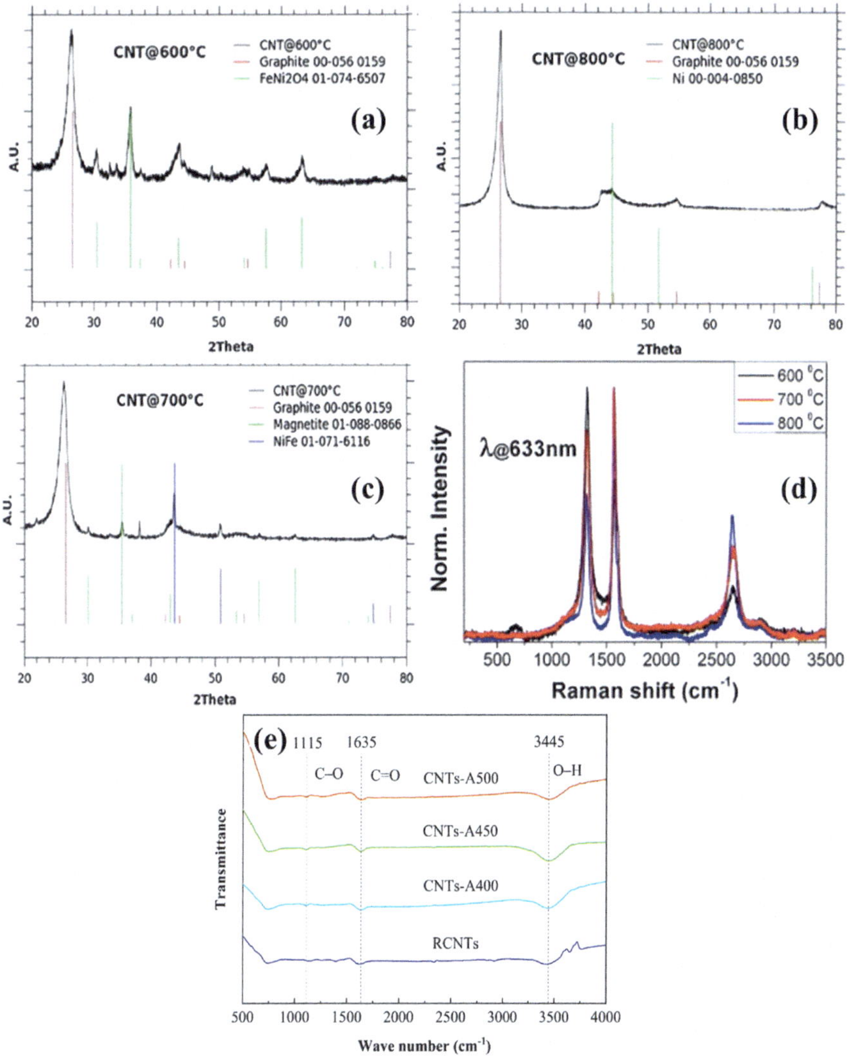

Fig. 1.6 XRD patterns of CNTs prepared at **a** 600 °C, **b** 800 °C **c** 700 °C, and **d** Raman spectra corresponding to prepared CNTs, respectively, reproduced with permission from ref. [111], **e** FTIR spectrum of air-treated plastic waste-derived CNTs, reproduced with permission from ref. [149]

sheets is disordered due to the carbon–carbon bond in-plane vibrations. Seo et al. [148] reported the formation of different layered regions of graphene nanosheets as shown in Fig. 1.5d. Mishra et al. [111] have investigated the vibrational properties of the CNTs derived from PP waste at different temperatures as shown in Fig. 1.6d. The presence of a sharp peak at around 1567 cm^{-1} confirmed the formation of high purity of MWCNTs.

1.3.2.3 Fourier Transform Infrared (FTIR) Spectroscopy

FTIR identifies the functional groups and chemical bonds on the surface of the CNMs. FTIR spectra measure the absorption of infrared radiation by the sample, which corresponds to the vibrational frequencies of chemical bonds. The resulting spectra provide information about the presence of specific functional groups (e.g., hydroxyl, carboxyl, carbonyl) introduced during functionalization. Garg et al. [147] determined the presence of functional groups of the graphene nanosheets derived from WPs. FTIR results shown in Fig. 1.5b confirmed the formation of graphene nanosheets in the final product. Wang et al. [149] have synthesized the CNTs and studied the impact of air and CO_2 treatment on the structural properties of CNTs. FTIR results (Fig. 1.6e) showed the impact of air treatment as the IR active bands varied.

1.3.3 Chemical Compositional Properties

1.3.3.1 X-ray Photoelectron Spectroscopy (XPS)

XPS analyzes the surface chemistry and elemental composition which provide insights into the types of functional groups and the oxidation state of carbon. XPS measures the binding energies of core electrons, revealing information about the elements present and their chemical states. This technique is used to study the surface modifications and functionalization of CNMs. Gan et al. [150] have synthesized the GO and reduced GO nanosheets using bagasse. Results confirmed the formation of GO and rGO nanosheets (Fig. 1.7a and b).

On the other hand, Wang et al. [149] have studied the chemical compositional studies of HNO_3 and HCl-treated CNTs using XPS. Figure 1.7c andd shows the C 1 s XPS data of the acid-treated CNTs. Results revealed the introduction of C–O/C = O bonds on CNTs through acid treatment.

1.3.4 Thermal Stability

1.3.4.1 Thermogravimetric Analysis (TGA)

TGA assesses the thermal stability and compositional changes of the CNMs as a function of temperature, indicating the presence of volatile components and residual catalysts. TGA measures the weight loss of the sample as it is heated in a controlled atmosphere. The resulting thermogram reveals information about the decomposition temperatures, thermal stability, and the content of inorganic residues. Pandey et al. [151] have synthesized the graphene nanosheets using WPs and studied their thermal properties (Fig. 1.8a). Slight weight loss confirmed the high thermal stability and

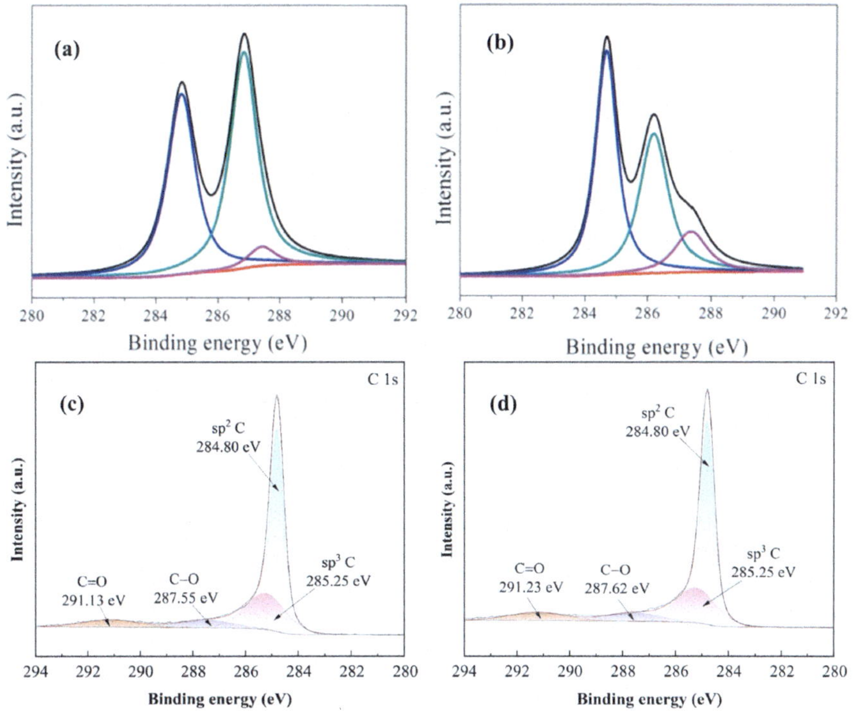

Fig. 1.7 XPS spectra of **a** GO, **b** B-rGO, reproduced with permission from ref. [150], C 1 s XPS spectra of **c** HNO_3-4 h **d** HCl-4 h treated CNTs, reproduced with permission from ref. [149]

high purity of the synthesized graphene nanosheets. Wang et al. [149] investigated the impact of CO_2 treatment at different temperatures (as shown in Fig. 1.8b). Results revealed the high stability of the CNTs under air conditions.

1.4 Conclusions

The existence and progress of human culture are at risk due to the global energy crisis and changes in climate conditions; consequently, there is a pronounced requirement for the development of novel technologies that produce NMs including CNMs with tunable characteristics. The scientific community has discovered that nanotechnology is a fantastic tool for the production of sophisticated, high-yield designed NMs including CNMs. Numerous innovative technologies have been developed for the synthesis of CNMs, but green synthesis approaches that utilize WMs are gaining more attention due to simultaneously managing the burden of WMs, tunable ability, insignificant toxicity, biodegradability, and renewability. As a result, researchers are seeking to develop contemporary technologies for the synthesis of CNMs using

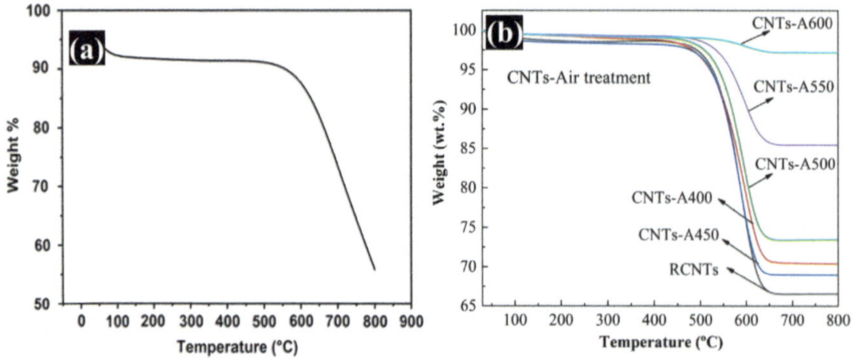

Fig. 1.8 TGA curves of **a** Graphene nanosheets, reproduced with permission from ref. [151] under creative commons license, **b** Air-treated CNTs, respectively, reproduced with permission from ref. [149]

simple, cost-effective, and eco-friendly routes through the utilization of WMs, particularly agricultural residues. This chapter focuses on the synthesis of CNMs (graphene and CNTs) from different types of WMs. These versatile CNMs synthesized from WMs could be used in different ways to provide novel systems with excellent mechanical, electrical, optical, physicochemical, and biocompatibility characteristics. Waste, plentiful, and naturally occurring renewable resources must be used to fabricate high-value-added CNMs at low cost, which are effectively used in various applications and developing more effective, convenient, economical, and environmentally friendly innovative technologies to address environmental problems to ensure a sustainable future.

References

1. Ripa M, Fiorentino G, Giani H, Clausen A, Ulgiati S (2017) Refuse recovered biomass fuel from municipal solid waste. A Life Cycle Assess Appl Energy 186:211–225
2. Solid waste management. Available online at: https://www.unep.org/explore-topics/resource-efficiency/what-we-do/cities/solid-waste-management (accessed on June 10, 2024).
3. Kaza S, Yao LC, B-Tata P, Van WF (2018) What a Waste 2.0: a global snapshot of solid waste management to 2050. Urban Development. © Washington, DC: World Bank. http://hdl.handle.net/10986/30317 License: CC BY 3.0 IGO
4. Dell J (2019) 157,000 Shipping containers of U.S. plastic waste exported to countries with poor waste management in 2018—plastic pollution coalition. Available online at: https://www.plasticpollutioncoalition.org/blog/2019/3/6/157000-shipping-containers-of-us-plastic-waste-exported-to-countrieswith-poorwastemanagement-in-2018(accessed June 10, 2024)
5. Sukanan D (2020) Waste imports are flooding Asian countries like Thailand. Available at: https://www.sustainability-times.com/environmental-protection/asian-countries-like-thailand-are-flooded-by-waste-imports (accessed June 10, 2024)
6. Ouyang J, Zhou L, Liu Z, Heng JY, Chen W (2020) Biomass-derived activated carbons for the removal of pharmaceutical mircopollutants from wastewater: a review. Sep Purif Technol 253:117536

7. Abbas A, Mariana LT, Phan AN (2018) Biomass-waste derived graphene quantum dots and their applications. Carbon 140:77–99
8. Hu B, Wang K, Wu L, Yu SH, Antonietti M, Titirici MM (2010) Engineering carbon materials from the hydrothermal carbonization process of biomass. Adv Mater 22:813–828
9. Raja S, Mattoso LHC, Moreira FKV. Biomass-derived nanomaterials. In: Rajendran S, Naushad M, Balakumar S (eds) Nanostructured materials for energy related applications. Environmental chemistry for a sustainable world, vol 24. Springer, Cham, pp 243–70. https://doi.org/10.1007/978-3-030-04500-5_10
10. Wang Y, Sun J, He B, Feng M (2020) Synthesis and modification of biomass derived carbon dots in ionic liquids and their application: a mini review. Green Chem Eng 1:94–108
11. Kral U, Morf LS, Vyzinkarova D, Brunner PH (2019) Cycles and sinks: two key elements of a circular economy. J Mate Cycles Waste Manag 21:1–9
12. Dhingra R, Naidu S, Upreti G, Sawhney R (2010) Sustainable nanotechnology: through green methods and life-cycle thinking. Sustainability 2:3323–3338
13. Falinski MM, Plata DL, Chopra SS, Theis TL, Gilbertson LM, Zimmerman JB (2018) A framework for sustainable nanomaterial selection and design based on performance, hazard, and economic considerations. Nat Nanotechnol 13:708–714
14. Silva GA (2008) Nanotechnology approaches to crossing the blood-brain barrier and drug delivery to the CNS. BMC Neurosci 9:S4
15. Nasir S, Hussein MZ, Zainal Z, Yusof NA (2018) Carbon-based nanomaterials/allotropes: A glimpse of their synthesis, properties and some applications. Materials 11:295
16. Magrez A, Kasas S, Salicio V, Pasquier N, Seo JW, Celio M, Catsicas S, Schwaller B, Forró L (2006) Cellular toxicity of carbon-based nanomaterials. Nano Lett 6:1121–1125
17. Rauti R, Musto M, Bosi S, Prato M, Ballerini L (2019) Properties and behavior of carbon nanomaterials when interfacing neuronal cells: How far have we come? Carbon 143:430–446
18. Villarreal CC, Pham T, Ramnani P, Mulchandani A (2017) Carbon allotropes as sensors for environmental monitoring. Curr Opin Electrochem 3:106–113
19. Siqueira JR, Oliveira ON (2017) 9—Carbon-Based Nanomaterials. In: Da Róz AL, Ferreira M, de Lima Leite F, Oliveira ON (eds) Nanostructures. William Andrew Publishing, Norwich, UK, pp 233–249
20. Novoselov KS, Geim AK, Morozov SV, Jiang DE, Zhang Y, Dubonos SV, Grigorieva IV, Firsov AA (2004) Electric field effect in atomically thin carbon films. Science 306:666–669
21. Novoselov KS, Geim AK, Morozov SV, Jiang D, Katsnelson MI, Grigorieva IV, Dubonos SV, Firsov AA (2005) Two-dimensional gas of massless Dirac fermions in graphene. Nature 438:197–200
22. Lee C, Wei X, Kysar JW, Hone J (2008) Measurement of the elastic properties and intrinsic strength of monolayer graphene. Science 321:385–388
23. Balandin AA, Ghosh S, Bao W, Calizo I, Teweldebrhan D, Miao F, Lau CN (2008) Superior thermal conductivity of single-layer graphene. Nano Lett 8:902–907
24. Vashist SK, Luong JH (2015) Recent advances in electrochemical biosensing schemes using graphene and graphene-based nanocomposites. Carbon 84:519–550
25. Wu X, Yang D, Wang C, Jiang Y, Wei T, Fan Z (2015) Functionalized three-dimensional graphene networks for high performance supercapacitors. Carbon 92:26–30
26. Xue Y, Liu Y, Lu F, Qu J, Chen H, Dai L (2012) Functionalization of graphene oxide with polyhedral oligomeric silsesquioxane (POSS) for multifunctional applications. J Phys Chem Lett 3:1607–1612
27. Low T, Avouris P (2014) Graphene plasmonics for terahertz to mid-infrared applications. ACS Nano 8:1086–1101
28. Essig S, Marquardt CW, Vijayaraghavan A, Ganzhorn M, Dehm S, Hennrich F, Ou F, Green AA, Sciascia C, Bonaccorso F, Bohnen KP (2010) Phonon-assisted electroluminescence from metallic carbon nanotubes and graphene. Nano Lett 10:1589–1594
29. De M, Chou SS, Dravid VP (2011) Graphene oxide as an enzyme inhibitor: modulation of activity of α-chymotrypsin. J Am Chem Soc 133:17524–17527

30. Guo CX, Yang HB, Sheng ZM, Lu ZS, Song QL, Li CM (2010) Layered graphene/quantum dots for photovoltaic devices. Angew Chem Int Ed 49:3014–3017
31. Liu J, Cui L, Losic D (2013) Graphene and graphene oxide as new nanocarriers for drug delivery applications. Acta Biomater 9:9243–9257
32. Pumera M (2011) Graphene in biosensing. Mater Today 14:308–315
33. Yang Y, Asiri AM, Tang Z, Du D, Lin Y (2013) Graphene based materials for biomedical applications. Mater Today 16:365–373
34. Li WZ, Xie SS, Qian LX, Chang BH, Zou BS, Zhou WY, Zhao RA, Wang G (1996) Large-scale synthesis of aligned carbon nanotubes. Science 274:1701–1703
35. Bronikowski MJ (2006) CVD growth of carbon nanotube bundle arrays. Carbon 44:2822–2832
36. Mayne M, Grobert N, Terrones M, Kamalakaran R, Rühle M, Kroto HW, Walton DR (2001) Pyrolytic production of aligned carbon nanotubes from homogeneously dispersed benzene-based aerosols. Chem Phys Lett 338:101–107
37. Zhang ZJ, Wei BQ, Ramanath G, Ajayan PM (2000) Substrate-site selective growth of aligned carbon nanotubes. Appl Phys Lett 77:3764–3766
38. Sen R, Govindaraj A, Rao CN (1997) Carbon nanotubes by the metallocene route. Chem Phys Lett 267:276–280
39. Kumar R, Singh RK, Singh DP (2016) Natural and waste hydrocarbon precursors for the synthesis of carbon based nanomaterials: graphene and CNTs. Renew Sustain Energy Rev 58:976–1006
40. Boruah A, Saikia M, Das T, Goswamee RL, Saikia BK (2020) Blue-emitting fluorescent carbon quantum dots from waste biomass sources and their application in fluoride ion detection in water. J Photochem Photobiol B: Biol 209:111940
41. Wang Z, Shen D, Wu C, Gu S (2018) State-of-the-art on the production and application of carbon nanomaterials from biomass. Green Chem 20:5031–5057
42. Geim AK, Novoselov KS (2007) The rise of graphene. Nat Mater 6:183–191
43. Abergel DS, Apalkov V, Berashevich J, Ziegler K, Chakraborty T (2010) Properties of graphene: a theoretical perspective. Adv Phys 59:261–482
44. Zhu Y, Murali S, Cai W, Li X, Suk JW, Potts JR, Ruoff RS (2010) Graphene and graphene oxide: synthesis, properties, and applications. Adv Mater 22:3906–3924
45. Avouris P, Dimitrakopoulos C (2012) Graphene: synthesis and applications. Mater Today 15:86–97
46. Allen MJ, Tung VC, Kaner RB (2010) Honeycomb carbon: a review of graphene. Chem Rev 110:132–145
47. Choi W, Lahiri I, Seelaboyina R, Kang YS (2010) Synthesis of graphene and its applications: a review. Crit Rev Solid State Mater Sci 35:52–71
48. Ni ZH, Wang HM, Kasim J, Fan HM, Yu T, Wu YH, Feng YP, Shen ZX (2007) Graphene thickness determination using reflection and contrast spectroscopy. Nano Lett 7:2758–2763
49. Ferrari AC, Meyer JC, Scardaci V, Casiraghi C, Lazzeri M, Mauri F, Piscanec S, Jiang D, Novoselov KS, Roth S, Geim AK (2006) Raman spectrum of graphene and graphene layers. Phys Rev Lett 97:187401
50. Nair RR, Blake P, Grigorenko AN, Novoselov KS, Booth TJ, Stauber T, Peres NM, Geim AK (2008) Fine structure constant defines visual transparency of graphene. Science 320:1308
51. Zhang P, Ma L, Fan F, Zeng Z, Peng C, Loya PE, Liu Z, Gong Y, Zhang J, Zhang X, Ajayan PM (2014) Fracture toughness of graphene. Nat Commun 5:3782
52. Castro Neto AH, Guinea F, Peres NM, Novoselov KS, Geim AK (2009) The electronic properties of graphene. Rev Mod Phys 81:109–162
53. Deng J, You Y, Sahajwalla V, Joshi RK (2016) Transforming waste into carbon-based nanomaterials. Carbon 96:105–115
54. Balandin AA (2011) Thermal properties of graphene and nanostructured carbon materials. Nat Mater 10:569–581
55. Gong J, Liu J, Wen X, Jiang Z, Chen X, Mijowska E, Tang T (2014) Upcycling waste polypropylene into graphene flakes on organically modified montmorillonite. Ind Eng Chem Res 53:4173–4181

56. Ruan G, Sun Z, Peng Z, Tour JM (2011) Growth of graphene from food, insects, and waste. ACS Nano 5:7601–7607
57. Sharma S, Kalita G, Hirano R, Shinde SM, Papon R, Ohtani H, Tanemura M (2014) Synthesis of graphene crystals from solid waste plastic by chemical vapor deposition. Carbon 72:66–73
58. Manukyan KV, Rouvimov S, Wolf EE, Mukasyan AS (2013) Combustion synthesis of graphene materials. Carbon 62:302–311
59. Takami T, Seino R, Yamazaki K, Ogino T (2014) Graphene film formation on insulating substrates using polymer films as carbon source. J Phys D Appl Phys 47:094015
60. El Essawy NA, Ali SM, Farag HA, Konsowa AH, Elnouby M, Hamad HA (2017) Green synthesis of graphene from recycled PET bottle wastes for use in the adsorption of dyes in aqueous solution. Ecotoxicol Environ Saf 145:57–68
61. Kamali AR, Yang J, Sun Q (2019) Molten salt conversion of polyethylene terephthalate waste into graphene nanostructures with high surface area and ultra-high electrical conductivity. Appl Surf Sci 476:539–551
62. Wen Y, Kierzek K, Chen X, Gong J, Liu J, Niu R, Mijowska E, Tang T (2019) Mass production of hierarchically porous carbon nanosheets by carbonizing "real-world" mixed waste plastics toward excellent-performance supercapacitors. Waste Manage 87:691–700
63. Pandey S, Karakoti M, Dhali S, Karki N, SanthiBhushan B, Tewari C, Rana S, Srivastava A, Melkani AB, Sahoo NG (2019) Bulk synthesis of graphene nanosheets from plastic waste: an invincible method of solid waste management for better tomorrow. Waste Manage 88:48–55
64. Sun L, Tian C, Li M, Meng X, Wang L, Wang R, Yin J, Fu H (2013) From coconut shell to porous graphene-like nanosheets for high-power supercapacitors. J Mater Chem A 1:6462–6470
65. Primo A, Atienzar P, Sanchez E, Delgado JM, García H (2012) From biomass wastes to large-area, high-quality, N-doped graphene: catalyst-free carbonization of chitosan coatings on arbitrary substrates. Chem Comm 48:9254–9256
66. Suryawanshi A, Biswal M, Mhamane D, Gokhale R, Patil S, Guin D, Ogale S (2014) Large scale synthesis of graphene quantum dots (GQDs) from waste biomass and their use as an efficient and selective photoluminescence on–off–on probe for Ag^+ ions. Nanoscale 6:11664–11670
67. Wang H, Xu Z, Kohandehghan A, Li Z, Cui K, Tan X, Stephenson TJ, King'Ondu CK, Holt CM, Olsen BC, Tak JK (2013) Interconnected carbon nanosheets derived from hemp for ultrafast supercapacitors with high energy. ACS Nano 7:5131–5141
68. Muramatsu H, Kim YA, Yang KS, Cruz-Silva R, Toda I, Yamada T, Terrones M, Endo M, Hayashi T, Saitoh H (2014) Rice husk-derived graphene with nano-sized domains and clean edges. Small 10:2766–2770
69. Rhee I, Kim YA, Shin GO, Kim JH, Muramatsu H (2015) Compressive strength sensitivity of cement mortar using rice husk-derived graphene with a high specific surface area. Constr Build Mater 96:189–197
70. Rhee I, Lee JS, Kim JH, Kim YA (2017) Thermal performance, freeze-and-thaw resistance, and bond strength of cement mortar using rice husk-derived graphene. Constr Build Mater 146:350–359
71. Wang Z, Ogata H, Morimoto S, Ortiz-Medina J, Fujishige M, Takeuchi K, Muramatsu H, Hayashi T, Terrones M, Hashimoto Y, Endo M (2015) Nanocarbons from rice husk by microwave plasma irradiation: From graphene and carbon nanotubes to graphenated carbon nanotube hybrids. Carbon 94:479–484
72. Singh P, Bahadur J, Pal K (2017) One-step one chemical synthesis process of graphene from rice husk for energy storage applications. Graphene 6:61–71
73. Uda MN, Gopinath SC, Hashim U, Uda MA, hulwani Ibrahim N, Parmin NA, Halim NH, Anbu P (2020) Simple and green approach strategy to synthesis graphene using rice straw ash. IOP Conf Ser Mater Sci Eng 864:012181
74. Seitzhanova MA, Mansurov ZA, Yeleuov M, Roviello V, Di Capua R (2019) The characteristics of graphene obtained from rice husk and graphite. Euras Chem Tech J 21:149–156

75. Che Othman FE, Ismail MS, Yusof N, Samitsu S, Yusop MZ, Tajul Arifin NF, Alias NH, Jaafar J, Aziz F, Wan Salleh WN, Ismail AF (2020) Methane adsorption by porous graphene derived from rice husk ashes under various stabilization temperatures. Carbon Lett 30:535–543
76. Hashmi A, Singh AK, Jain B, Singh A (2020) Muffle atmosphere promoted fabrication of graphene oxide nanoparticle by agricultural waste. Fuller Nanotub Car Nanostructures 28:627–636
77. Tohamy HA, Anis B, Youssef MA, Abdallah AE, El-Sakhawy M, Kamel S (2020) Preparation of eco-friendly graphene oxide from agricultural wastes for water treatment. Desalin Water Treat 191:250–262
78. Naik MJ, Debbarma J, Saha M, Bhargava A (2020) Graphene oxide nanoflakes from various agrowastes. Materialwiss Werkstofftech 51:368–374
79. Tewari C, Tatrari G, Karakoti M, Pandey S, Pal M, Rana S, SanthiBhushan B, Melkani AB, Srivastava A, Sahoo NG (2019) A simple, eco-friendly and green approach to synthesis of blue photoluminescent potassium-doped graphene oxide from agriculture waste for bio-imaging applications. Mater Sci Eng: C 104:109970
80. Akhavan O, Bijanzad K, Mirsepah A (2014) Synthesis of graphene from natural and industrial carbonaceous wastes. RSC Adv 4:20441–20448
81. Adolfsson KH, Hassanzadeh S, Hakkarainen M (2015) Valorization of cellulose and waste paper to graphene oxide quantum dots. RSC Adv 5:26550–26558
82. Gu S, Zhou J, Yu C, Luo Z, Wang Q, Shi Z (2015) A novel two-staged thermal synthesis method of generating nanosilica from rice husk via pre-pyrolysis combined with calcination. Ind Crops Prod 65:1–6
83. Pode R (2016) Potential applications of rice husk ash waste from rice husk biomass power plant. Renew Sustain Energy Rev 53:1468–1485
84. Umeda J, Kondoh K (2010) High-purification of amorphous silica originated from rice husks by combination of polysaccharide hydrolysis and metallic impurities removal. Ind Crops Prod 32:539–544
85. Letcher TM (2020) Future energy: improved, sustainable and clean options for our planet. Elsevier, AE Amsterdam, Netherlands
86. Azizovna SM (2020) Synthesis and application of membrane technology for desalination of water. University of Naples Federico II: Naples, Italy
87. Kumar M, Sachdeva A, Garg RK, Singh S (2020) Synthesis and characterization of graphene prepared from rice husk by a simple microwave process. Nano Hybrids and Compos 29:74–83
88. Hosseinalipour SM, Jabbari E, Madadelahi M, Fardad A (2014) Gas mixing simulation in a t-shape micro channel using the DSMC Method. Chall Nano Micro Scale Sci Tech 2:132–139
89. Hillert M, Lange N (1959) The structure of graphite filaments. Z für Krist-Cryst Mater 111:24–34
90. Lukyanovich VM (1952) The structure of carbon forming in thermal decomposition of carbon monoxide on an iron catalyst. Russ J Phys Chem 26:88–95
91. Oberlin A, Endo M, Koyama T (1976) Filamentous growth of carbon through benzene decomposition. J Cryst Growth 32:335–349
92. Iijima S (1991) Helical microtubules of graphitic carbon. Nature 354:56–58
93. Ebbesen TW, Ajayan PM (1992) Large-scale synthesis of carbon nanotubes. Nature 358:220–222
94. Greil P (2015) Perspectives of nano-carbon based engineering materials. Adv Eng Mater 17:124–137
95. Baughman RH, Zakhidov AA, De Heer WA (2002) Carbon nanotubes–the route toward applications. Science 297:787–792
96. Liu WW, Aziz A, Chai SP, Mohamed AR, Tye CT (2011) The effect of carbon precursors (methane, benzene and camphor) on the quality of carbon nanotubes synthesised by the chemical vapour decomposition. Phys E Low-Dimen Syst Nanostruct 43:1535–1542
97. De Volder MF, Tawfick SH, Baughman RH, Hart AJ (2013) Carbon nanotubes: present and future commercial applications. Science 339:535–539

98. Prasek J, Drbohlavova J, Chomoucka J, Hubalek J, Jasek O, Adam V, Kizek R (2011) Methods for carbon nanotubes synthesis. J Mater Chem 21:15872–15884
99. Dai H (2002) Carbon nanotubes: synthesis, integration, and properties. Acc Chem Res 35:1035–1044
100. Hong EH, Lee KH, Oh SH, Park CG (2003) Synthesis of carbon nanotubes using microwave radiation. Adv Funct Mater 13:961–966
101. Qiu H, Yang J (2017) Structure and properties of carbon nanotubes. In: Peng H, Li Q, Chen T (eds) In micro and nano technologies, Industrial Applications of Carbon Nanotubes. Elsevier, pp 47–69
102. Hu Z, Lu X (2014) Mechanical properties of carbon nanotubes and graphene. In: Iijima KT (ed) Carbon nanotubes and graphene, 2nd edn. Elsevier, Oxford, pp 165–200
103. Corral EL, Shofner ML, Barrera EV (2004) Applications. In: Carbon nanotubes. CRC Press, PP 253–275
104. Yanagi K (2014) Differentiation of carbon nanotubes with different chirality. In: Iijima KT (ed) Carbon nanotubes and graphene, 2nd edn. Elsevier, Oxford, pp 19–38
105. Bandaru PR (2007) Electrical properties and applications of carbon nanotube structures. J Nanosci Nanotechnol 7:1239–1267
106. Toshishige Y (2004) Nanoelectronics applications. In: Meyyappan M (ed) Carbon nanotubes: science and applications. CRC Press, pp 163–93
107. Saito R, Dresselhaus MS (2014) Optical properties of carbon nanotubes. In: Iijima KT (ed) Carbon nanotubes and graphene. 2nd ed. Elsevier, Oxford, pp 77–98
108. Titirici MM, White RJ, Brun N, Budarin VL, Su DS, Del Monte F, Clark JH, MacLachlan MJ (2015) Sustainable carbon materials. Chem Soc Rev 44:250–290
109. Bazargan A, McKay G (2012) A review–synthesis of carbon nanotubes from plastic wastes. Chem Eng J 195:377–391
110. Gong J, Liu J, Wan D, Chen X, Wen X, Mijowska E, Jiang Z, Wang Y, Tang T (2012) Catalytic carbonization of polypropylene by the combined catalysis of activated carbon with Ni_2O_3 into carbon nanotubes and its mechanism. Appl Catal A Gen 449:112–120
111. Mishra N, Das G, Ansaldo A, Genovese A, Malerba M, Povia M, Ricci D, Di Fabrizio E, Di Zitti E, Sharon M, Sharon M (2012) Pyrolysis of waste polypropylene for the synthesis of carbon nanotubes. J Anal Appl Pyrol 94:91–98
112. Gong J, Liu J, Jiang Z, Feng J, Chen X, Wang L, Mijowska E, Wen X, Tang T (2014) Striking influence of chain structure of polyethylene on the formation of cup-stacked carbon nanotubes/carbon nanofibers under the combined catalysis of CuBr and NiO. Appl Catal B Environ 147:592–601
113. Zhuo C, Hall B, Richter H, Levendis Y (2010) Synthesis of carbon nanotubes by sequential pyrolysis and combustion of polyethylene. Carbon 48:4024–4034
114. Zhang J, Li J, Cao J, Qian Y (2008) Synthesis and characterization of larger diameter carbon nanotubes from catalytic pyrolysis of polypropylene. Mater Lett 62:1839–1842
115. Bajad GS, Tiwari SK, Vijayakumar RP (2015) Synthesis and characterization of CNTs using polypropylene waste as precursor. Mater Sci Eng B 194:68–77
116. Tang T, Chen X, Meng X, Chen H, Ding Y (2005) Synthesis of multiwalled carbon nanotubes by catalytic combustion of polypropylene. Angew Chem Int Ed 117:1541–1544
117. Nath DCD, Sahajwalla V (2012) Analysis of carbon nanotubes produced by pyrolysis of composite film of poly (vinyl alcohol) and modified fly ash. Mater Sci Appl 3:103–109
118. Berkmans AJ, Jagannatham M, Priyanka S, Haridoss P (2014) Synthesis of branched, nano channeled, ultrafine and nano carbon tubes from PET wastes using the arc discharge method. Waste Manage 34:2139–2145
119. Bajad G, Guguloth V, Vijayakumar RP, Bose S (2016) Conversion of plastic waste into CNTs using Ni/Mo/MgO catalyst—an optimization approach by mixture experiment. Fuller Nanotub Carbon Nanostruct 24:162–169
120. Pol VG, Thiyagarajan P (2010) Remediating plastic waste into carbon nanotubes. J Environ Monit 12:455–459

121. Wu C, Wang Z, Wang L, Williams PT, Huang J (2012) Sustainable processing of waste plastics to produce high yield hydrogen-rich synthesis gas and high quality carbon nanotubes. RSC Adv 2:4045–4047
122. Jiang Z, Song R, Bi W, Lu J, Tang T (2007) Polypropylene as a carbon source for the synthesis of multi-walled carbon nanotubes via catalytic combustion. Carbon 45:449–458
123. Gong J, Liu J, Ma L, Wen X, Chen X, Wan D, Yu H, Jiang Z, Borowiak-Palen E, Tang T (2012) Effect of Cl/Ni molar ratio on the catalytic conversion of polypropylene into Cu–Ni/C composites and their application in catalyzing "Click" reaction. Appl Catal B Environ 117:185–193
124. Zhuo C, Alves JO, Tenorio JA, Levendis YA (2012) Synthesis of carbon nanomaterials through up-cycling agricultural and municipal solid wastes. Ind Eng Chem Res 51:2922–2930
125. Liu J, Jiang Z, Yu H, Tang T (2011) Catalytic pyrolysis of polypropylene to synthesize carbon nanotubes and hydrogen through a two-stage process. Polym Degrad Stab 96:1711–1719
126. Suriani AB, Dalila AR, Mohamed A, Mamat MH, Salina M, Rosmi MS, Rosly J, Nor RM, Rusop M (2013) Vertically aligned carbon nanotubes synthesized from waste chicken fat. Mater Lett 101:61–64
127. Suriani AB, Nor RM, Rusop M (2010) Vertically aligned carbon nanotubes synthesized from waste cooking palm oil. J Ceram Soc Jpn 118:963–968
128. Fathy NA (2017) Carbon nanotubes synthesis using carbonization of pretreated rice straw through chemical vapor deposition of camphor. RSC Adv 7:28535–28541
129. Lotfy VF, Fathy NA, Basta AH (2018) Novel approach for synthesizing different shapes of carbon nanotubes from rice straw residue. J Environ Chem Eng 6:6263–6274
130. Suriani AB, Alfarisa S, Mohamed A, Isa IM, Kamari A, Hashim N, Mamat MH, Mohamed AR, Rusop M (2015) Quasi-aligned carbon nanotubes synthesised from waste engine oil. Mater Lett 139:220–223
131. Dunens OM, MacKenzie KJ, Harris AT (2010) Synthesis of multi-walled carbon nanotubes on 'red mud' catalysts. Carbon 48:2375–2377
132. Quan C, Li A, Gao N (2010) Synthesis of carbon nanotubes and porous carbons from printed circuit board waste pyrolysis oil. J Hazard Mater 179:911–917
133. Mis-Fernández R, Rios-Soberanis CR, Arenas-Alatorre J, Azamar-Barrios JA (2012) Synthesis of carbon nanostructures from residual solids waste tires. J Appl Polym Sci 123:1960–1967
134. Mittal H, Mishra SB, Mishra AK, Kaith BS, Jindal R (2013) Water-soluble carbon nanotubes from bitumen waste: synthesis, functionalisation and derivatisation for its use as superabsorbent. J Inorg Organomet Polym Mater 23:1128–1137
135. Hu C, Yu C, Li M, Fan X, Yang J, Zhang P, Wang S, Zhao Z, Qiu J (2014) Preparation of single-walled carbon nanotubes from fullerene waste soot. ACS Sustain Chem Eng. 2:14–18
136. TermehYousefi A, Bagheri S, Shinji K, Rouhi J, Rusop Mahmood M, Ikeda S (2014) Fast synthesis of multilayer carbon nanotubes from camphor oil as an energy storage material. BioMed Res Int 2014:691537
137. Wu C, Wang Z, Williams PT, Huang J (2013) Renewable hydrogen and carbon nanotubes from biodiesel waste glycerol. Sci Rep 3:2742
138. Nath DC, Sahajwalla V (2011) Application of fly ash as a catalyst for synthesis of carbon nanotube ribbons. J Hazard Mater 192:691–697
139. Nath DC, Sahajwalla V (2011) Growth mechanism of carbon nanotubes produced by pyrolysis of a composite film of poly (vinyl alcohol) and fly ash. Appl Phys A 104:539–544
140. Dalila AR, Abu Bakar S, Rosmi MS, Azmina MS, Rosazley R, Supian FL, Rosli J, Rusop M (2014) The effect of synthesis temperature on the growth of carbon nanotubes from waste chicken fat precursor. Adv Mater Res 832:798–803
141. Hidalgo P, Navia R, Hunter R, Coronado G, Gonzalez M (2019) Synthesis of carbon nanotubes using biochar as precursor material under microwave irradiation. J Environ Manage 244:83–91
142. Debalina B, Reddy RB, Vinu R (2017) Production of carbon nanostructures in biochar, bio-oil and gases from bagasse via microwave assisted pyrolysis using Fe and Co as susceptors. J Anal Appl Pyrol 124:310–318

143. Thompson E, Danks AE, Bourgeois L, Schnepp Z (2015) Iron-catalyzed graphitization of biomass. Green Chem 17:551–556
144. Asnawi M, Azhari S, Hamidon MN, Ismail I, Helina I (2018) Synthesis of carbon nanomaterials from rice husk via microwave oven. J Nanomater 2018:2898326
145. Sharma S, Kalita G, Hirano R, Hayashi Y, Tanemura M (2013) Influence of gas composition on the formation of graphene domain synthesized from camphor. Mater Lett 93:258–262
146. Acomb JC, Wu C, Williams PT (2014) Control of steam input to the pyrolysis-gasification of waste plastics for improved production of hydrogen or carbon nanotubes. Appl Catal B Environ 147:571–584
147. Garg KK, Pandey S, Kumar A, Rana A, Sahoo NG, Singh RK (2022) Graphene nanosheets derived from waste plastic for cost-effective thermoelectric applications. Results Mater 13:100260
148. Seo DH, Pineda S, Fang J, Gozukara Y, Yick S, Bendavid A, Lam SK, Murdock AT, Murphy AB, Han ZJ, Ostrikov K (2017) Single-step ambient-air synthesis of graphene from renewable precursors as electrochemical genosensor. Nat Commun 8:14217
149. Wang C, Zhang W, Yao D, Hu Q, Yang H (2024) Gas-phase and liquid-phase purification of plastic waste derived carbon nanotubes: effect of purification agent and operational parameters. Chem Eng J 496:153725
150. Gan L, Li B, Chen Y, Yu B, Chen Z (2019) Green synthesis of reduced graphene oxide using bagasse and its application in dye removal: a waste-to-resource supply chain. Chemosphere 219:148–154
151. Pandey S, Karakoti M, Surana K, Dhapola PS, SanthiBhushan B, Ganguly S, Singh PK, Abbas A, Srivastava A, Sahoo NG (2021) Graphene nanosheets derived from plastic waste for the application of DSSCs and supercapacitors. Sci Rep 11:3916

Chapter 2
Agrowaste-Derived ‘Natural’ Carbon Nanomaterials (NCNM) with Versatile Applications: Bacterial Cellulose

Jyutika M. Rajwade, Snehal S. Kulkarni, Aditya V. Wadekar, and Aniket S. Khandagale

Abstract Bacterial Cellulose (BC), a natural Carbon Nanomaterial (NCNM) is an extracellular polysaccharide produced fermentatively by specific bacterial genera. In static conditions, BC is produced at the air–liquid interface, as an easily visible moist hydrogel pellicle in a chemically ‘pure’ form. The pellicle contains individual fibers ~70 nm in width, forming bundles ~130 nm in thickness, which are crystalline in nature. Thus, the carbohydrate polymer (containing only C, H, and O as its constituents) per se qualifies as a ‘natural’ nanomaterial obtained by fermentation of agro wastes, which serve as carbon sources. The peculiar physicochemical and mechanical properties of bacterial cellulose, such as its purity, water-holding capacity, high tensile strength, flexibility, mouldability, sterilizable nature, high Young’s modulus, and biological properties such as non-immunogenic and non-toxic nature, biocompatibility make it an appropriate material for biomedical and non-biomedical applications. Bacterial cellulose is easily convertible into carbon nanotubes and aerogels, which can have several applications such as biosensing, manufacture of reinforced materials, materials with thermal resistance characteristics, etc. The chapter overviews bacterial cellulose production using agro wastes and its applications.

Keywords Bacterial cellulose · ‘Natural’ biopolymer · Nanoscale material · Biomedical applications · Non-biomedical applications

J. M. Rajwade (✉) · S. S. Kulkarni
Nanobioscience Group, Agharkar Research Institute, G. G. Agharkar Road, Pune 411 004, India
e-mail: jrajwade@aripune.org

A. V. Wadekar · A. S. Khandagale
Rajiv Gandhi Institute of I.T and Biotechnology, Bharati Vidyapeeth, Katraj, Pune 411045, India

N. Talreja et al. (eds.), *Waste-Derived Carbon Nanostructures*, Nanostructure Science and Technology, https://doi.org/10.1007/978-3-031-75247-6_2

2.1 Introduction

Globally, the population is growing at a fast pace, and to satisfy the need for food, agriculture including aquaculture, is practiced on a very large scale [1, 2]. Research is focused on increasing land under cultivation, improving crops, introducing newer techniques for fertilizer applications, and controlling pests and pathogens [3, 4]. Moreover, climate-resilient, sustainable agriculture is the need of the hour [5]. With increased agricultural produce, 'waste' is generated. Agricultural waste can be defined as materials that are undesired or non-marketable, originating exclusively from agricultural activities directly associated with crop cultivation or animal husbandry, with the primary intention of generating profit or sustaining a livelihood. Huge quantities (~ 998 million tons) of agricultural residue (waste) are generated annually [6]. Rapid changes in the various farming practices and technological improvements have led to the generation of agricultural residues, which, if disposed of inappropriately, pose significant environmental risks. Agriculture is one of the biggest contributors to the economy of India. With the huge quantities of agricultural waste generated, efforts at its utilization as a substrate for producing valued products need to be maximized. Agricultural residues are usually categorized into crop residues (examples of which are rice husk, wheat straws, and sugarcane bagasse), animal by-products (including animal excreta and deceased animals), processing leftovers (such as packaging materials and fertilizer containers), and hazardous waste (e.g., pesticides, insecticides) [7]. Agrowastes, typically crop residues, contain elevated levels of intricate carbohydrates, proteins, fiber, polyphenolic constituents, biologically active compounds, and other components [8]. Agro industries contribute to three major types of wastes viz., a) livestock excreta b) residual farm produce c) agricultural industry by-products [9].

(a) **Livestock excreta**—Agricultural land is typically allocated for three primary categories of farm animals: cattle, pigs, and poultry, with land usage of ~15.3, 0.10, and 1.3 million hectares, respectively [10]. Annually, around 120 million tons of manure is generated collectively by the livestock sector.
(b) **Residual farm produce**—These materials fall under the classification of primary agricultural residues, primarily comprising straw, husk, and stalks of cultivated crops that remain post-harvest. They can be utilized as substrates, as these remnants are abundant in cellulosic fibers [11].
(c) **Agricultural industry by-products**—These belong to the classification of secondary agricultural residues generated after the processing of crops into diverse bioproduct forms. This category encompasses materials like husk, molasses, bagasse, peels, hulls, and similar by-products [10].

Agricultural by-products derived from plant biomass exhibit a potential to function as sustainable, renewable, and cost-effective raw materials for producing polymers at an industrial scale [12]. Biopolymers are polymers obtained from biological entities, i.e., plants, animals, and microorganisms. Agro wastes and fossil fuels are also a source of biopolymers. Bacteria possess the ability to produce a diverse array of biopolymers with varied biological functions, exhibiting material properties that are suited for a multitude of uses in industrial and medical settings. Biopolymers are primarily categorized based on their origins into three groups: a) Natural biopolymers b) Synthetic biopolymers c) Microbial biopolymers [13].

(a) **Natural Biopolymer**—A natural biopolymer originates organically from living organisms. Within the realm of natural biopolymers, proteins, and polysaccharides are examples of biopolymers falling under this category [14], including collagen, cellulose, silk, chitosan, etc. [15].
(b) **Synthetic biopolymers**—These are obtained by modification of natural polymers or by chemical process with synthetic monomers as their building blocks. These modifications are designed to enable the biopolymers to naturally degrade without producing residues and intermediates that pose hazards to fauna and the ecosystem by and large [16]. These are further categorized into biodegradable and non-biodegradable biopolymers, examples include polylactic acid (PLA), polycaprolactone (PCL), polyglycolic acid (PGA), polyvinyl alcohol (PVOH), and polybutylene succinate (PBS) [16].
(c) **Microbial biopolymers**—Bacteria generate several biopolymers, such as polyamides, polysaccharides, poly-phosphates, and polyesters, because of their diverse biochemical capabilities [17]. When synthesized by pathogenic bacteria, the identical biopolymers serve as significant virulence factors, whereas in the case of non-pathogenic bacteria, they are repurposed as ingredients in food or materials for biological applications [18]. Examples of such materials include curdlan, xanthan, bacterial cellulose, gellan, K30 antigen, etc. Some of the microbial polymers are described in detail. Figure 2.1 shows the categorization of biopolymers.

Biopolymers have attracted researchers from all parts of the world because they possess unique characteristics. Furthermore, the understanding of the reactions involved in the biosynthesis of polymer and the control mechanisms governing

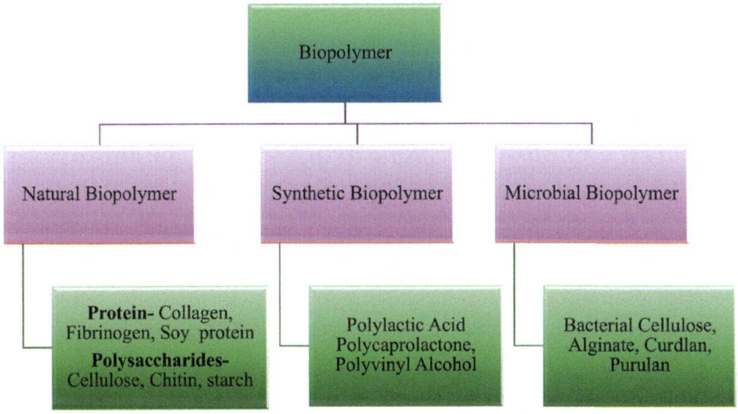

Fig. 2.1 Classification of biopolymers

these processes laid the groundwork for employing metabolic and protein engineering strategies aimed at enhancing their production. These approaches also aim at enhancing economic production efficiency and crafting customized polymers endowed with highly desirable material properties [19]. Among the biopolymers, microbial polysaccharides are produced for several diverse applications.

Polysaccharides consist of extended sequences of monosaccharides joined by glycosidic bonds, typically utilized by our bodies either for energy production or to support cellular structure. D-glucose stands out as the primary compound among them. Polysaccharides also contain various derivatives of monosaccharides such as glucuronic and iduronic acid (simple sugar acids), D-galactosamine, and D-glucosamine (amino sugars), along with their derivatives such as N-acetylneuraminic acid and N-acetylmuramic acid [20]. Naturally occurring polysaccharides include chitosan, chitin, alginic acid, starch, cellulose, konjac, and various gums (guar gum, gum Arabic, and gum karaya). Bacterial polysaccharides stand out for their high purity, hydrophilic characteristics, and finely structured three-dimensional fibers. Mainly, these polymers furnish microorganisms with protection, structural reinforcement, and energy. Nevertheless, recent discoveries have unveiled additional functionalities of these biopolymers [21]. Cellulosic fiber and its derivatives, when isolated, have garnered interest for a variety of applications spanning paper and packaging to sensors, water purification, textiles, biomedical devices, and drug delivery. These applications are driven by their favorable characteristics, namely, low density, high surface area, and robust mechanical strength.

Cellulose $(C_6H_{10}O_5)_n$ is an organic compound found widely in plant cell walls. Anselme Payen, for the first time in 1839 coined the term 'cellulose' [22] to describe the fibrous substance formed when wood was mixed with nitric acid. Payen described cellulose as a strong and durable fiber-like substance that is left behind after following the treatment of plant tissues by acids, and ammonia, followed by extraction with hot water, alcohol, and other solvents [23]. Among all other known biopolymers cellulose per se is considered the most abundant macromolecule found on our planet, which is biodegradable and can be obtained from a variety of natural sources such as plants, algae, and bacteria. Lignin and hemicellulose are the macromolecules associated with plant-based cellulose.

2.2 Bacterial Cellulose (BC)/Natural Carbon Nanomaterial (NCNM): A Historical Perspective

Cellulose is the basic and structural unit in plant cell walls, but it can also be produced with the help of different bacteria such as *Acetobacter*, *Sarcina ventriculi,* and *Agrobacterium.* The discovery of bacterial cellulose produced specifically from the bacterium *Acetobacter xylinum* was accredited to Prof. A. J Brown in 1886, who reported the synthesis of bacterial cellulose (BC) or natural carbon nanomaterial (NCNM) as a gelatinous mat [24].

Figure 2.2 depicts the notable contributions related to BC/NCNM. More extensive studies related to BC/NCNM started in the twentieth century, several years after the discovery by Prof. A.J. Brown. The findings of Prof. A.J. Brown were confirmed by C. A Browne, studied BC production using Louisiana sugarcane [25]. Later in 1947, Hestrin et al. proved that in the presence of oxygen and glucose, resting as well as lyophilized bacteria produced cellulose and *Acetobacter xylinum* was the typical bacterium for BC production [26]. Further, in 1957, Colvin demonstrated in vitro cellulose synthesis using cell-free extracts of *Acetobacter xylinum* in the presence of glucose and ATP [27]. In 1954 Schramm and Hestrin observed the presence of cellulose-negative mutants in *A. xylinum.* It was reported that upon cultivation and successive transfers under shaking conditions, cellulose-producing strains formed cellulose-negative mutants spontaneously, which reduced the BC yield. In 1957, Steel and Walker discovered cellulose-deficient isolates in agitated and aerated conditions. Coucheron in 1991 highlighted the association between the insertion of sequence elements in genes which prevented the biosynthesis of cellulose in an *A. xylinum* strain, suggesting that cellulose deficiency in mutants could be attributed to the insertion of sequence elements [28].

Some of the prominent bacterial genera known for extracellular BC/NCNM formation are *Acetobacter*, *Achromobacter*, *Pseudomonas*, *Rhizobium*, *Salmonella, Azotobacter,* and species of *Gluconacetobacter,* viz., *G. xylinum, G. hansenii,* and *G. pasteurianus* [22, 29]. Most commonly *Gluconacetobacter xylinum* is used to produce BC/NCNM. As it synthesizes acetic acid it was later named *Acetobacter xylinum*. Presently, the bacterium is referred to as *Gluconacetobacter xylinus* which belongs to the genus *Gluconacetobacter* and is one of the 29 genera of the *Acetobacteriaceae* family [30]. Bacteria belonging to the genus (*Agrobacterium, Rhizobium,*

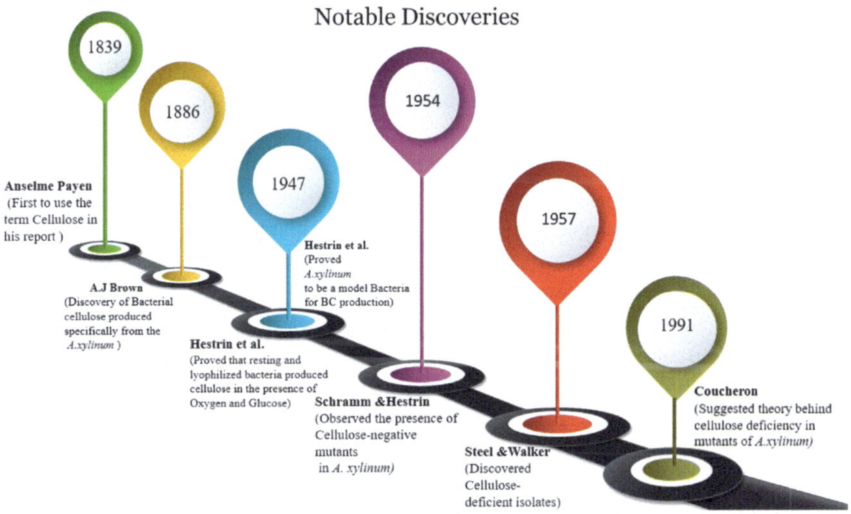

Fig. 2.2 A schematical illustration of the contributions related to BC/NCNM

Aerobacter, etc.) produce extracellular fibrils made up of cellulose which help them to adhere to the plant tissues. Some species like *Sarcina* produce cellulose in an amorphous form [31].

2.3 BC/NCNM: An Exceptional Carbohydrate Polymer

Typically, carbohydrates, proteins, fats, and nucleic acids are polymers of biological origin derived from natural sources. Several biopolymers represent sustainable resources and have been explored in numerous applications. Considering the scope of the present chapter, the characteristics of a few carbohydrate polymers of microbial origin are compared (Table 2.1). Natural polymers provide unique characteristics and above all ensure biocompatibility and are preferred because of environmental safety and biodegradable nature [32]. Various carbohydrate biopolymers like cellulose, chitosan, alginate, dextran, glycogen, etc. are widely used in many industries.

2.3.1 Physiochemical Characteristics of BC/NCNM

BC/NCNM and plant cellulose are similar in chemical structure [33]. Cellulose is a linear homopolymer of glucose monomers linked by β-1,4 glycosidic linkage with the chemical formula $(C_6H_{10}O_5)_n$ [34]. BC/NCNM possesses high integral purity, without the presence of lignin and hemicellulose, which contributes to minimizing post-treatments for the removal of impurities [35]. However, in comparison to plant-derived cellulose, BC/NCNM differs in certain structural and chemical properties. BC/NCNM has a degree of polymerization (2000 to 6000), which is much less than plant-derived cellulose (13,000 to 14,000) [29]. Furthermore, fibrils of BC/NCNM are much smaller (~100 times smaller than plant cellulose [36]), which makes it an important nanomaterial of natural origin. Figure 2.3 is a pictorial representation of the properties of BC/NCNM and Fig. 2.4 indicates its chemical structure.

BC/NCNM has repeated glucose monomers forming a long unbranched chain, which, in turn, are held together by intra- and inter-molecular hydrogen bonds that help to form a ribbon-like assembly. Bacterial cellulose is generally formed at the interface of air and the medium, and its shape is maintained by hydrophilic interactions [37, 38]. Bacterial cellulose fibers are arranged in a nanofibrous 3D network. The size and structural properties of the filaments are governed by the growth conditions. The crystalline structure of BC/NCNM is a result of the hydrogen bonding between the cellulose sheets. The thickness of BC/NCNM fibers depends on the bacterial species used. For example, the BC/NCNM fibers obtained from *Gluconacetobacter xylinum* are 3 to 4 nm thick and 70 to 80 nm wide [39], which are thus carbon-containing nanostructures.

During synthesis, the bacterial cellulose fibrils have loose arrangements with lots of empty spaces, which increases the surface area, creating a highly porous

Table 2.1 Types of biopolymers, their source of origin, and characteristics

Biopolymer	Bacterial cellulose	Chitosan	Alginate	Hyaluronic acid	Dextran
Producer strains	*Acetobacter xylinum* [50], *Acetobacter sp. V6* [51], *Gluconacetobacter sp RKY5* [52], *Gluconacetobacter swingsii* [53]	*Aspergillus Niger and Salmonella typhi, Salmonella paratyphi A, Escherichia coli, Proteus vulgaris and Pseudomonas aeruginosa, Bacillus sp. and Serratia sp.*[54]	*Azotobacter sp. and Pseudomonas sp* [55]	*Streptococcus zooepidemicus* [56]	*Aerococcus, Alloiococcus, Carnobacterium, Dolosigranulum, Enterococcus, Globicatella, Lactococcus, Lactobacillus, Leuconostoc, Oenococcus, Pediococcus, Streptococcus, Vaiscoccus*, and *Weiscoccus* [57]
Tensile strength (of Films)	70 to 300 MPa [58]	122.6 MPa [59]	33.6 ± 3.1 MPa [60]	-	-
GRAS +	Yes [61]	Yes [62]	Yes [63]	Yes (Sodium Hyaluronate) [64]	Yes [65]

(continued)

Table 2.1 (continued)

Biopolymer	Bacterial cellulose	Chitosan	Alginate	Hyaluronic acid	Dextran
Solubility	Insoluble [66]	Soluble in acidic pH [67]	Soluble in water [68]	Soluble in water [69]	Water soluble [70]
Chemical composition	Polymer of glucose molecules-individual glucose units are connected via acetal linkages between the C1 and C4 carbons of the glucopyranose ring [71]	Linear polysaccharides composed of varying amounts of (β1-4) linked residues of N-acetyl glucosamine, residues [25]	Two monomeric units: β-(1,4) linked d-mannuronic acid with 4C1 ring conformation and α-(1,4)-linked l-guluronic acid with 1C4 ring conformation [72]	Repeating polymeric disaccharides of D-glucuronic acid and N-acetyl-D-glucosamine linked by a glucuronidic β (1 → 3) bond [73]	(1,6)-linked glucan with side chains attached to the C-3 position of the backbone [74]

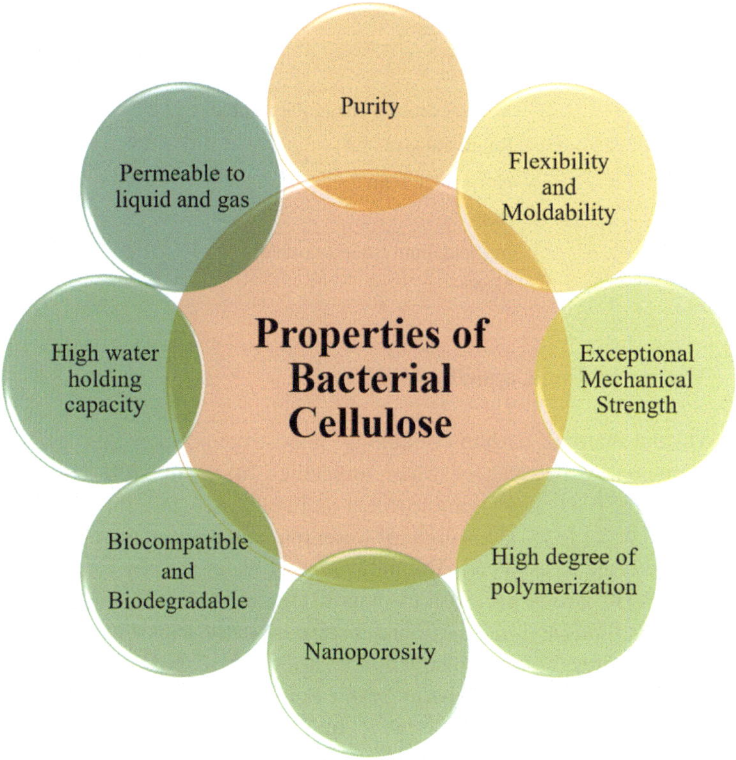

Fig. 2.3 Characteristics of BC/NCNM

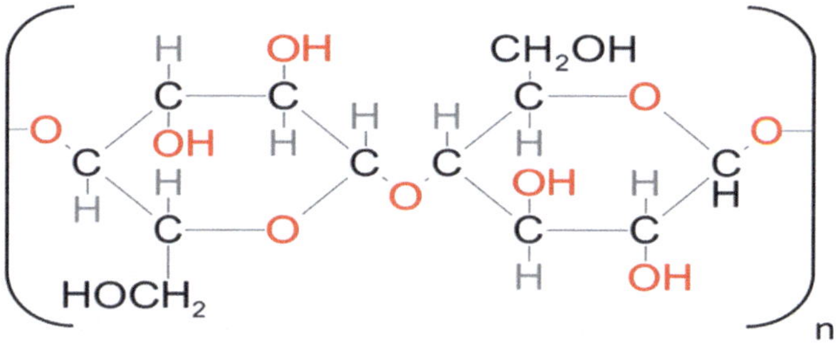

Fig. 2.4 Chemical structure of bacterial cellulose

matrix [40, 41]. It is reported that upon supplementation with an additional carbon source, the density and strength of the microbial cellulose fiber can be enhanced [42]. The physical, mechanical, and thermal characteristics of microbial cellulose are significantly influenced by structural changes [43]. Structurally strong and stable fibrils can provide superior resistance to applied stress [44].

Table 2.2 Water-holding capacity of bacterial cellulose analyzed using different methods

Method used	Average WHC (g/g)
Vacuum	148
Vertical drum (rectangular sample)	285
Rapid transfer	309
Horizontal drain	204
Wipe	249
Vertical drain (round sample)	209
Shake	291

2.3.1.1 Water-Holding Capacity (WHC)

BC/NCNM, the natural carbon-containing nanostructure, appears to have an expanded appearance because of water molecules fitting into the empty spaces between the pores. It has been demonstrated that cellulose can hold 100–200 times its dry weight in water [45]. The weight of water retained per unit weight of cellulose nanofibrils is referred to as WHC. The molecules of water are physically bonded to cellulose. The WHC will be greater if the nanofibrils are loose [46, 47]. Additionally, water-immiscible fluids and tiny particles can also fill in the gaps which results in the creation of composites. This has resulted in a variety of BC/NCNM-based composites incorporated with nano-and biocompatible polymeric materials [48]. According to a study conducted by Mazar-Ulf-Islam et al. 2012, the WHC of BC/NCNM was found to be 106.43 (±5.23) g [42]. Another study conducted by Mani Illa et al. 2019 revealed the wet weight of BC/NCNM produced by *Komagataeibacter hansenii* (ATCC 3769) to be 11.2 ± 0.3 (g/100 mL) with a water content of 97% [49]. In an interesting study that was conducted by Schrecker and Gostomski (2005), the authors used different methods for calculating WHC. According to their findings, WHC varied according to the method of analysis Table 2.2 [45].

2.3.1.2 Degree of Polymerization (DP)

It is defined as the number of monomeric units in a macromolecule or oligomer molecule [75]). Sharma et al. (2019) in his work stated the degree of polymerization of cellulose produced by *Komagataeibacter hansenii* ATCC 23,769 strain to be 876 ± 169[49]. Tahera, et al. (1999) studied the effect of cellulase production on DP of cellulose produced by the same species. The result of this study indicated a pH-dependent decrease in the DP of cellulose corresponding to an increase in cellulase concentration [76].

2.3.1.3 Elasticity (Young's Modulus)

The mechanical characteristics of BC, the natural carbon nanomaterial, are determined by examining the material in sheet/pellicle form. Yamanaka et al. (1989) carried out some of the early work in which they discovered that BC sheets may have Young's modulus greater than 15 GPa [77]. Nishi et al. (1990) further reported improvement in the modulus up to 30 GPa [78] and Young's modulus in the range of 30–40 GPa was also recorded by Tajima et al. [79]. According to a study conducted by Potivaea et al. (2011), the natural rubber-BC/NCNM films had a ~ twofold increase in Young's modulus values than the unmodified BC/NCNM film (with Young's modulus of 9.14 GPa) [80].

2.3.1.4 Microporosity

Tissue engineering requires a 3D structure with fine hollow spaces. The porosity of BC/NCNM can be changed by altering the carbon source or the culture used in the fermentative production. A report by Tang et al. 2010 suggests that freeze-dried BC/NCNM membranes have more uniform pores as compared to air-dried BC/NCNM membranes [81]. Another study found that grinding BC/NCNM and then freezing and lyophilizing it produced porous bacterial cellulose sponges [82]. Additionally, the freezing and lyophilization procedure was used to produce chitosan and BC/NCNM/agarose composites [83]. By adding potato starch to the growth media, bacterial cellulose potato starch composites were produced, resulting in a 40 μm pore size [84]. In BC/NCNM, pore diameters as large as 20 μm were achieved with the use of surfactant-assisted foaming treatment [85]. Additionally, porogens such as nylon, paraffin wax, and starch particles have been added to create microporous BC/NCNM [86–89]. Due to this characteristic, BC/NCNM the natural carbon-containing nanomaterial finds application as scaffolding suited for tissue engineering.

2.3.1.5 Moldability

Moldability is the ability of any soft material to take the desired shape. BC/NCNM, a natural nanomaterial of microbial origin with unique characteristics like biocompatibility and moldability, has led to applications of BC/NCNM in tissue engineering, biomedical implants, and wound dressings [90]. A report by Salinas et al. (2021) addresses the conversion of BC in tubes, which can be used for many biomedical applications [91]. Another study by Martina Pepicelli et al. shows the use of BC/NCNM to create and modify cellulose to prepare capsules. Agrochemicals, nutrition, flavor, aroma, and medication encapsulation are just a few of the many uses for the resulting cellulose capsules [92].

Table 2.3 The tensile stress of BC/NCNM film soaked for 24 h in different concentrations of NaOH

NaOH concentration (%)	Tensile stress (MPa)	Breaking strain (%)
0	208	3.08
1	201	2.96
5	182	3.64
10	165	3.00

2.3.1.6 Tensile Strength

According to Suratago et al. (2016), the tensile strength of BC/NCNM sheets is in the range of 70–300 MPa ([93]). Another study by Suryanto et al. (2019) reports (Table 2.3) that the tensile stress of BC/NCNM film soaked in NaOH for 24 h decreased in a concentration-dependent manner [94]. In the presence of 10% NaOH, the stress was only 165 compared to 208 MPa for the native BC/NCNM. According to data published by Santmarti, the tensile strength of BC/NCNM nano-papers was 113–125 MPa, (under quasi-static testing condition) [95].

2.3.1.7 Biocompatibility

The capacity of a material to function with an expected host response in a particular application is known as biocompatibility. Additionally, the material should not have any harmful effects on the biological systems. Moreover, biocompatibility also refers to the requirement for a suitable host response to guarantee good performance on a particular application [96]. According to Petersen and Gatenholm (2003), BC/NCNM structural resemblance to extracellular matrix elements like collagen may contribute to its biocompatibility for use in tissue engineering applications. In comparison to native BC/NCNM, certain composites, such as BC/NCNM/Poly(ethylene glycol) [97], BC/NCNM/Chitosan [98], and BC/NCNM/Gelatine [99], demonstrated higher NIH3T3 bioactivity [100].

Wound dressings have been made using BC/NCNM gels. As per the findings of Czaja et al. [101], biocompatibility and the unique carbon-containing nano-fibrillar structure of BC/NCNM provide an ideal environment for wound healing. When compared to traditional wound dressings like moist gauze and ointments, BC/NCNM speeds up the skin's healing process. Studies have found that BC/NCNM-based dressings lessen wound inflammation, accelerate the process of re-epithelization, and lower the risk of wound infections and scarring [102]. Legeza et al. [103] developed a BC/NCNM wound dressing impregnated with poviargol (Procel-Super and Procel-PA) or the antioxidant superoxide dismutase for the treatment of third-degree burns. Xcell®, Bioprocess®, and Biofill® are the three BC/NCNM-based coverings that are marketed commercially. A few other uses of BC/NCNM include the manufacture of contact lenses [104], electroconductive composite hydrogel biosensors [105], topical lidocaine delivery membranes [106], and nutritional supplements to treat diabetes [107]. Due to its excellent characteristics, BC/NCNM the natural carbon nanomaterial has several applications. However, commercial applications would require economic media and conditions that favor fermentative production. All such media and types of reactors used to synthesize BC/NCNM and strategies to overproduce it are discussed in the following paragraphs.

HS (Hestrin–Schramm) medium is notable because it supports the growth of microbes that produce copious amounts of BC/NCNM extracellularly. Hestrin and Schramm used the medium for the first time in 1954. Typically, the Hestrin–Schramm (HS) medium is composed of Glucose (2.0 g%), Peptone (0.5 g%), Yeast extract (0.27 g%), Na_2HPO_4(0.27 g%) and pH adjusted to 6.0 [108]. HS medium is a popular choice among researchers and manufacturers interested in producing BC/NCNM for various applications, including biomedical, food technology, and materials science. Additionally, modifications to the original medium formulation have been made over time to optimize BC/NCNM production efficiency and quality for specific purposes. Instead of glucose as the carbon source, several authors reported substrates such as mannitol, sucrose, food-grade glucose, and date syrup in the production of BC/NCNM [109]. Since BC/NCNM overproducing microorganisms can utilize diverse organic compounds, several researchers employed complex agricultural wastes as carbon sources in the form of a cheaper medium to produce BC/NCNM, the natural carbon-containing nanomaterial for commercial applications. Figure 2.5 is a pictorial representation of various agricultural wastes that were reported in BC/NCNM production. Generally, wastes containing simple as well as complex sugars are reported in the synthesis of BC/NCNM. The substrates were chosen as a source of carbon. Attempts were made to minimize the pretreatments to economize BC/NCNM production (Table 2.4).

In a few studies, the agricultural wastes were also supplemented with other polymers for obtaining BC/NCNM nanocomposites during fermentation. Table 2.5 enlists these studies.

2.4 Types of Reactors Employed to Produce BC/NCNM

Bioreactors are described as vessels that furnish the necessary biological, biochemical, and biomechanical conditions to support the optimal growth of fermenting microorganisms and/or biochemical reactions on an industrial scale to produce desired products. Different types of reactor configurations are tested to improve the quality and quantity (yield) of the product. Each of these traits relies also on the culture's conditions, which include oxygen levels, carbon sources, and temperature control [74, 138]. Bioreactors that are mostly used for the production of BC/NCNM are broadly classified into (a) static, (b) agitated, and (c) stirred systems and their combinations. These conditions are reported to influence the production of BC/NCNM, the natural nanomaterial.

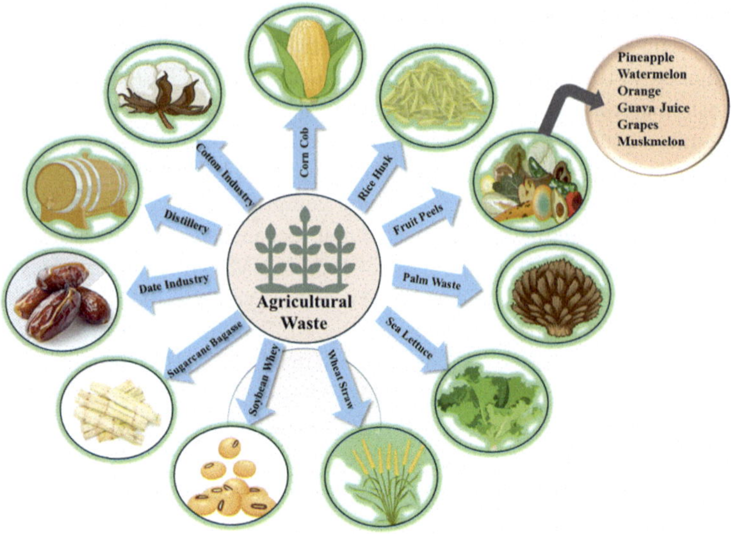

Fig. 2.5 Different types of agricultural wastes used for BC/NCNM production

2.4.1 *Static Reactors*

The static culture technique is a commonly utilized technique for manufacturing bacterial cellulose, which is a straightforward approach suitable for small-scale laboratory production of BC/NCNM [139]. In static environments, cellulose microfibrils emerge from the bacterial surface and aggregate to create a pellicle at the air–liquid interface. Biocellulose nanofibers generated under static conditions exhibited an average size of 478.0 μm [140]. The aggregation of cellulose fibers is a consequence of oxygen depletion within the bulk, prompting bacteria to migrate to the interface, where growth and polymer synthesis take place [141]. In static conditions, a larger area and workforce are needed, leading to increased expenses for industrial-scale production. The first demerit of static reactors is that as the culture progresses, the depletion of the oxygen and carbon sources occurs, leading to a gradual decrease in BC/NCNM production over time. The second demerit is that the structural integrity of BC/NCNM relies on its uninterrupted formation process, requiring it to remain still and undisturbed throughout [142]. Research has confirmed that *Acetobacter xylinum* synthesizes BC/NCNM when incubated statically [143] in Schramm–Hestrin (SH) medium supplemented with glucose (carbon source) [144]. The yield in the static fermenter was reported to be 3.0 g/100 ml [112]. Also, the reported yield using HS medium in static conditions was 3.2 g/l [145].

Table 2.4 Agricultural wastes used for the fermentative production of BC/NCNM

Agricultural waste (substrate)	Bacterial species	BC/NCNM yield	Reference
Oil palm frond (OPF) juice	*Acetobacter xylinum*	0.339 g/L	[110]
Oil palm frond juice + coconut water	*Acetobacter xylinum*	4.50 g/L	[111]
Watermelon peels	*Komagataeibacter hansenii*	3.0 g/100 ml (wet weight)	[112]
Fermented rice water (noodles)	*Komagataeibacter sp. PAP1*	11.76 g/L	[113]
Guava Juice(overripe)	*Acetobacter Xylinum*	–	[114]
Baijiu (Chinese liquor)	*Gluconacetobacter xylinus*	7.42 g/L	[115]
Pineapple peel	*Komagataeibacter hansenii MCM B-967*	–	[116]
Date syrup	*Acetobacter xylinum 0416*	–	[117]
White and red grape bagasse	*Gluconacetobacter xylinus*	–	[118]
Oil palm frond juice	*Acetobacter Xylinum 0416*	2.33 g/L	[119]
Date syrup and cheese whey 50:50 conc	*Komagataeibacter xylinus PTCC 1734*	Dry weight 1.88 g/ 100 mL	[120]
Waste fiber sludge	*Gluconacetobacter xylinus* ATCC 23,770	10 g/L	[121]
Wheat straw acid hydrolysate	*G. xylinus (formerly A. xylinum) ATCC 23,770*	–	[122]
Corn steep liquor	*Gluconacetobacter hansenii,*	–	[123]
Corn cob	*Gluconacetobacter xylinus*	4 g/L	[124]
Muskmelon/cantaloupe juice	*Komagataeibacter*	3.49 g/L	[125]
Discarded potatoes	*Gluconacetobacter xylinus*	4.7 g/L	[126]
Sugarcane bagasse	*Gluconacetobacter hansenii*	–	[127]
Tobacco waste extract	*K. xylinum* ATCC 23,767	5.2 g/L	[128]
Wastewater of candied jujube processing industry	*K. xylinum* CGMCC 2955	2.25 g/L	[129]

(continued)

Table 2.4 (continued)

Agricultural waste (substrate)	Bacterial species	BC/NCNM yield	Reference
Orange peels	*Komagataeibacter sucrofermentans DSM15*	11.6 g/L	[130]
Prickly pear peels	*Acetobacter xylinus*	2.94 g/L	[131]
Tomato juice	*Acetobacter pasteurianus RSV-4*	7.8 g/L	[132]

Table 2.5 List of additives and media used for the production of BC/NCNM

Additive	Media used	Bacterial strain	BC/NCNM yield (g/L)	Ref.
Sodium alginate	Corn steep liquor	*Acetobacter xylinum NUST4.1*	6.0	[133]
Polyacrylamide-co-acrylic acid (3 g L^{-1})	Fructose + CSL	*A. xylinum BPR2001*	3.7	[134]
Molasses (H_2SO_4 heat treatment)	Molasses	*A. xylinum BPR2001*	7.8	[135]
Agar (0.1 w/w)	CSL-Fructose	*A. xylinum subsp sucrofermentans BPR 2001*	8.7	[136]
Lactate (12.5 g L^{-1})	Fructose-CSL	*A. xylinum subsp. BPR3001A*	–	[137]

2.4.2 Agitated Reactor

It is widely recognized that the growth of microbes and the production of their metabolites in bioreactors are significantly impacted by both the constituents of the medium and various physical factors. These factors include agitation, aeration, temperature, fermentation duration, and the concentration of dissolved oxygen [146]. Both agitation and aeration play crucial roles in all aerobic processes, exerting a notable influence on the production of the majority of biopolymers. Agitation is an important aspect of fermentation procedures because it contributes to both mixing and shearing. It enhances the transfer of mass and oxygen across various phases and ensures continuous mixing to maintain homogeneous chemical and physical conditions within the medium [147]. It is observed that more BC/NCNM is produced in static conditions compared to agitated conditions. However, static fermentation requires greater surface area, making them less preferred systems in the commercial production of this natural nanomaterial.

Agitated biocellulose displayed an average bundle diameter of 21.2 μm [140]. Various culture conditions, including shake flasks, stirred tanks, airlift reactors, rotating disk reactors, and modified airlift or gas-lift bioreactors, have undergone testing for BC/NCNM production under agitated conditions [148]. Three types of cellulose 9fibrous suspensions, spheres, and pellets) are usually generated during agitated cultivation and stirred fermentation processes. Another variation of the agitated system is the rotary biological contactor. In this particular design, the circular disks can rotate while they interact with the air-medium interface. While BC/NCNM generated through this method achieved a uniform structure, its yield did not show a significant increase compared to static culture methods [149]. The strain *Gluconacetobacter* sp. RKY5 was used in a rotating biofilm reactor in which the amount of bacterial cellulose produced, and cell concentration were 5.52 and 4.98 g/L, respectively [27].

2.4.3 Stirred Tank Reactor

Stirred tank bioreactors represent a widely adopted technology that provides a valuable alternative to conventional culture methods. The production of BC/NCNM is closely linked to volumetric oxygen transfer coefficients, enabling stirred tank bioreactors to enhance production rates [142]. These bioreactors exhibit a high volumetric mass-transfer coefficient facilitating oxygen transfer [150]. The agitation system and baffles within these bioreactors guide the transfer of nutrients and oxygen into the bacterial cells [150]. The strain *A. xylinum* Nust4.1 was used in a stirred tank bioreactor operating at 200 rpm, which reported a yield of 1.89 g/l [133]. *A. xylinum* subsp *sucrofermentas* BPR 2001 when used in stirred tank bioreactor gave a yield of 6.9 g/l [136].

2.4.4 Airlift Bioreactor

Airlift bioreactors resemble bubble-column reactors, but they feature a draft tube, which distinguishes them slightly. When an airlift bioreactor includes an inner draft tube, it is referred to as an airlift bioreactor with an internal loop. Conversely, when an airlift bioreactor contains an external draft tube, it is termed an airlift bioreactor with an external loop. The presence of the draft tube enhances circulation and oxygen transfer and helps balance shear forces within the reactor [151]. BC/NCNM exhibited a distinct morphology viz., unique elliptical pellets when grown in an airlift reactor, while the fibrous form was typically produced in an agitated stirred tank fermenter [152]. This method ensures a steady supply of nutrients to the system with minimal disturbance. Small, controlled amounts of oxygen are introduced, allowing them to dissolve into the medium over time gently. The rising gas creates a gentle internal flow that mixes the liquid below the surface, avoiding disruption to the productive culture layer at the top [153]. *Acetobacter xylinum* subsp. *sucrofermentans* BPR2001 in a 50-L internal-loop airlift reactor produced 3.8 g/L of BC/NCNM [152] and the

Table 2.6 Overall comparison of reactors reported for BC/NCNM production

Reactor type	Merits	Demerits
Static	• Thick pellicles with less significant branching are formed • Requires minimal equipment and technical expertise • Accessible method for research and small-scale production • High Young Modules	• Large surface area needed • Long culture periods • Inefficient for large-scale manufacturing • Low production yield
Agitated	• High production yield • **Scalability** • **Uniform pellicle formation**	• Shear forces that can damage fragile bacterial cells • High Cost
Stirred tank	• **Higher BC/NCNM production** • Homogeneity • **Controlled Environment**	• **Shear Stress** • **Increased complexity** • **BC/NCNM produced has low mechanical strength**
Air lift	• **Reduced Shear Stress** • Good oxygen transfer • Manipulative Design	• Operational Complexities • High power demands • Foam formation

same culture produced 8.7 g/L of bacterial cellulose in a 50-L airlift reactor [136]. The overall comparison of reactors reported for BC/NCNM production is provided in Table 2.6.

Krystynowicz and colleagues observed that under conditions of intense vibration and air exposure, there is a notable decrease in bacterial cellulose production. These conditions can prompt the emergence of spontaneous cellulose mutants, reducing the strain's production capacity [154].

2.5 Strategies to Improve the Yield of BC/NCNM the Natural Carbon Nanomaterial

BC/NCNM, an exceptional microbial polymer, is used and studied widely. The biosynthesis of cellulose involves a sequence of successive processes involving enzymes. Studying the process is essential to understanding the role of regulatory proteins that may be the likely candidates for genetic alteration to improve or control BC/NCNM production and change BC/NCNM characteristics. Through genetic modification, it is possible to target the genes involved in cellulose biosynthesis. Such studies provide a means of upregulating or controlling the synthesis of cellulose in the bacterium, which is desirable for the industrial production of BC/NCNM [155].

Komagataeibacter xylinus, (formerly known as *Gluconobacter/Acetobacter xylinus)* is often used as a model bacterium for studying cellulose manufacturing. The extracellular cellulose produced by this microbe has some excellent qualities,

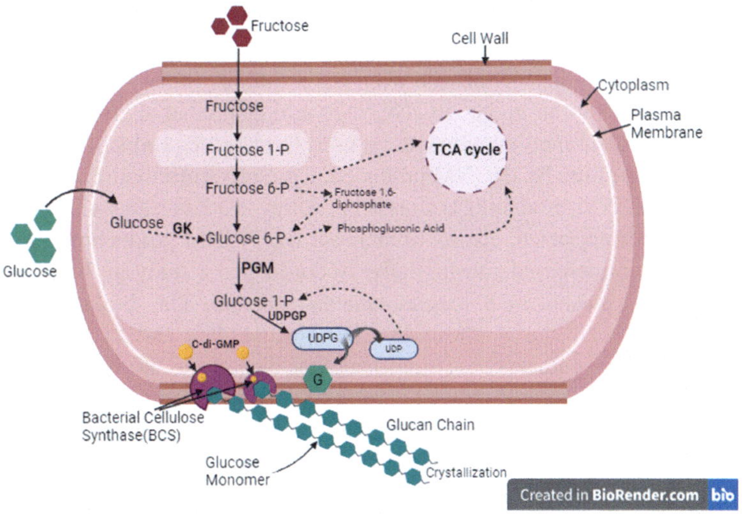

Fig. 2.6 Diagrammatic representation of cellulose synthesis pathway in *K. xylinus* adapted from Singhania et al. [155]. GK = Glucokinase, PGM = Phosphoglucomutase, UDP = Uridyl diphosphate, UDPG = Uridyl diphosphate glucose, UDPGP = UDPG phosphorylase, C-di-GMP = cyclic diguanylic acid

such as strength, flexibility, water retention capacity, and malleability. The cellulose biosynthesis process is a well-controlled series of reactions in a pathway involving multiple quantities of individual enzymes as well as sets of regulatory proteins [156] (Fig. 2.6).

The inner membrane-associated bacterial cellulose synthase *BcsA* and *BcsB* subunits catalyze the production and transportation of cellulose in bacteria [157]. The production of BC can be increased via many methods, such as implementing metabolic and biochemical strategies or genetic manipulation. A brief overview of the strategies to increase BC/NCNM yield is presented in the following paragraphs. Anguluri et al. (2022) reported an improvement in BC/NCNM of about 38% in the HS mannitol compared to the HS glucose, followed by an increase in mannitol utilization for BC/NCNM production. This information is significant because BC/NCNM is produced from mannitol-containing waste. Furthermore, the mannitol adaptation did not significantly impact the strain's capacity to use glucose, demonstrating the adaptability of *Komagataeibacter xylinus* K2G30 in generating BC/NCNM from various carbon sources [158].

Sharma et al. reported improved production of BC/NCNM by *Gluconacetobacter kombuchae* upon supplementing the medium with lactic acid and ethanol. The most suitable concentration of lactic acid was found to be 0.6%, which improved BC/NCNM yield by 4.11 g. The optimum concentration of ethanol was found to be 1%, which increased the BC/NCNM yield by 2.97 g [159]. The work done by Tian et al. suggests a 7% increase in BC/NCNM yield after ethanol and sodium citrate were added to the medium for fermentation [160]. Another study by Cielecka et al.

using *Komagataeibacter xylinus E25* reports that the addition of 0.6% lactic acid and 1.0% ethanol in media gave the highest BC/NCNM yield. The media supplemented with 0.1%–0.2% of either of the additives gave similar results. Subsequently, up to 0.8% lactate resulted in higher BC/NCNM yield than ethanol [161]. It has been suggested that genetic modification can be used to increase production or improve the characteristics of the BC/NCNM produced. Currently, information on the whole genome of BC/NCNM-producing strains is becoming available, and with the help of the bioinformatics approach, efforts to increase BC/NCNM production by genetic manipulation have increased substantially. According to a study by Floreaa et al., the *UGPase* gene present in *Komagataeibacter rhaeticus* may be knocked down using sRNA to prevent the synthesis of UDP glucose. It is essential to target the *UGPase* gene for the synthesis of cellulose because it catalyzes the synthesis of UDP glucose [162]. Another study done on *K. xylinus* AY201 suggests the introduction and expression of the *crdS* gene in *K. xylinus*, which helps in the simultaneous synthesis of cellulose/curdlan [163]. A study conducted by Yadav et al. suggests the insertion of three gene operons (NAG5, AGM1, UAP1) isolated from yeast *Candida albicans* in *K. xylinum*, helped in the synthesis of cellulose with enhanced biodegradability and low crystallinity [164].

Modifications of BC/NCNM post-synthesis have been performed to diversify the end-user. The surface-OH groups are modified chemically, and several reports on the grafting of metal ions, biomolecules, chemical ligands, linkers, etc., are presented in Table 2.7.

2.6 Application of the Natural Carbon-Containing Nanomaterial BC/NCNM

Numerous non-medical applications of BC/NCNM are reported. The following summarizes these reports. Applications in food, packaging, biosensors, bioremediation, etc. (Table 2.8).

Since BC/NCNM synthesis was demonstrated over a century ago, products containing BC/NCNM are available commercially. However, all commercial applications were concentrated around the health of an individual, and hence, BC/NCNM obtained from chemically defined media was reported.

Products containing BC/NCNM are available commercially (Table 2.9). Prima Cel™ is a form of cellulose made in the USA by Xylos Corp., utilized as a wound dressing in clinical studies to treat ulcers [156]. Biofill™ and Bioprocess™ are commercial preparations of *A. xylinum* cellulose. Germany's Fzmb GmbH (formerly fzmb e.V.) is one of the largest manufacturers of cellulose sheets which are used in the medical and cosmetic industries [248]. In the Philippines, cellulose is utilized in the culinary sector as Nata de coco (Shyu et al. 2004, Budhiono et al. 1999) [197, 249]. Using microbial cellulose, Sony Corporation and Ajinimoto of Japan created the first audio speaker diaphragms [250].

Table 2.7 Different BC/NCNM-based composites and their applications

BC/NCNM modified/additives	Applications	Ref.
BC/NCNM-chitosan-Ag-NPs	**Dermal restoration**	[165]
BC/NCNM-Montmorillonite (MMT)	Wound healing	[40]
BC/NCNM-Hydroxyapatite (HAP)	**Repair and creation of musculoskeletal tissues**	[166]
BC/NCNM-Ag triethanolamine (TEA)	Wound healing	[167]
Ag-nanofibrillated BC/NCNM	Antibacterial activity	[168]
BNC-GO-polyacrylic acid	Wound healing	[169]
BC/NCNM-PVA	Cardiovascular soft tissue replacement	[170]
BC/NCNM nanofibers and Chitin nanocrystals	Bactericidal activity	[171]
BC/NCNM-PEG	Wound healing	[97]
Cu-Cellulose	Antimicrobial activity	[172]
BC/NCNM-Silver sulfadiazine (SSD)	Antimicrobial activity	[173]
BC/NCNM-kaolin	Wound healing	[174]
BC/NCNM-GO-CuO	Antimicrobial activity	[175]
BC/NCNM-SiO_2-Cu	Antimicrobial activity	[176]
BC/NCNM-Vaccarin	Wound healing	[177]
BC/NCNM-Methylglyoxal	Antimicrobial wound dressing	[178]
BC/NCNM-ZnO	Antibacterial properties	[179]
BC/NCNM-Benzalkonium Chloride	Antibacterial properties	[180]
BC/NCNM-silk fibroin	Tissue regeneration	[181]
BC/NCNM-TiO_2	Wound healing and tissue regeneration	[182]
BC/NCNM nanocrystals-Regenerated chitin	Wound healing	[183]
BC/NCNM-FeO	Blood vessels	[184]
BC/NCNM-AgNPs	Antibacterial application	[185]
BC/NCNM-glycerine	Skin disease treatment and moisturizer	[186]
BC/NCNM-[EDA][DLA-Tyr] (ethylene diamine, Di linoleic acid and Tyrosine)	Wound healing and antimicrobial Activity	[187]
BC/NCNM-Goat bone apatite (GBA)	Bone defects repair and construction	[188]
BC/NCNM-2,3-dialdehyde	Tissue engineering scaffolds	[189]
BC/NCNM-Hap-Fe_3O_4	Bone tissue engineering	[190]
BC/NCNM-Paraffin Wax	Bone tissue engineering applications	[89]

(continued)

Table 2.7 (continued)

BC/NCNM modified/additives	Applications	Ref.
BC/NCNM-PVA	**Synthetic cornea and bioengineering**	[191]
BC/NCNM-Chitosan-carboxymethyl cellulose	**Corneal regeneration**	[192]
BC/NCNM-poly(ethylene glycol) (PEG)	Tissue-engineering scaffolds	[97]
BC/NCNM-Serum Albumin	Drug delivery	[193]
BC/NCNM membrane	**Repair dural defect in mice**	[194]
BC/NCNM-human urine-derived stem cells (USC)	**Urinary conduits**	[87]
BC/NCNM-lingual keratinocytes	Urethral reconstruction	[195]
BC/NCNM-Silicate cement	**Dentistry**	[196]

BC/NCNM is a non-toxic, biocompatible, biodegradable, organic polysaccharide which makes BC/NCNM a novel material for many uses and industrial applications [251]. BC/NCNM fibers are typically 70–130 nm in width and are classified as waste-derived carbon nanomaterial (WDCNs). BC/NCNM as nanomaterials (NMs) possesses a multitude of advantageous characteristics, including copious production from agro wastes, high tolerance to environmental conditions, thermal stability, non-toxicity, antibacterial and antifungal activities upon subtle modifications, biocompatibility, biodegradability, and other unique features like strong adsorption capacities and easy functionalization [252]. Some important uses of BC/NCNM are described in the following paragraphs.

2.6.1 BC/NCNM as a Conductor of Electricity

BC/NCNM itself is a dielectric material, but when modified with other conductive elements, it acts as a good conductor of electricity [253]. BC/NCNM is reported to have a resistance of $10^8 \Omega$ [254]. Considering these aspects, BC/NCNM proves to be a novel insulating material that can be used in designing electronic implants [255].

2.6.2 BC/NCNM in Biomedical Applications

BC/NCNM finds application as an appropriate material for bone tissue engineering, drug delivery, artificial blood vessel construction, wound dressing, etc. Furthermore, BC/NCNM can be easily modified to create derivatives or composites with improved functional and physicochemical characteristics. To create hybrid materials with better

Table 2.8 Applications of BC/NCNM

Modification/additives (If any)	Specific application	Reference
BC/NCNM in food		
Nata-de-coco	Food material	[197]
BC/NCNM	Vegetarian meat	[198]
Biocellulose	Low calorie bulking agent	[199]
BC/NCNM-soy protein isolate (SPI)	Emulsifier	[200]
BC/NCNM	Fat replacer in mayonnaise	[201]
BC/NCNM	Meat analogue	[202]
BC/NCNM	Thickener and gelling agent	[203]
Nata-De-Coco (BC/NCNM)	Fat replacer in meat balls	[204]
BC/NCNM + Lactic Acid bacteria	Immobilizer of probiotics and enzymes	[205]
BC/NCNM	Increase viscosity, gelation agent	[206]
BC/NCNM in packaging		
BC/NCNM-pectin-blackberry pomace powder (BPP)-carboxymethylcellulose (CMC)	Antioxidant and antimicrobial edible coatings	[207]
BC/NCNM-cashew tree gum (CG)	Food packaging or coating purposes	[208]
BC/NCNM-pectin added with fruit purees	Food wrapping or coating	[209]
BC/NCNM-lysozyme	Active packaging	[210]
BC/NCNM (tapioca starch/chitosan-based films	Food packaging	[211]
BC/NCNM-ε-polylysine (ε-PL)	Bacteriostatic sausage casing	[212]
BC/NCNM-ZnO	Controlled release antimicrobial food active packaging	[213]
BC/NCNM-*Echium amoenum* extract (EAE)	Intelligent packaging of protein-rich food	[214]
BC/NCNM in biosensors		
BC/NCNM immobilizing lactate oxidase	Electrochemical biosensor	[215]
BC/NCNM-novel gold nanoparticle	Bioelectroanalysis and bioelectrocatalysis	[216]
BC/NCNM sponges-Ag	Bioanalysis	[217]
BC/NCNM membrane	Biosensor for whole blood measurements	[218]
BC/NCNM nanopaper	Optical sensor	[219]
AuNPs-modified BC/NCNM	Glucose biosensor	[220]
BC/NCNM-CNTs	Biodetector	[221]
BC/NCNM-TiO_2	Biosensor (photocatalysis)	[222]

(continued)

Table 2.8 (continued)

Modification/additives (If any)	Specific application	Reference
BC/NCNM-ZnO	Metal Oxides (MOX) gas sensor	[223]
BC/NCNM/polyaniline/SWCNTs	Biological sensor	[224]
BC/NCNM	Deformation sensor	[225]
BC/NCNM in bioremediation		
BC/NCNM-GO	Absorption for organic liquids	[226]
BC/NCNM	Magnetic carbon nanofiber adsorbent	[227]
BC/NCNM-derived activated carbon (BC/NCNM-AC500)	Waste water treatment	[228]
Soy protein isolate/BC/NCNM	Air filtration	[229]
BC/NCNM aerogels (BCAs)	Filter membrane	[230]
BC/NCNM	Treatment of waste water	[231]
BC/NCNM membranes	Industrial water remediation	[232]
BC/NCNM-pandan	Heavy metal removal in wastewater treatment	[233]
BC/NCNM-CMC cellulose stabilized-Cu-NPs	Pollutants degradation	[234]
Immobilized BC/NCNM-TiO_2	Textile dye degradation	[235]
BC/NCNM in cosmetics		
BC/NCNM	Facial mask	[236]
BC/NCNM-glycerine	Skin compatibility	[186]
BC/NCNM-silk sericin	Facial treatment application	[237]
BC/NCNM-aloe vera-ascorbic acid	Natural facial scrub	[238]
BC/NCNM-cationic polymer	Personal cleansing composition	[239]
BC/NCNM	Contact lens	[104]
BC/NCNM	Skin moisturization, skin exfoliation and sebum Absorption	[240]
BC/NCNM	Skincare	[241]

functional capabilities, several polymers, CBNs, and metal nanoparticles (NPs) have been added to BC/NCNM using both in-situ and ex-situ techniques [256].

Table 2.9 Commercialized products based on BC/NCNM

Application	Product name	Company	Country
Bandages	Bionext (BENNETT) [242]	bionext®	Brazil
	MTA protective tissue [243]	Xylos Corporation, USA	U.S.A
Wound dressing	Biofill [139]	Bioprocess®	Brazil
	Biofill [244]	Gengiflex®	Brazil
	Xcell [245]	XCell®	Pennsylvania U.S.A
	Dermafill [246]	Dermafill™	Colquitt Georgia U.S. A
Ulcer treatment	Prima [247, 248]	Prima Cel™ (produced by Xylos Corp.)	U.S. A

2.6.3 BC/NCNM in the Food Industry

As native BC/NCNM is a hydrogel, its rheological qualities have been used in the food industry to create dessert cuisines like *Nata de coco* [206]. *Nata de coco*, a native dietary fiber of Asian countries is served as a gelatinous cube. The texture of *Nata* is smooth, chewy, and soft on the surface. It is low in fat, calories, and cholesterol. During the manufacture of *Nata de coco*, under static culture conditions, BC/NCNM forms at the air–liquid interface. The bacterium *Acetobacter xylinum* was cultivated with coconut water as a carbon source. BC/NCNM can also be used in enzyme immobilization [48, 257].

A recent advancement in BC/NCNM research is the conversion of BC/NCNM into aerogels. Aerogels can be defined as a highly porous material that can be made of any substance including metals, semiconductors, polymers, biopolymers, etc. [258]. Aerogels are lightweight and have many interesting characteristic features such as high specific surface area, incredibly low heat conductivity, sound propagation, or superior shock absorption [259]. Aerogels have mostly been studied as filter materials, gas storage media (particularly for fuel cells), catalyst carriers, heat, and sound insulators, or even as dust scavengers because of their high porosity and easily accessible interconnected open pore system [260, 261]. Due to their various physicochemical properties cellulose aerogels would gradually replace silica and synthetic polymer-based aerogels. This advance is because it is a naturally occurring pure material comprising glucose units forming a 3D nanoscale structure.

2.6.4 BC/NCNM Aerogel: Preparation, Recent Developments and Applications

Cellulose aerogels have a 3D network structure with very high porosity and are biocompatible because of the high specific surface area [6, 7, 262]. BC/NCNM aerogel can be prepared by two methods, viz. (a) freeze-drying method in which the BC/NCNM hydrogel obtained during growth under static conditions is pulverized and then frozen in a foil to attain the required shape [263] and the (b) sol–gel method in which gelation of the sol occurs in a continuous liquid phase (gel) to form a 3D network structure [264]. During the sol–gel reaction, the substance changes from the liquid sol phase to the solid gel phase. Currently, this method is used to create all aerogels [265, 266]. For preparing BC/NCNM sol-gels, the cellulose is dissolved in a solution, and it turns into a gel either by clumping together its polymers or through a separation process when a different substance is added. When the solution is treated to remove the solvent, causing a shift in its composition, the BC/NCNM molecules lose their solvent and form bonds with each other again. For example, when cellulose is dissolved in N-methyl morpholine-N-oxide (NMMO) and then mixed with water, it forms a gel as the solution separates into regions rich in cellulose and regions with less cellulose [267, 268]. In general, by adding chemical crosslinking agents such as epichlorohydrin (ECH) and N, N'- methylene bis acrylamide (MBA) to the liquid sol or changing physical conditions (temperature, pH, ultrasonic treatment, etc.), colloidal particle aggregation can be induced to form a 3D interconnected network structure. This converts the material into a solid gel [269, 270].

Wu et al. (2003) reported the production of ultralight, flexible, and fire-resistant carbon nanofiber (CNF) aerogels on a large scale from BC/NCNM pellicles. It was also reported that the CNF can be used as a good absorbent for absorbing a wide range of organic solvents and oils. Apart from absorption and fire resistance, electric conductivity was observed in BC/NCNM-CNF aerogels, and it was also found to be sensitive to compressive strain, which makes it a good pressure-sensing material [271]. The synthesis of nanocomposites using BC/NCNM aerogels was reported. Hosseini et al. (2018) prepared BC/NCNM-rGO composite (bacterial cellulose reduced graphene oxide), which can be used as an electrical conductor to develop strain sensors [272]. Mousavi (2021) prepared polyaniline bacterial cellulose aerogel nanocomposites (BC/NCNM/PANI) [273]. BC/NCNM aerogels can also be modified with carbon nanotubes and carboxylic carbon nanotubes (BCA/CNT and BCA/CCNT) which can be used in oil absorption and remediation of the environment [274]. The principal uses of BC/NCNM aerogel are adsorption and separation, insulation, biomedical applications, preparing carbon aerogels, and as carriers of metal NPs or metal oxides.

Aerogels prepared from BC are mainly used for adsorption of oil which occurs due to oil spill incidents and cause damage to aquatic life. Materials like polypropylene (PP), Zeolite, and activated carbon were used, but due to one-time usage, insufficient ability of absorption, and less biodegradability, BC/NCNM aerogels were preferred [275, 276]. BC/NCNM aerogels can also be coated with titanium dioxide on the surface using the chemical vapor deposition method and used as absorption matrices [277].

Nanocomposites of metal NPs and BC/NCNM aerogels were prepared and used for loading and targeted delivery of molecules. Au-NPs were used to develop a catalyst that can be used for catalyzing styrene epoxidation (96%, 1 h) [278]. The work done by Hiruvengadam and Vitta suggests the incorporation of Ni-BC/NCNM aerogel to prepare nanocomposites that are thermally sensitive and magnetic [279].

BC/NCNM aerogels find applications in drug transport, cell culture, biosensors, and various biomedical fields. 3D cell culture is a crucial technique in cell biology, cell therapy, and drug development. Antibacterial materials can be fixed to the surface of cellulose aerogels to induce bacterial adsorption or growth inhibition. Henschen et al. adsorbed polyvinyl amine and polyacrylic acid onto the surface of cellulose aerogels using a layer-by-layer self-assembly technique. In a bacterial suspension, this substance was able to absorb almost 99.9 percent of the live bacteria [280].

Edward et al. suggested the use of polypeptide chemistry to implant tripeptide molecules onto a nanocellulose aerogel. The inclusion of fluorenyl methoxycarbonyl peptide resulted in a polypeptide–nanocellulose aerogel (PepNA) which could be used to track the amount of protease in chronic wounds because of its excellent detection sensitivity for human neutrophil elastase [281]. Carbon aerogels are made by carbonization of BC/NCNM aerogel in nitrogen or argon by heating them to 500 to 1000 C at a fixed rate [282, 283]. The aerogel derived from BC/NCNM proved to have good hydrophobicity flame retardancy good conductivity and adsorption [284–286].

2.7 Conclusion

The chapter overviews BC/NCNM production using agrowastes and their applications. BC/NCNM is an extracellular polysaccharide produced fermentatively from various agro wastes by specific bacterial genera. With its excellent characteristics, such as biocompatibility, mechanical strength, water-holding capacity, degree of polymerization, microporosity, elasticity, and moldability, BC/NCNM has several applications. BC/NCNM is a natural nanoscale material, and this chapter focuses on optimizing the production of bacterial cellulose from different types of agrowaste and different types of reactors. Various modifications of BC/NCNM post-synthesis are performed to improve the quantity and quality of the product. BC/NCNM can be applied for various uses, including biomedical devices, the food industry, the cosmetic industry, and environmental protection. Bacterial cellulose is easily convertible into carbon nanotubes and aerogels, which can have several applications such as biosensing, manufacture of reinforced materials, materials with thermal resistance characteristics, etc. With the advent of Nanotechnology, this versatile natural nanoscale polymer will provide a sustainable alternative to many more products.

References

1. Hemathilake DM, Gunathilake DM (2022) Agricultural productivity and food supply to meet increased demands. In: Future foods, Academic Press, pp 539–553
2. Agriculture Organization of the United Nations (2000) Fisheries department the state of world fisheries and aquaculture, 2000. Food & Agriculture Org
3. Akramovna ON (2021) Scientific basis for increasing the efficiency of cultivation of crops on the lands of farms and the population. Academicia: Int Multidisciplin Res J 11(2):1297–1304
4. Lucas JA (2011) Advances in plant disease and pest management. J Agricult Sci 149(S1):91–114
5. Ghag SB, Alok A, Rajam MV, Penna S (2023) Designing climate-resilient crops for sustainable agriculture: a silent approach. J Plant Growth Regul 42(10):6503–6522
6. Agamuthu P, Khidzir KM, Hamid FS (2009) Drivers of sustainable waste management in Asia. 27(7):625–633
7. Raut NA, Kokare DM, Bhanvase BA, Randive KR, Dhoble SJ (2023) 360-degree waste management, volume 1: fundamentals, agricultural and domestic waste, and remediation. Elsevier
8. Dwivedi S, Tanveer A, Yadav S, Anand G, Yadav D (2022) Agro-wastes for cost effective production of industrially important microbial enzymes: an overview. Microb Biotechnol Role Ecolo Sustain Res 26:435–460
9. Ogbu CC, Okechukwu SN Agro-industrial waste management: the circular and bioeconomic perspective
10. Iqbal N, Agrawal A, Dubey S, Kumar J (2020) Role of decomposers in agricultural waste management. In: Biotechnological applications of biomass IntechOpen
11. Clancy-Hepturn M (1998) Agricultural residues: a promising alternative to virgin wood fiber. Issue Resour Conserv Brief Ser
12. Biswas A, Saha BC, Lawton JW, Shogren RL, Willett JL (2006) Process for obtaining cellulose acetate from agricultural by-products. Carbohyd Polym 64(1):134–137
13. Ibrahim S, Riahi O, Said SM, Sabri MFM, Rozali S (2019) Biopolymers from crop plants. Reference module in materials science and materials engineering
14. Ekiert M, Mlyniec A, Uhl T (2015) The influence of degradation on the viscosity and molecular mass of poly (lactide acid) biopolymer. Diagnostyka 16(4):63–70
15. Reddy N, Yang Y (2011) Potential of plant proteins for medical applications. Trends Biotechnol 29(10):490–498
16. Das A, Ringu T, Ghosh S, Pramanik N (2023) A comprehensive review on recent advances in preparation, physicochemical characterization, and bioengineering applications of biopolymers. Polym Bull 80(7):7247–7312
17. Ghosh S, Lahiri D, Nag M, Dey A, Sarkar T, Pathak SK, Atan Edinur H, Pati S, Ray RR (2021) Bacterial biopolymer: its role in pathogenesis to effective biomaterials. Polymers 13(8):1242
18. Moradali MF, Rehm BH (2020) Bacterial biopolymers: from pathogenesis to advanced materials. Nat Rev Microbiol 18(4):195–210
19. Rehm BH (2010) Bacterial polymers: biosynthesis, modifications and applications. Nat Rev Microbiol 8(8):578–592
20. Roy A, Bulut O, Some S, Mandal AK, Yilmaz MD (2019) Green synthesis of silver nanoparticles: biomolecule-nanoparticle organizations targeting antimicrobial activity. RSC Adv 9(5):2673–2702
21. Khan R, Shah MD, Shah L, Lee PC, Khan I (2022) Bacterial polysaccharides—a big source for prebiotics and therapeutics. Front Nutr 9:1031935
22. Klemm D, Heublein B, Fink HP, Bohn A (2005) Cellulose: fascinating biopolymer and sustainable raw material. Angew Chem Int Ed 44(22):3358–3393
23. Chen P, Cho SY, Jin HJ (2010) Modification and applications of bacterial celluloses in polymer science. Macromol Res 18:309–320
24. Brown AJ (1886) XLIII—On an acetic ferment which forms cellulose. J Chem Soc Trans 49:432–439

25. Browne CA Jr (1906) The fermentation of sugar-cane products. J Am Chem Soc 28(4):453–469
26. Hestrin S, Aschner M, Mager J (1947) Synthesis of cellulose by resting cells of Acetobacter xylinum. Nature 159(4028):64–65
27. Kim YJ, Kim JN, Wee YJ, Park DH, Ryu HW (2007) Bacterial cellulose production by Gluconacetobacter sp. PKY5 in a rotary biofilm contactor. Appl Biochemistr Biotechnol 137:529–537
28. Park JK, Shehzad O, Khan T, Khan S (2008) Production of bacterial cellulose by Gluconacetobacter hasenii using a new bioreactor equipped with centrifugal impeller. J Biotechnol 136:S485
29. Jonas R, Farah LF (1998) Production and application of microbial cellulose. Polym Degrad Stab 59(1–3):101–106
30. Lisdiyanti P, Navarro RR, Uchimura T, Komagata K (2006) Reclassification of Gluconacetobacter hansenii strains and proposals of Gluconacetobacter saccharivorans sp. nov. and Gluconacetobacter nataicola sp. nov. Int J Systemat Evolution Microbiol 56(9):2101–2111
31. @.Cannon RE, Anderson SM. Biogenesis of bacterial cellulose. Critical reviews in microbiology. 1991;17(6):435–47.
32. Vroman I, Tighzert L (2009) Biodegradable polymers. Materials 2(2):307–344
33. Jozala AF, de Lencastre-Novaes LC, Lopes AM, de Carvalho S-E, Mazzola PG, Pessoa-Jr A, Grotto D, Gerenutti M, Chaud MV (2016) Bacterial nanocellulose production and application: a 10-year overview. Appl Microbiol Biotechnol 100:2063–2072
34. Lahiri D, Nag M, Dutta B, Dey A, Sarkar T, Pati S, Edinur HA, Abdul Kari Z, Mohd Noor NH, Ray RR (2021) Bacterial cellulose: Production, characterization, and application as antimicrobial agent. Int J Mol Sci 22(23):12984
35. Reiniati I, Hrymak AN, Margaritis A (2017) Recent developments in the production and applications of bacterial cellulose fibers and nanocrystals. Crit Rev Biotechnol 37(4):510–524
36. Klemm D, Schumann D, Udhardt U, Marsch S (2001) Bacterial synthesized cellulose—artificial blood vessels for microsurgery. Prog Polym Sci 26(9):1561–1603
37. Ul-Islam M, Khattak WA, Kang M, Kim SM, Khan T, Park JK (2013) Effect of post-synthetic processing conditions on structural variations and applications of bacterial cellulose. Cellulose 20:253–263
38. Ullah MW, Ul-Islam M, Khan S, Kim Y, Park JK (2016) Structural and physico-mechanical characterization of bio-cellulose produced by a cell-free system. Carbohyd Polym 136:908–916
39. Ullah MW, Manan S, Kiprono SJ, Ul-Islam M, Yang G (2019) Synthesis, structure, and properties of bacterial cellulose. Nanocell Fundament Adv Mater 6:81–113
40. Ul-Islam M, Khan T, Khattak WA, Park JK (2013) Bacterial cellulose-MMTs nanoreinforced composite films: novel wound dressing material with antibacterial properties. Cellulose 20:589–596
41. Dahman Y (2009) Nanostructured biomaterials and biocomposites from bacterial cellulose nanofibers. J Nanosci Nanotechnol 9(9):5105–5122
42. Ul-Islam M, Khan T, Park JK (2012) Water holding and release properties of bacterial cellulose obtained by in situ and ex situ modification. Carbohyd Polym 88(2):596–603
43. Shezad O, Khan S, Khan T, Park JK (2010) Physicochemical and mechanical characterization of bacterial cellulose produced with an excellent productivity in static conditions using a simple fed-batch cultivation strategy. Carbohyd Polym 82(1):173–180
44. Ul-Islam M, Ha JH, Khan T, Park JK (2013) Effects of glucuronic acid oligomers on the production, structure and properties of bacterial cellulose. Carbohyd Polym 92(1):360–366
45. Schrecker ST, Gostomski PA (2005) Determining the water holding capacity of microbial cellulose. Biotech Lett 27:1435–1438
46. Wu JM, Liu RH (2013) Cost-effective production of bacterial cellulose in static cultures using distillery wastewater. J Biosci Bioeng 115(3):284–290
47. Cao Y, Lu S, Yang Y (2018) Production of bacterial cellulose from byproduct of citrus juice processing (citrus pulp) by Gluconacetobacter hansenii. Cellulose 25:6977–6988

48. Clasen C, Sultanova B, Wilhelms T, Heisig P, Kulicke WM (2006) Effects of different drying processes on the material properties of bacterial cellulose membranes. In: Macromolecular symposia, vol 244, No. 1. Weinheim, WILEY-VCH Verlag, pp 48–58
49. Illa MP, Sharma CS, Khandelwal M (2019) Tuning the physiochemical properties of bacterial cellulose: effect of drying conditions. J Mater Sci 54(18):12024–12035
50. Singhsa P, Narain R, Manuspiya H (2018) Physical structure variations of bacterial cellulose produced by different Komagataeibacter xylinus strains and carbon sources in static and agitated conditions. Cellulose. 25(3):1571–1581
51. Jung HI, Jeong JH, Lee OM, Park GT, Kim KK, Park HC, Lee SM, Kim YG, Son HJ (2010) Influence of glycerol on production and structural–physical properties of cellulose from Acetobacter sp. V6 cultured in shake flasks. Bioresource Technol 101(10):3602–3608
52. Kim SY, Kim JN, Wee YJ, Park DH, Ryu HW (2006) Production of bacterial cellulose by Gluconacetobacter sp. RKY5 isolated from persimmon vinegar. In: Twenty-seventh symposium on biotechnology for fuels and chemicals 2006. Humana Press, pp 705–715
53. Castro C, Zuluaga R, Putaux JL, Caro G, Mondragon I, Gañán P (2011) Structural characterization of bacterial cellulose produced by Gluconacetobacter swingsii sp. from Colombian agroindustrial wastes. Carbohydrate Polym 84(1):96–102
54. Kumaresapillai N, Basha RA, Sathish R (2011) Production and evaluation of chitosan from Aspergillus niger MTCC strains. Iranian J Pharmaceutic Res IJPR 10(3):553
55. Hay ID, Rehman ZU, Moradali MF, Wang Y, Rehm BH (2013) Microbial alginate production, modification and its applications. Microbial Biotechnol 6(6):637–650
56. Liu L, Liu Y, Li J, Du G, Chen J (2011) Microbial production of hyaluronic acid: current state, challenges, and perspectives. Microbial Cell Fact 10(1):1–9
57. Holzapfel WH, Wood BJ (2014) Lactic acid bacteria: biodiversity and taxonomy. John Wiley & Sons
58. Elsacker E, Vandelook S, Damsin B, Van Wylick A, Peeters E, De Laet L (2021) Mechanical characteristics of bacterial cellulose-reinforced mycelium composite materials. Fungal Biol Biotechnol 8(1):1–4
59. Gadgey KK, Sharma GS (2020) Investigation of mechanical properties of chitosan based films prepared from narmada riverside crab shells. Int J Mech Eng Tech 11(6):21–28
60. Rhim JW (2004) Physical and mechanical properties of water resistant sodium alginate films. LWT-Food Sci Technol 37(3):323–330
61. Shi Z, Zhang Y, Phillips GO, Yang G (2014) Utilization of bacterial cellulose in food. Food Hydrocoll 35:539–545
62. Le Tien C, Lacroix M, Ispas-Szabo P, Mateescu MA (2003) N-acylated chitosan: hydrophobic matrices for controlled drug release. J Controll Release 93(1):1–3
63. Draget KI Alginates: Fundamental properties and food applications
64. Oe M, Ito S, Kaneda H, Takeda Y, Matsuoka R, Takara T (2023) Safety evaluation of the overintake of sodium hyaluronate. Oyo Yakuri (Pharmacometrics)(0300–8533) 105
65. De Bellis P, Ferrara M, Bavaro AR, Linsalata V, Di Biase M, Musio B, Gallo V, Mulè G, Valerio F (2022) Characterization of dextran produced by the food-related strain weissella cibaria c43–11 and of the relevant dextransucrase gene. Foods 11(18):2819
66. Lima GD, Sierakowski MR, Faria-Tischer P, Tischer CA Characterization of the bacterial cellulose dissolved on dimethylacetamide/lithium chloride
67. Aranaz I, Alcántara AR, Civera MC, Arias C, Elorza B, Heras Caballero A, Acosta N (2021) Chitosan: an overview of its properties and applications. Polymers 13(19):3256
68. Lee KY, Mooney DJ (2012) Alginate: properties and biomedical applications. Progr Polym Sci 37(1):106–126
69. Snetkov P, Zakharova K, Morozkina S, Olekhnovich R, Uspenskaya M (2020) Hyaluronic acid: The influence of molecular weight on structural, physical, physico-chemical, and degradable properties of biopolymer. Polymers 12(8):1800
70. Güner A (2004) The algorithmic calculations of solubility parameter for the determination of interactions in dextran/certain polar solvent systems. Euro Polym J 40(7):1587–1594

71. McNamara JT, Morgan JL, Zimmer J (2015) A molecular description of cellulose biosynthesis. Ann Rev Biochemistr 84:895–921
72. Abka-Khajouei R, Tounsi L, Shahabi N, Patel AK, Abdelkafi S, Michaud P (2022) Structures, properties and applications of alginates. Mar Drugs 20(6):364
73. Weissmann B, Meyer K, Sampson P, Linker A (1954) Isolation of oligosaccharides enzymatically produced from hyaluronic acid. J Biologic Chemistr 208(1):417–429
74. Sajna KV, Gottumukkala LD, Sukumaran RK, Pandey A (2015) White biotechnology in cosmetics. In: Industrial biorefineries & white biotechnology. Elsevier, pp 607–652
75. Cowie JM, Arrighi V (2007) Polymers: chemistry and physics of modern materials. CRC press
76. Tahara N, Tabuchi M, Watanabe K, Yano H, Morinaga Y, Yoshinaga F (1997) Degree of polymerization of cellulose from Acetobacter xylinum BPR2001 decreased by cellulase produced by the strain. Biosci Biotechnol Biochem 61(11):1862–1865
77. Yamanaka S, Watanabe K, Kitamura N, Iguchi M, Mitsuhashi S, Nishi Y, Uryu M (1989) The structure and mechanical properties of sheets prepared from bacterial cellulose. J Mater Sci 24:3141–3145
78. Nishi Y, Uryu M, Yamanaka S, Watanabe K, Kitamura N, Iguchi M, Mitsuhashi S (1990) The structure and mechanical properties of sheets prepared from bacterial cellulose: part 2 Improvement of the mechanical properties of sheets and their applicability to diaphragms of electroacoustic transducers. J Mater Sci 25:2997–3001
79. He W, Wu J, Xu J, Mosselhy DA, Zheng Y, Yang S (2021) Bacterial cellulose: functional modification and wound healing applications. Adv Wound Care 10(11):623–640
80. Potivara K, Phisalaphong M (2019) Development and characterization of bacterial cellulose reinforced with natural rubber. Materials 12(14):2323
81. Tang W, Jia S, Jia Y, Yang H (2010) The influence of fermentation conditions and post-treatment methods on porosity of bacterial cellulose membrane. World J Microbiol Biotechnol 26:125–131
82. Gao C, Wan Y, Lei X, Qu J, Yan T, Dai K (2011) Polylysine coated bacterial cellulose nanofibers as novel templates for bone-like apatite deposition. Cellulose 18:1555–1561
83. Nge TT, Nogi M, Yano H, Sugiyama J (2010) Microstructure and mechanical properties of bacterial cellulose/chitosan porous scaffold. Cellulose 17:349–363
84. Yang C, Gao C, Wan Y, Tang T, Zhang S, Dai K (2011) Preparation and characterization of three-dimensional nanostructured macroporous bacterial cellulose/agarose scaffold for tissue engineering. J Porous Mater 18:545–552
85. Yin N, Chen S, Li Z, Ouyang Y, Hu W, Tang L, Zhang W, Zhou B, Yang J, Xu Q, Wang H (2012) Porous bacterial cellulose prepared by a facile surfactant-assisted foaming method in azodicarbonamide-NaOH aqueous solution. Mater Lett 81:131–134
86. Bäckdahl H, Esguerra M, Delbro D, Risberg B, Gatenholm P (2008) Engineering microporosity in bacterial cellulose scaffolds. J Tissue Eng Regen Med 2(6):320–330
87. Bodin A, Bharadwaj S, Wu S, Gatenholm P, Atala A, Zhang Y (2010) Tissue-engineered conduit using urine-derived stem cells seeded bacterial cellulose polymer in urinary reconstruction and diversion. Biomaterials 31(34):8889–8901
88. Andersson J, Stenhamre H, Bäckdahl H, Gatenholm P (2010) Behavior of human chondrocytes in engineered porous bacterial cellulose scaffolds. J Biomed Mater Res Part A. 94(4):1124–1132
89. Zaborowska M, Bodin A, Bäckdahl H, Popp J, Goldstein A, Gatenholm P (2010) Microporous bacterial cellulose as a potential scaffold for bone regeneration. Acta Biomater 6(7):2540–2547
90. Haimer E, Wendland M, Schlufter K, Frankenfeld K, Miethe P, Potthast A, Rosenau T, Liebner F (2010) Loading of bacterial cellulose aerogels with bioactive compounds by antisolvent precipitation with supercritical carbon dioxide. In: Macromolecular symposia, vol 294, no 2. WILEY-VCH Verlag, Weinheim, pp 64–74
91. Salinas DC, Sordelli A, Martínez LA, Villoldo G, Bernal C, Perez MS, Cerrutti P, Foresti ML (2021) Production of bacterial cellulose tubes for biomedical applications: analysis of the effect of fermentation time on selected properties. Int J Biol Macromol 189:1

92. Pepicelli M, Binelli MR, Studart AR, Ruhs PA, Fischer P (2021) Self-grown bacterial cellulose capsules made through emulsion templating. ACS Biomater Sci Eng 7(7):3221–3228
93. Suratago T, Taokaew S, Kanjanamosit N, Kanjanaprapakul K, Burapatana V, Phisalaphong M (2015) Development of bacterial cellulose/alginate nanocomposite membrane for separation of ethanol–water mixtures. J Ind Eng Chem 32:305–312
94. Suryanto H, Muhajir M, Sutrisno TA, Mudjiono, Zakia N, Yanuhar U (2019) The mechanical strength and morphology of bacterial cellulose films: the effect of NaOH concentration. In: IOP conference series: materials science and engineering, vol 515, p 012053. IOP Publishing
95. Santmarti A, Liu HW, Herrera N, Lee KY (2020) Anomalous tensile response of bacterial cellulose nanopaper at intermediate strain rates. Sci Rep 10(1):15260
96. Williams DF (1999) Dictionary of biomaterials. Liverpool University Press, Liverpool, p 42
97. Cai Z, Kim J (2010) Bacterial cellulose/poly (ethylene glycol) composite: characterization and first evaluation of biocompatibility. Cellulose 17(1):83–91
98. Kim J, Cai Z, Lee HS, Choi GS, Lee DH, Jo C (2011) Preparation and characterization of a bacterial cellulose/chitosan composite for potential biomedical application. J Polym Res 18:739–744
99. Zhijiang C, Guang Y (2011) Bacterial cellulose/collagen composite: Characterization and first evaluation of cytocompatibility. J Appl Polym Sci 120(5):2938–2944
100. Mormino R, Bungay H (2003) Composites of bacterial cellulose and paper made with a rotating disk bioreactor. Appl Microbiol Biotechnol 62:503–506
101. Czaja W, Krystynowicz A, Bielecki S, Brown RM Jr (2006) Microbial cellulose—the natural power to heal wounds. Biomaterials 27(2):145–151
102. Fontana JD, De Souza AM, Fontana CK, Torriani IL, Moreschi JC, Gallotti BJ, De Souza SJ, Narcisco GP, Bichara JA, Farah LF (1990) Acetobacter cellulose pellicle as a temporary skin substitute. Appl Biochem Biotechnol 24:253–264
103. Legeza VI, Galenko-Yaroshevskii VP, Zinov'ev EV, Paramonov BA, Kreichman GS, Turkovskii II, Gumenyuk ES, Karnovich AG, Khripunov AK (2004) Effects of new wound dressings on healing of thermal burns of the skin in acute radiation disease. Bull Experiment Biol Med 138:311–315
104. Levinson DJ, Glonek T, inventors; Levinson Dennis J, assignee (2010) Microbial cellulose contact lens. United States patent US 7,832,857
105. Shi Z, Zang S, Jiang F, Huang L, Lu D, Ma Y, Yang G (2012) In situ nano-assembly of bacterial cellulose–polyaniline composites. RSC Adv 2(3):1040–1046
106. Trovatti E, Silva NH, Duarte IF, Rosado CF, Almeida IF, Costa P, Freire CS, Silvestre AJ, Neto CP (2011) Biocellulose membranes as supports for dermal release of lidocaine. Biomacromol 12(11):4162–4168
107. Park HO, Bang YB, Joung HJ, Kim BC, Kim HR, inventors; Bioneer Corp, assignee (2004) Lactobacillus KCTC 0774BP and acetobacter KCTC 0773BP for treatment or prevention of obesity and diabetes mellitus. United States patent US 6,808,703
108. Schramm M, Hestrin S (1954) Factors affecting production of cellulose at the air/liquid interface of a culture of Acetobacter xylinum. Microbiology 11(1):123–129
109. Mohammadkazemi F, Azin M, Ashori A (2015) Production of bacterial cellulose using different carbon sources and culture media. Carbohyd Polym 117:518–523
110. Supian NN, Zakaria J, Amin KN, Mohamad S, Mohamad SF (2021) Effect of fermentation period on bacterial cellulose production from oil palm frond (OPF) juice. In: IOP conference series: materials science and engineering, vol 1092, no 1. IOP Publishing, p 012048
111. Fathiyah SS, Shahril M, Junaidi Z (2021) Oil palm frond juice and coconut water as alternative fermentation substrate for bacterial cellulose production. In: IOP conference series: materials science and engineering, vol 1092, no 1. IOP Publishing, p 012055
112. Kumbhar JV, Rajwade JM, Paknikar KM (2015) Fruit peels support higher yield and superior quality bacterial cellulose production. Appl Microbiol Biotechnol 99:6677–6691
113. Sutthiphatkul T, Suwanposri A, Ochaikul D (2020) Optimization of bacterial cellulose production from wastewater of noodle processing by Komagataeibacter sp. PAP1 and bio-cellulose paper production. Walailak J Sci Technol (WJST) 17(11):1241–1251

114. Lathifah AN, Nuryana I (2021) Bacterial cellulose production in the overripe guava juice by acetobacter xylinum as a solution to reduce organic waste. J Kimia Terapan Indonesia 23(2):41–48
115. He F, Yang H, Zeng L, Hu H, Hu C (2020) Production and characterization of bacterial cellulose obtained by Gluconacetobacter xylinus utilizing the by-products from Baijiu production. Bioprocess Biosyst Eng 43:927–936
116. Kumbhar JV, Jadhav SH, Bodas DS, Barhanpurkar-Naik A, Wani MR, Paknikar KM, Rajwade JM (2017) In vitro and in vivo studies of a novel bacterial cellulose-based acellular bilayer nanocomposite scaffold for the repair of osteochondral defects. Int J Nanomed 2017:6437–6459
117. Lotfiman S, Awang Biak DR, Ti TB, Kamarudin S, Nikbin S (2018) Influence of date syrup as a carbon source on bacterial cellulose production by Acetobacter xylinum 0416. Adv Polym Technol 37(4):1085–1091
118. Ogrizek L, Lamovšek J, Čuš F, Leskovšek M, Gorjanc M (2021) Properties of bacterial cellulose produced using white and red grape bagasse as a nutrient source. Processes 9(7):1088
119. Mohamad S, Abdullah LC, Jamari SS, Osman Al Edrus SS, Aung MM, Sy Mohamad SF (2022) Production and characterization of bacterial cellulose nanofiber by Acetobacter Xylinum 0416 using only oil palm frond juice as fermentation medium. J Nat Fibers 19(17):16005–16016.
120. Raiszadeh-Jahromi Y, Rezazadeh-Bari M, Almasi H, Amiri S (2020) Optimization of bacterial cellulose production by Komagataeibacter xylinus PTCC 1734 in a low-cost medium using optimal combined design. J Food Sci Technol 57:2524–2533
121. Cavka A, Guo X, Tang SJ, Winestrand S, Jönsson LJ, Hong F (2013) Production of bacterial cellulose and enzyme from waste fiber sludge. Biotechnol Biofuels 6:1
122. Chen L, Hong F, Yang XX, Han SF (2013) Biotransformation of wheat straw to bacterial cellulose and its mechanism. Biores Technol 135:464–468
123. Costa AF, Almeida FC, Vinhas GM, Sarubbo LA (2017) Production of bacterial cellulose by Gluconacetobacter hansenii using corn steep liquor as nutrient sources. Front Microbiol 8:2027
124. Huang C, Yang XY, Xiong L, Guo HJ, Luo J, Wang B, Zhang HR, Lin XQ, Chen XD (2015) Utilization of corncob acid hydrolysate for bacterial cellulose production by Gluconacetobacter xylinus. Appl Biochem Biotechnol 175:1678–1688
125. Saleh AK, El-Gendi H, El-Fakharany EM, Owda ME, Awad MA, Kamoun EA (2022) Exploitation of cantaloupe peels for bacterial cellulose production and functionalization with green synthesized Copper oxide nanoparticles for diverse biological applications. Sci Rep 12(1):19241
126. Abdelraof M, Hasanin MS, El-Saied H (2019) Ecofriendly green conversion of potato peel wastes to high productivity bacterial cellulose. Carbohyd Polym 211:75–83
127. Costa AF, Galdino Jr CJ, Meira HM, Júlia DP, Siva ID, Gomes EA, Sarubbo LA (2020) Production of paper using bacterial cellulose and residue from the sugar and alcohol industry. CET J Chem Eng Trans 79.
128. Ye J, Zheng S, Zhang Z, Yang F, Ma K, Feng Y, Zheng J, Mao D, Yang X (2019) Bacterial cellulose production by Acetobacter xylinum ATCC 23767 using tobacco waste extract as culture medium. Bioresour Technol 274:518–524
129. Li Z, Wang L, Hua J, Jia S, Zhang J, Liu H (2015) Production of nano bacterial cellulose from waste water of candied jujube-processing industry using Acetobacter xylinum. Carbohyd Polym 120:115–119
130. Tsouko E, Maina S, Ladakis D, Kookos IK, Koutinas A (2020) Integrated biorefinery development for the extraction of value-added components and bacterial cellulose production from orange peel waste streams. Renew Energy 160:944–954
131. El-Gendi H, Salama A, El-Fakharany EM, Saleh AK (2023) Optimization of bacterial cellulose production from prickly pear peels and its ex situ impregnation with fruit byproducts for antimicrobial and strawberry packaging applications. Carbohyd Polym 302:120383

132. Kumar V, Sharma DK, Bansal V, Mehta D, Sangwan RS, Yadav SK (2019) Efficient and economic process for the production of bacterial cellulose from isolated strain of Acetobacter pasteurianus of RSV-4 bacterium. Biores Technol 275:430–433
133. Zhou LL, Sun DP, Hu LY, Li YW, Yang JZ (2007) Effect of addition of sodium alginate on bacterial cellulose production by Acetobacter xylinum. J Ind Microbiol Biotechnol 34(7):483
134. Joseph G, Rowe GE, Margaritis A, Wan W (2003) Effects of polyacrylamide-co-acrylic acid on cellulose production by Acetobacter xylinum. J Chem Technol Biotechnol: Int Res Process Environ Clean Technol 78(9):964–970
135. Bae S, Shoda M (2004) Bacterial cellulose production by fed-batch fermentation in molasses medium. Biotechnol Prog 20(5):1366–1371
136. Chao Y, Mitarai M, Sugano Y, Shoda M (2001) Effect of addition of water-soluble polysaccharides on bacterial cellulose production in a 50-L airlift reactor. Biotechnol Prog 17(4):781–785
137. Naritomi T, Kouda T, Yano H, Yoshinaga F (1998) Effect of lactate on bacterial cellulose production from fructose in continuous culture. J Ferment Bioeng 85(1):89–95
138. Gupta RK, Guha P, Srivastav PP (2022) Natural polymers in bio-degradable/edible film: a review on environmental concerns, cold plasma technology and nanotechnology application on food packaging-A recent trends. Food Chemistr Adv 1:100135
139. Jamsheera CP, Pradeep BV (2021) Production of bacterial cellulose from acetobacter species and its applications-a review. J Pure Appl Microbiol 15(2)
140. Watanabe K, Tabuchi M, Morinaga Y, Yoshinaga F (1998) Structural features and properties of bacterial cellulose produced in agitated culture. Cellulose 5:187–200
141. Augimeri RV, Varley AJ, Strap JL (2015) Establishing a role for bacterial cellulose in environmental interactions: lessons learned from diverse biofilm-producing Proteobacteria. Front Microbiol 6:1282
142. Claude I (2022) Bioreactor design for bacterial cellulose production. Diss Worces Polytech Inst
143. Brown JD (1986) Evaluations of self and others: self-enhancement biases in social judgments. Soc Cogn 4(4):353–376
144. Hestrin S, Schramm MJ (1954) Synthesis of cellulose by Acetobacter xylinum. 2. Preparation of freeze-dried cells capable of polymerizing glucose to cellulose. Biochem J 58(2):345
145. Machado RT, Meneguin AB, Sábio RM, Franco DF, Antonio SG, Gutierrez J, Tercjak A, Berretta AA, Ribeiro SJ, Lazarini SC, Lustri WR (2018) Komagataeibacter rhaeticus grown in sugarcane molasses-supplemented culture medium as a strategy for enhancing bacterial cellulose production. Ind Crops Prod 122:637–646
146. Ravichandra Potumarthi SC, Jetty A (2007) Alkaline protease production by submerged fermentation in stirred tank reactor using Bacillus licheniformis NCIM-2042: effect of aeration and agitation regimes. Biochem Eng J 34(2):185–192
147. Zhou Y, Han LR, He HW, Sang B, Yu DL, Feng JT, Zhang X (2018) Effects of agitation, aeration and temperature on production of a novel glycoprotein GP-1 by Streptomyces kanasenisi ZX01 and scale-up based on volumetric oxygen transfer coefficient. Molecules 23(1):125
148. Sani A, Dahman Y (2010) Improvements in the production of bacterial synthesized biocellulose nanofibres using different culture methods. J Chem Technol Biotechnol 85(2):151–164
149. Wang J, Tavakoli J, Tang Y (2019) Bacterial cellulose production, properties and applications with different culture methods–a review. Carbohyd Polym 219:63–76
150. Reiniati I, Hrymak AN, Margaritis A (2017) Kinetics of cell growth and crystalline nanocellulose production by Komagataeibacter xylinus. Biochem Eng J 127:21–31
151. Sakairi N, Asano H, Ogawa M, Nishi N, Tokura S (1998) A method for direct harvest of bacterial cellulose filaments during continuous cultivation of Acetobacter xylinum. Carbohyd Polym 35(3–4):233–237
152. Chao Y, Ishida T, Sugano Y, Shoda M (2000) Bacterial cellulose production by Acetobacter xylinum in a 50-L internal-loop airlift reactor. Biotechnol Bioeng 68(3):345–352
153. Shoda M, Sugano Y (2005) Recent advances in bacterial cellulose production. Biotechnol Bioprocess Eng 10:1–8

154. Krystynowicz A, Czaja W, Wiktorowska-Jezierska A, Gonçalves-Miśkiewicz M, Turkiewicz M, Bielecki S (2002) Factors affecting the yield and properties of bacterial cellulose. J Ind Microbiol Biotechnol 29(4):189–195
155. Singhania RR, Patel AK, Tsai ML, Chen CW, Di Dong C (2021) Genetic modification for enhancing bacterial cellulose production and its applications. Bioengineered 12(1):6793–6807
156. Bielecki S, Krystynowicz A, Turkiewicz M, Kalinowska H (2005) Bacterial cellulose. In: Polysaccharides and polyamides in the food industry: properties, production, and patents, pp 31–84
157. Omadjela O, Narahari A, Strumillo J, Mélida H, Mazur O, Bulone V, Zimmer J (2013) BcsA and BcsB form the catalytically active core of bacterial cellulose synthase sufficient for in vitro cellulose synthesis. Proc Natl Acad Sci 110(44):17856–17861
158. Anguluri K, La China S, Brugnoli M, Cassanelli S, Gullo M (2022) Better under stress: improving bacterial cellulose production by Komagataeibacter xylinus K2G30 (UMCC 2756) using adaptive laboratory evolution. Front Microbiol 13:994097
159. Sharma P, Sharma R, Ahuja S, Yadav A, Arora S, Aggarwal NK (2023) Enhancement of bacterial cellulose production by ethanol and lactic acid by using Gluconacetobacter kombuchae. Preparat Biochemistr Biotechnol 6:1–9
160. Li Y, Tian C, Tian H, Zhang J, He X, Ping W, Lei H (2012) Improvement of bacterial cellulose production by manipulating the metabolic pathways in which ethanol and sodium citrate involved. Appl Microbiol Biotechnol 96:1479–1487
161. Cielecka I, Ryngajłło M, Maniukiewicz W, Bielecki S (2021) Response surface methodology-based improvement of the yield and differentiation of properties of bacterial cellulose by metabolic enhancers. Int J Biol Macromol 187:584–593
162. Florea M, Hagemann H, Santosa G, Abbott J, Micklem CN, Spencer-Milnes X, de Arroyo GL, Paschou D, Lazenbatt C, Kong D, Chughtai H (2016) Engineering control of bacterial cellulose production using a genetic toolkit and a new cellulose-producing strain. Proc Natl Acad Sci 113(24):E3431–E3440
163. Fang J, Kawano S, Tajima K, Kondo T (2015) In vivo curdlan/cellulose bionanocomposite synthesis by genetically modified Gluconacetobacter xylinus. Biomacromol 16(10):3154–3160
164. Yadav V, Paniliatis BJ, Shi H, Lee K, Cebe P, Kaplan DL (2010) Novel in vivo-degradable cellulose-chitin copolymer from metabolically engineered Gluconacetobacter xylinus. Appl Environ Microbiol 76(18):6257–6265
165. Legeza VI, Galenko-Yaroshevskii VP, Zinov'ev EV, Paramonov BA, Kreichman GS, Turkovskii II, Gumenyuk ES, Karnovich AG, Khripunov AK (2004) Effects of new wound dressings on healing of thermal burns of the skin in acute radiation disease. Bull Experiment Biol Med 138:311–315
166. Hong L, Wang YL, Jia SR, Huang Y, Gao C, Wan YZ (2006) Hydroxyapatite/bacterial cellulose composites synthesized via a biomimetic route. Mater Lett 60(13–14):1710–1713
167. Barud HS, de Araújo Júnior AM, Santos DB, de Assunção RM, Meireles CS, Cerqueira DA, Rodrigues Filho G, Ribeiro CA, Messaddeq Y, Ribeiro SJ (2008) Thermal behavior of cellulose acetate produced from homogeneous acetylation of bacterial cellulose. Thermochim Acta 471(1–2):61–69
168. Pal S, Nisi R, Stoppa M, Licciulli A Silver-functionalized bacterial cellulose as antibacterial membrane for wound-healing applications. ACS Omega 2:3632–3639
169. Chen XY, Low HR, Loi XY, Merel L, Mohd Cairul Iqbal MA (2019) Fabrication and evaluation of bacterial nanocellulose/poly (acrylic acid)/graphene oxide composite hydrogel: Characterizations and biocompatibility studies for wound dressing. J Biomed Mater Res Part B: Appl Biomater 107(6):2140–2151
170. Millon LE, Wan WK (2006) The polyvinyl alcohol–bacterial cellulose system as a new nanocomposite for biomedical applications. J Biomed Mater Res Part B: Appl Biomater Offic J Soc Biomater Japanese Soc Biomater Austral Soc Biomater Korean Soc Biomater 79(2):245–253

171. Butchosa N, Brown C, Larsson PT, Berglund LA, Bulone V, Zhou Q (2013) Nanocomposites of bacterial cellulose nanofibers and chitin nanocrystals: fabrication, characterization and bactericidal activity. Green Chem 15(12):3404–3413
172. Pinto RJ, Daina S, Sadocco P, Neto CP, Trindade T (2013) Antibacterial activity of nanocomposites of copper and cellulose. BioMed Res Int
173. Luan J, Wu J, Zheng Y, Song W, Wang G, Guo J, Ding X (2012) Impregnation of silver sulfadiazine into bacterial cellulose for antimicrobial and biocompatible wound dressing. Biomed Mater 7(6):065006
174. Wanna D, Alam C, Toivola DM, Alam P (2013) Bacterial cellulose–kaolin nanocomposites for application as biomedical wound healing materials. Adv Nat Sci Nanosci Nanotechnol 4(4):045002
175. Xie YY, Hu XH, Zhang YW, Wahid F, Chu LQ, Jia SR, Zhong C (2020) Development and antibacterial activities of bacterial cellulose/graphene oxide-CuO nanocomposite films. Carbohyd Polym 229:115456
176. Ma B, Huang Y, Zhu C, Chen C, Chen X, Fan M, Sun D (2016) Novel Cu@ SiO2/bacterial cellulose nanofibers: preparation and excellent performance in antibacterial activity. Mater Sci Eng, C 62:656–661
177. Qiu Y, Qiu L, Cui J, Wei Q (2016) Bacterial cellulose and bacterial cellulose-vaccarin membranes for wound healing. Mater Sci Eng, C 59:303–309
178. Yang M, Ward J, Choy KL (2020) Nature-inspired bacterial cellulose/methylglyoxal (BC/MGO) nanocomposite for broad-spectrum antimicrobial wound dressing. Macromol Biosci 20(8):2000070
179. Wahid F, Duan YX, Hu XH, Chu LQ, Jia SR, Cui JD, Zhong C (2019) A facile construction of bacterial cellulose/ZnO nanocomposite films and their photocatalytic and antibacterial properties. Int J Biol Macromol 132:692–700
180. Wei B, Yang G, Hong F (2011) Preparation and evaluation of a kind of bacterial cellulose dry films with antibacterial properties. Carbohyd Polym 84(1):533–538
181. Barud HO, Barud HD, Cavicchioli M, do Amaral TS, de Oliveira Junior OB, Santos DM, Petersen AL, Celes F, Borges VM, de Oliveira CI, de Oliveira PF (2015) Preparation and characterization of a bacterial cellulose/silk fibroin sponge scaffold for tissue regeneration. Carbohydr Polym 128:41–51
182. Khalid A, Ullah H, Ul-Islam M, Khan R, Khan S, Ahmad F, Khan T, Wahid F (2017) Bacterial cellulose–TiO2 nanocomposites promote healing and tissue regeneration in burn mice model. RSC Adv 7(75):47662–47668
183. Wu H, Williams GR, Wu J, Wu J, Niu S, Li H, Wang H, Zhu L (2018) Regenerated chitin fibers reinforced with bacterial cellulose nanocrystals as suture biomaterials. Carbohyd Polym 180:304–313
184. Arias SL, Shetty AR, Senpan A, Echeverry-Rendón M, Reece LM, Allain JP (2016) Fabrication of a functionalized magnetic bacterial nanocellulose with iron oxide nanoparticles. JoVE (J Visualiz Experim) 111:e52951
185. Yang G, Xie J, Deng Y, Bian Y, Hong F (2012) Hydrothermal synthesis of bacterial cellulose/AgNPs composite: a "green" route for antibacterial application. Carbohyd Polym 87(4):2482–2487
186. Almeida IF, Pereira T, Silva NH, Gomes FP, Silvestre AJ, Freire CS, Lobo JS, Costa PC (2014) Bacterial cellulose membranes as drug delivery systems: An in vivo skin compatibility study. Eur J Pharm Biopharm 86(3):332–336
187. Żywicka A, Fijałkowski K, Junka AF, Grzesiak J, El Fray M (2018) Modification of bacterial cellulose with quaternary ammonium compounds based on fatty acids and amino acids and the effect on antimicrobial activity. Biomacromol 19(5):1528–1538
188. Fan X, Zhang T, Zhao Z, Ren H, Zhang Q, Yan Y, Lv G (2013) Preparation and characterization of bacterial cellulose microfiber/goat bone apatite composites for bone repair. J Appl Polym Sci 129(2):595–603
189. Li J, Wan Y, Li L, Liang H, Wang J (2009) Preparation and characterization of 2, 3-dialdehyde bacterial cellulose for potential biodegradable tissue engineering scaffolds. Mater Sci Eng, C 29(5):1635–1642

190. Torgbo S, Sukyai P (2019) Fabrication of microporous bacterial cellulose embedded with magnetite and hydroxyapatite nanocomposite scaffold for bone tissue engineering. Mater Chem Phys 237:121868
191. Wang HY, Wei RH, Zhao SZ (2013) Evaluation of corneal cell growth on tissue engineering materials as artificial cornea scaffolds. Int J Ophthalmol 6(6):873
192. Goncalves S, Padrao J, Rodrigues IP, Silva JP, Sencadas V, Lanceros-Mendez S, Girao H, Dourado F, Rodrigues LR (2015) Bacterial cellulose as a support for the growth of retinal pigment epithelium. Biomacromol 16(4):1341–1351
193. Müller A, Ni Z, Hessler N, Wesarg F, Müller FA, Kralisch D, Fischer D (2013) The biopolymer bacterial nanocellulose as drug delivery system: investigation of drug loading and release using the model protein albumin. J Pharm Sci 102(2):579–592
194. Xu C, Ma X, Chen S, Tao M, Yuan L, Jing Y (2014) Bacterial cellulose membranes used as artificial substitutes for dural defection in rabbits. Int J Mol Sci 15(6):10855–10867
195. Huang JW, Lv XG, Li Z, Song LJ, Feng C, Xie MK, Li C, Li HB, Wang JH, Zhu WD, Chen SY (2015) Urethral reconstruction with a 3D porous bacterial cellulose scaffold seeded with lingual keratinocytes in a rabbit model. Biomed Mater 10(5):055005
196. Voicu G, Jinga SI, Drosu BG, Busuioc C (2017) Improvement of silicate cement properties with bacterial cellulose powder addition for applications in dentistry. Carbohyd Polym 174:160–170
197. Budhiono A, Rosidi B, Taher H, Iguchi M (1999) Kinetic aspects of bacterial cellulose formation in nata-de-coco culture system. Carbohyd Polym 40(2):137–143
198. Oliveira AA, Mesquita Ed, Furtado AA (2021) Use of bacterial cellulose as a fat replacer in emulsified meat products. Food Sci Technol 42:e42621
199. Chau CF, Yang P, Yu CM, Yen GC (2008) Investigation on the lipid-and cholesterol-lowering abilities of biocellulose. J Agric Food Chem 56(6):2291–2295
200. Guo Y, Zhang X, Hao W, Xie Y, Chen L, Li Z, Zhu B, Feng X (2018) Nano-bacterial cellulose/soy protein isolate complex gel as fat substitutes in ice cream model. Carbohyd Polym 198:620–630
201. Akoğlu A, Çakır İ, Karahan AG, Çakmakçı ML (2018) Effects of bacterial cellulose as a fat replacer on some properties of fat-reduced mayonnaise. Roman Biotechnol Lett
202. Marchetti L, Muzzio B, Cerrutti P, Andrés SC, Califano AN (2017) Bacterial nanocellulose as novel additive in low-lipid low-sodium meat sausages. effect on quality and stability. Food Struct 14:52–59
203. Okiyama A, Motoki M, Yamanaka S (1993) Bacterial cellulose IV application to processed foods. Food Hydrocolloid 6(6):503–511
204. Lin KW, Lin HY (2004) Quality characteristics of Chinese-style meatball containing bacterial cellulose (Nata). J Food Sci 69(3):SNQ107–11
205. Fijałkowski K, Peitler D, Rakoczy R, Żywicka A (2016) Survival of probiotic lactic acid bacteria immobilized in different forms of bacterial cellulose in simulated gastric juices and bile salt solution. LWT-Food Sci Technol 68:322–328
206. Shi Z, Zhang Y, Phillips GO, Yang G (2014) Utilization of bacterial cellulose in food. Food Hydrocolloids 35:539–545
207. Isopencu GO, Stoica-Guzun A, Busuioc C, Stroescu M, Deleanu IM (2021) Development of antioxidant and antimicrobial edible coatings incorporating bacterial cellulose, pectin, and blackberry pomace. Carbohydr Polym Technol Appl 2:100057
208. Silva SM, Ribeiro HL, Mattos AL, Borges MD, Rosa MD, de Azeredo HM (2021) Films from cashew byproducts: cashew gum and bacterial cellulose from cashew apple juice. J Food Sci Technol 58:1979–1986
209. Viana RM, Sá NM, Barros MO, de Fátima BM, Azeredo HM (2018) Nanofibrillated bacterial cellulose and pectin edible films added with fruit purees. Carbohyd Polym 196:27–32
210. Buruaga-Ramiro C, Valenzuela SV, Valls C, Roncero MB, Pastor FJ, Díaz P, Martinez J (2020) Development of an antimicrobial bioactive paper made from bacterial cellulose. Int J Biol Macromol 158:587–594

211. Abral H, Pratama AB, Handayani D, Mahardika M, Aminah I, Sandrawati N, Sugiarti E, Muslimin AN, Sapuan SM, Ilyas RA (2021) Antimicrobial edible film prepared from bacterial cellulose nanofibers/starch/chitosan for a food packaging alternative. Int J Polym Sci 2021:1–1
212. Zhu H, Jia S, Yang H, Tang W, Jia Y, Tan Z (2010) Characterization of bacteriostatic sausage casing: a composite of bacterial cellulose embedded with ε-polylysine. Food Sci Biotechnol 19:1479–1484
213. Jebel FS, Almasi H (2016) Morphological, physical, antimicrobial and release properties of ZnO nanoparticles-loaded bacterial cellulose films. Carbohyd Polym 149:8–19
214. Mohammadalinejhad S, Almasi H, Moradi M (2020) Immobilization of Echium amoenum anthocyanins into bacterial cellulose film: a novel colorimetric pH indicator for freshness/spoilage monitoring of shrimp. Food Control 113:107169
215. Gomes NO, Carrilho E, Machado SA, Sgobbi LF (2020) Bacterial cellulose-based electrochemical sensing platform: a smart material for miniaturized biosensors. Electrochim Acta 349:136341
216. Zhang T, Wang W, Zhang D, Zhang X, Ma Y, Zhou Y, Qi L (2010) Biotemplated synthesis of gold nanoparticle–bacteria cellulose nanofiber nanocomposites and their application in biosensing. Adv Func Mater 20(7):1152–1160
217. Marques PA, Nogueira HI, Pinto RJ, Neto CP, Trindade T (2008) Silver-bacterial cellulosic sponges as active SERS substrates. J Raman Spectroscop Int J Origin Work Aspect Raman Spectroscop Includ High Order Process Brill Rayleigh Scatter 39(4):439–443
218. Eisele S, Ammon HP, Kindervater R, Gröbe A, Göpel W (1994) Optimized biosensor for whole blood measurements using a new cellulose based membrane. Biosens Bioelectron 9(2):119–124
219. Abbasi-Moayed S, Golmohammadi H, Hormozi-Nezhad MR (2018) A nanopaper-based artificial tongue: a ratiometric fluorescent sensor array on bacterial nanocellulose for chemical discrimination applications. Nanoscale 10(5):2492–2502
220. Lv P, Zhou H, Mensah A, Feng Q, Wang D, Hu X, Cai Y, Lucia LA, Li D, Wei Q (2018) A highly flexible self-powered biosensor for glucose detection by epitaxial deposition of gold nanoparticles on conductive bacterial cellulose. Chem Eng J 351:177–188
221. Kim YH, Park S, Won K, Kim HJ, Lee SH (2013) Bacterial cellulose–carbon nanotube composite as a biocompatible electrode for the direct electron transfer of glucose oxidase. J Chem Technol Biotechnol 88(6):1067–1070
222. Gutierrez J, Tercjak A, Algar I, Retegi A, Mondragon I (2012) Conductive properties of TiO_2/bacterial cellulose hybrid fibres. J Colloid Interface Sci 377(1):88–93
223. Núñez-Carmona E, Bertuna A, Abbatangelo M, Sberveglieri V, Comini E, Sberveglieri G (2019) BC-MOS: the novel bacterial cellulose based MOS gas sensors. Mater Lett 237:69–71
224. Jasim A, Ullah MW, Shi Z, Lin X, Yang G (2017) Fabrication of bacterial cellulose/polyaniline/single-walled carbon nanotubes membrane for potential application as biosensor. Carbohyd Polym 163:62–69
225. Di Pasquale G, Graziani S, Pollicino A, Trigona C (2019) Performance characterization of a biodegradable deformation sensor based on bacterial cellulose. IEEE Trans Instrum Meas 69(5):2561–2569
226. Wang Y, Yadav S, Heinlein T, Konjik V, Breitzke H, Buntkowsky G, Schneider JJ, Zhang K (2014) Ultra-light nanocomposite aerogels of bacterial cellulose and reduced graphene oxide for specific absorption and separation of organic liquids. RSC Adv 4(41):21553–21558
227. Thaveemas P, Chuenchom L, Kaowphong S, Techasakul S, Saparpakorn P, Dechtrirat D (2021) Magnetic carbon nanofiber composite adsorbent through green in-situ conversion of bacterial cellulose for highly efficient removal of bisphenol A. Biores Technol 333:125184
228. Khamkeaw A, Phisalaphong M, Jongsomjit B, Lin KY, Yip AC (2020) Synthesis of mesoporous MFI zeolite via bacterial cellulose-derived carbon templating for fast adsorption of formaldehyde. J Hazard Mater 384:121161
229. Liu X, Souzandeh H, Zheng Y, Xie Y, Zhong WH, Wang C (2017) Soy protein isolate/bacterial cellulose composite membranes for high efficiency particulate air filtration. Composites Sci Technol 138:124–133

230. Sai H, Fu R, Xing L, Xiang J, Li Z, Li F, Zhang T (2015) Surface modification of bacterial cellulose aerogels' web-like skeleton for oil/water separation. ACS Appl Mater Interfaces 7(13):7373–7381
231. Galdino CJ Jr, Maia AD, Meira HM, Souza TC, Amorim JD, Almeida FC, Costa AF, Sarubbo LA (2020) Use of a bacterial cellulose filter for the removal of oil from wastewater. Process Biochem 91:288–296
232. Alves AA, Silva WE, Belian MF, Lins LS, Galembeck A (2020) Bacterial cellulose membranes for environmental water remediation and industrial wastewater treatment. Int J Environ Sci Technol 17:3997–4008
233. Muhamad II, Muhamad SN, Salehudin MH, Zahan KA, Tong WY, Pa'e N (2020) Effect of pandan extract concentration to chromium (IV) removal using bacterial cellulose-pandan composites prepared by in-situ modification technique. Mater Today: Proceed 31:89–95
234. Kamal T, Ahmad I, Khan SB, Ul-Islam M, Asiri AM (2019) Microwave assisted synthesis and carboxymethyl cellulose stabilized copper nanoparticles on bacterial cellulose nanofibers support for pollutants degradation. J Polym Environ 27:2867–2877
235. Li G, Nandgaonkar AG, Wang Q, Zhang J, Krause WE, Wei Q, Lucia LA (2017) Laccase-immobilized bacterial cellulose/TiO_2 functionalized composite membranes: evaluation for photo-and bio-catalytic dye degradation. J Membr Sci 525:89–98
236. Amnuaikit T, Chusuit T, Raknam P, Boonme P (2011) Effects of a cellulose mask synthesized by a bacterium on facial skin characteristics and user satisfaction. Med Dev: Evid Res 23:77–81
237. Aramwit P, Bang N (2014) The characteristics of bacterial nanocellulose gel releasing silk sericin for facial treatment. BMC Biotechnol 14:1–1
238. Hasan N, Biak DR, Kamarudin S (2012) Application of bacterial cellulose (BC) in natural facial scrub. Int J Adv Sci Eng Inf Technol 2(4):1–4
239. Heath BP, Coffindaffer TW, Kyte III KE, Smith III ED, McConaughy SD, inventors; Gillette Co LLC, assignee (2012) Personal cleansing compositions comprising a bacterial cellulose network and cationic polymer. United States patent US 8,097,574
240. Lin YC, Wey YC, Lee ML, Lin PC, inventors; NYMPHEAS INTERNATIONAL CORP, assignee (2015) Cosmetic composition containing fragments of bacterial cellulose film and method for manufacturing thereof. United States patent application US 14/614,855
241. Amorim JD, Costa AF, Galdino Jr CJ, Glória MV, Santos E, Sarubbo LA (2019) Bacterial cellulose production using fruit residues as substract to industrial application. CET J Chem Eng Trans 74
242. Chopra H, Gandhi S, Gautam RK, Kamal MA (2022) Bacterial nanocellulose based wound dressings: current and future prospects. Curr Pharm Des 28(7):570–580
243. de Amorim JD, da Silva Junior CJ, de Medeiros AD, do Nascimento HA, Sarubbo M, de Medeiros TP, Costa AF, Sarubbo LA (2022) Bacterial cellulose as a versatile biomaterial for wound dressing application. Molecules 27(17):5580
244. Johnson DC, Neogi AN (1989) Sheeted products formed from reticulated microbial cellulose. US patent 4863565
245. Frankel VH, Serafica GC, Damien CJ (2004) Development and testing of a novel biosynthesized XCell for treating chronic wounds. Surg Technol Int 12:27–33
246. Choi SM, Rao KM, Zo SM, Shin EJ, Han SS (2022) Bacterial cellulose and its applications. Polymers 14(6):1080
247. El-Saied H, Basta AH, Gobran RH (2004) Research progress in friendly environmental technology for the production of cellulose products (bacterial cellulose and its application). Polym-Plast Technol Eng 43(3):797–820
248. Chawla PR, Bajaj IB, Survase SA, Singhal RS (2009) Microbial cellulose: fermentative production and applications. Food Technol Biotechnol 47(2)
249. Ng CC, Shyu YT (2004) Development and production of cholesterol-lowering Monascus-nata complex. World J Microbiol Biotechnol 20:875–879
250. Iguchi M, Mitsuhashi S, Ichimura K, Nishi Y, Uryu M, Yamanaka S, Watanabe K (1988) Bacterial cellulose-containing molding material having high dynamic strength. Patent NumberUS Patent 4, 742, 164

251. Singhsa P, Narain R, Manuspiya H (2018) Physical structure variations of bacterial cellulose produced by different Komagataeibacter xylinus strains and carbon sources in static and agitated conditions. Cellulose 25(3):1571–1581
252. Pu W, Shen C, Wei B, Yang Y, Li Y (2018) A comprehensive review of polysaccharide biopolymers for enhanced oil recovery (EOR) from flask to field. J Ind Eng Chem 61:1–1
253. Kiziltas EE, Kiziltas A, Rhodes K, Emanetoglu NW, Blumentritt M, Gardner DJ (2016) Electrically conductive nano graphite-filled bacterial cellulose composites. Carbohyd Polym 136:1144–1151
254. Lin Z, Guan Z, Huang Z (2013) New bacterial cellulose/polyaniline nanocomposite film with one conductive side through constrained interfacial polymerization. Ind Eng Chem Res 52(8):2869–2874
255. Zhuravleva N, Reznik A, Kiesewetter D, Stolpner A, Khripunov A (2018) Possible applications of bacterial cellulose in the manufacture of electrical insulating paper. In: Journal of Physics: Conference Series, vol 1124. IOP Publishing, p 031008
256. Wahid F, Huang LH, Zhao XQ, Li WC, Wang YY, Jia SR, Zhong C (2021) Bacterial cellulose and its potential for biomedical applications. Biotechnol Adv 53:107856
257. Cannon RE, Anderson SM (1991) Biogenesis of bacterial cellulose. Crit Rev Microbiol 17(6):435–447
258. Montes S, Maleki H (2020) Aerogels and their applications. In: Colloidal metal oxide nanoparticle. Elsevier, pp 337–399
259. Hrubesh LW, Pekala RW (1994) Thermal properties of organic and inorganic aerogels. J Mater Res 9:731–738
260. Schwertfeger F, Frank D, Schmidt M (1998) Hydrophobic waterglass based aerogels without solvent exchange or supercritical drying. J Non-Cryst Solids 225:24–29
261. Hüsing N, Schubert U (1998) Aerogels—airy materials: chemistry, structure, and properties. Angew Chem Int Ed 37(1–2):22–45
262. Long LY, Weng YX, Wang YZ (2018) Cellulose aerogels: synthesis, applications, and prospects. Polymers 10(6):623
263. Revin VV, Nazarova NB, Tsareva EE, Liyaskina EV, Revin VD, Pestov NA (2020) Production of bacterial cellulose aerogels with improved physico-mechanical properties and antibacterial effect. Front Bioeng Biotechnol 8:603407
264. Asmatulu R (2012) Nanocoatings for corrosion protection of aerospace alloys. In: Corrosion protection and control using nanomaterials. Woodhead Publishing, pp 357–374
265. Gesser HD, Goswami PC (1989) Aerogels and related porous materials. Chem Rev 89(4):765–788
266. Gurav JL, Jung IK, Park HH, Kang ES, Nadargi DY (2010) Silica aerogel: synthesis and applications. J Nanomater 2010:1–1
267. Gavillon R, Budtova T (2008) Aerocellulose: new highly porous cellulose prepared from cellulose−NaOH aqueous solutions. Biomacromolecules 9(1):269–277
268. Biganska O, Navard P (2009) Morphology of cellulose objects regenerated from cellulose–N-methylmorpholine N-oxide–water solutions. Cellulose 16:179–188
269. Shen X, Shamshina JL, Berton P, Gurau G, Rogers RD (2016) Hydrogels based on cellulose and chitin: fabrication, properties, and applications. Green Chemistr 18(1):53–75
270. Sannino A, Demitri C, Madaghiele M (2009) Biodegradable cellulose-based hydrogels: design and applications. Materials 2(2):353–373
271. Wu ZY, Li C, Liang HW, Chen JF, Yu SH (2013) Ultralight, flexible, and fire-resistant carbon nanofiber aerogels from bacterial cellulose. Angewandte Chemie 125(10):2997–3001
272. Hosseini H, Kokabi M, Mousavi SM (2018) BC/rGO conductive nanocomposite aerogel as a strain sensor. Polymer 137:82–96
273. Hosseini H, Mousavi SM, Wurm FR, Goodarzi V (2021) Display of hidden properties of flexible aerogel based on bacterial cellulose/polyaniline nanocomposites with helping of multiscale modeling. Eur Polymer J 146:110251
274. Nie X, Lv P, Stanley SL, Wang D, Wu S, Wei Q (2019) Ultralight nanocomposite aerogels with interpenetrating network structure of bacterial cellulose for oil absorption. J Appl Polym Sci 136(39):48000

275. Zhao S, Malfait WJ, Guerrero-Alburquerque N, Koebel MM, Nyström G (2018) Biopolymer aerogels and foams: chemistry, properties, and applications. Angew Chem Int Ed 57(26):7580–7608
276. Adebajo MO, Frost RL, Kloprogge JT, Carmody O, Kokot S (2003) Porous materials for oil spill cleanup: a review of synthesis and absorbing properties. J Porous Mater 10:159–170
277. Kettunen M, Silvennoinen RJ, Houbenov N, Nykänen A, Ruokolainen J, Sainio J, Pore V, Kemell M, Ankerfors M, Lindström T, Ritala M (2011) Photoswitchable superabsorbency based on nanocellulose aerogels. Adv Func Mater 21(3):510–517
278. Keshipour S, Khezerloo M (2017) Gold nanoparticles supported on cellulose aerogel as a new efficient catalyst for epoxidation of styrene. J Iran Chem Soc 14:1107–1112
279. Thiruvengadam V, Vitta S (2016) Interparticle interactions mediated superspin glass to superferromagnetic transition in Ni-bacterial cellulose aerogel nanocomposites. J Appl Phys 119(24)
280. Henschen J, Illergård J, Larsson PA, Ek M, Wågberg L (2016) Contact-active antibacterial aerogels from cellulose nanofibrils. Colloids Surf B 146:415–422
281. Edwards JV, Fontenot KR, Prevost NT, Pircher N, Liebner F, Condon BD (2016) Preparation, characterization and activity of a peptide-cellulosic aerogel protease sensor from cotton. Sensors 16(11):1789
282. Wan C, Jiao Y, Qiang T, Li J (2017) Cellulose-derived carbon aerogels supported goethite (α-FeOOH) nanoneedles and nanoflowers for electromagnetic interference shielding. Carbohyd Polym 156:427–434
283. Chen Z, Peng X, Zhang X, Jing S, Zhong L, Sun R (2017) Facile synthesis of cellulose-based carbon with tunable N content for potential supercapacitor application. Carbohyd Polym 170:107–116
284. Meng Y, Wang X, Wu Z, Wang S, Young TM (2015) Optimization of cellulose nanofibrils carbon aerogel fabrication using response surface methodology. Eur Polymer J 73:137–148
285. Yang X, Fei B, Ma J, Liu X, Yang S, Tian G, Jiang Z (2018) Porous nanoplatelets wrapped carbon aerogels by pyrolysis of regenerated bamboo cellulose aerogels as supercapacitor electrodes. Carbohyd Polym 180:385–392
286. Wu ZY, Li C, Liang HW, Chen JF, Yu SH (2013) Inside back cover: ultralight, flexible, and fire-resistant carbon nanofiber aerogels from bacterial cellulose (Angew. Chem. Int. Ed. 10/2013). Angewandte Chemie International Edition. 52(10):3035

Chapter 3
Synthesis of Carbon Nanomaterials from Agro-Industrial Wastes and Their Extensive Applications

Smruti Ranjan Padhan, Sanjay Singh Rathore, Shankar Lal Jat, Sushmita Saini, Ayesha Mohanty, Kaushik Kumar Panigrahi, Shivani Ranjan, Sumit Sow, Pratikshya Mishra, Kirttiranjan Baral, Akanksha, and K. Swapna Rani

Abstract The synthesis of an array of carbon nanomaterials (CNMs) from agro-industrial wastes is a growing area of research, which offers great potential for sustainable nanotechnology and waste minimization. The novel approach employs a broad spectrum of agro-industrial wastes, including agro-residues, food processing by-products, and forestry residues as renewable feedstocks for CNMs production. Since there are several methods for synthesis, for example, pyrolysis, hydrothermal carbonization, and chemical vapor deposition, we can make these wastes into carbonaceous materials with the desired properties and functionalities by controlling the process. The characterization techniques, such as microscopy and spectroscopy, and even computational modeling, allow for the comprehensive study of nanomaterial structure, morphology, and efficacy, which can be used to further optimize the materials for unique applications. Biocompatibility and toxicity studies are the compliance tools to ensure the safety of CNMs in biomedical, environmental, and industrial sectors, and this in turn guides regulatory compliance and risk management goals. The widespread application of CNMs derived from agro-industrial wastes

S. R. Padhan (✉) · Akanksha · K. Swapna Rani
KVK-East Sikkim, ICAR-Research Complex for North Eastern Hill Region, Sikkim 737102, India
e-mail: smrutiranjaniari@gmail.com

S. R. Padhan · S. S. Rathore · S. Saini
Indian Agricultural Research Institute (IARI), New Delhi 110012, India

S. L. Jat
ICAR-Indian Institute of Maize Research, New Delhi 110012, India

A. Mohanty · K. K. Panigrahi · P. Mishra
Odisha University of Agriculture & Technology (OUAT), Bhubaneswar 751003, India

S. Ranjan · S. Sow
Dr. Rajendra Prasad Central Agricultural University, Pusa, Samastipur 848125, India

K. Baral
National Academy of Agricultural Research Management (NAARM), Hyderabad 500030, India

N. Talreja et al. (eds.), *Waste-Derived Carbon Nanostructures*, Nanostructure Science and Technology, https://doi.org/10.1007/978-3-031-75247-6_3

covers different fields, including energy storage and conversion, pollution treatment, catalysis, biomedical engineering, agriculture, and material science. They possess important features like high surface area, chemical stability, mechanical strength, and electrical conductivity. These features have resulted in innovative solutions in areas like supercapacitors, water purification, drug delivery, and soil improvement. Yet there are still some issues, such as upscaling, cost-effectiveness, and regulatory compliance, which the research is trying to solve by developing a variety of feedstocks, finding sustainable synthesis strategies, and advancing characterization methods to realize the full potential of CNMs. This area of study, however, offers good prospects by addressing such challenges as well as taking advantage of emerging opportunities. There is no doubt that this field of CNMs will redefine the way we manage waste, boost the development of new technologies, and build a circular economy model that is based on efficiency in resource utilization and sustainability.

Keywords Agro-industrial wastes · Biocompatibility · Carbon nanomaterials · Characterization · Environmental implications · Graphene · Nanomaterials synthesis

3.1 Introduction

Being a multi-disciplinary field, nanotechnology researches a vast spectrum of all the materials that are produced at the nanoscale level through various physical, chemical, and biological methods [1]. Nowadays, the production and use of carbon nanomaterials (CNMs) in materials science are a borderline between new directions in developments and discoveries. It is these new-generation materials and their unique attributes with meaningful capabilities that are getting the attention of researchers, engineers, and industries globally [2, 3]. On the one side, the exponential growth of the global issue of sustainable development goals and environmental conservation has given birth to an unsolicited re-evaluation of waste or biomass management strategies, particularly, those that are generated from agriculture and the industrial sectors. The interface between scientific research and societal demand for the sustainability of CNMs production has birthed a vibrant area of specialization focusing on the recycling of various agro-industrial wastes into these nanomaterials, a feat considering it presents a green way of recycling and reusing resources.

The history of CNMs revealed the significant progress that occurred in the nanomaterial's regime across the different time frames and dimensions (Fig. 3.1). The CNMs encompass a diverse array of structures and compositions, each exhibiting remarkable physical, chemical, and mechanical properties. Carbon atoms may arrange themselves into a variety of configurations. For instance, a stack of two-dimensional single sheets called the graphite crystal is created when C atoms are placed in a honeycomb lattice structure [4]. Among the most renowned CNMs are carbon nanotubes (CNTs), graphene, carbon nanofibers (CNFs), and carbon quantum dots. CNTs, with their exceptional tensile strength and electrical conductivity, have found applications in electronics, aerospace, and biomedical fields [5]. Graphene, a single layer of carbon atoms settled in a hexagonal lattice holds extraordinary mechanical strength, thermal conductivity, and electronic properties, rendering it promising for various applications, including energy storage, composites, and sensors [6, 7]. CNFs, characterized by their fibrous morphology, exhibit high aspect ratios and mechanical strength, making them suitable for reinforcing materials in composites and structural applications. Carbon quantum dots, on the other hand, offer unique optical and electronic properties, making them attractive for bioimaging, sensing, and optoelectronics [8]. The remarkable properties of these CNMs have fueled extensive research efforts aimed at understanding their synthesis, characterization, and application across diverse fields.

Agro-industrial wastes are generally generated from various sectors, viz agricultural, forestry, and food processing activities, and represent a significant environmental and economic challenge. These wastes including crop residues, animal manure, food processing by-products, and forestry residues, are often disposed of through landfilling, open burning, or indiscriminate dumping, posing environmental pollution risks and resource wastage [9]. Residue management holds immense importance in India due to the country's heavy reliance on agriculture as a primary livelihood source. With a vast agricultural sector, India is the second-largest producer of agricultural residues in the world, generating about 686 MT of crop residues each year. India produces 368 MT (54%) of its leftovers from cereals (rice, wheat, maize, pearl millet, barley, minor millets, and sorghum), with sugarcane contributing for

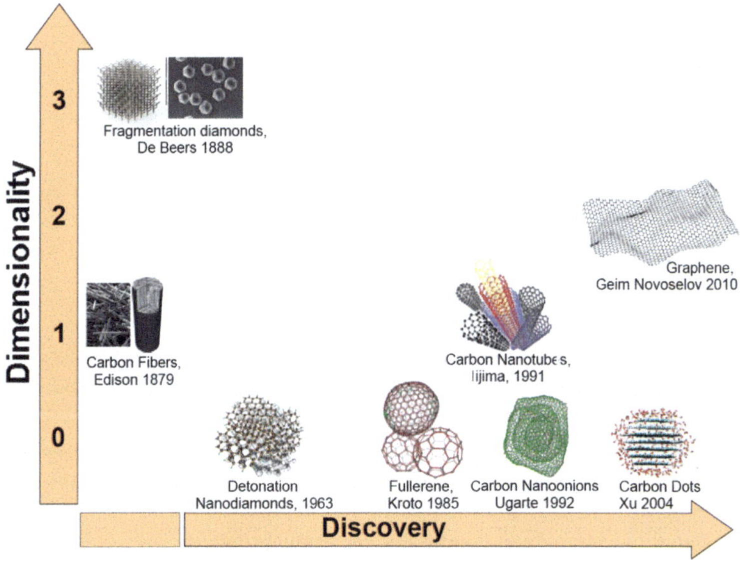

Fig. 3.1 History of various CNMs. The figure was reproduced with the permission [4]

the remaining 111 MT (16%) [10]. Addressing the management of agro-industrial wastes is crucial for mitigating environmental pollution, conserving resources, and promoting sustainable development. Moreover, the valorization of agro-industrial wastes presents significant prospects for maximizing resource recovery and use efficiency, energy generation, and vertical diversification through value-added products, thereby contributing to the circular economy model. Therefore, the CNMs obtained from agro-industrial wastes can be considered as an entity of circular economy because they provide a solution to both the environmental problems and the innovations in materials. Through the use of naturally occurring and inexpensive biomass resources as the initial precursors, this method provides a sustainable and low-cost substitute to the conventional sources of carbon found in fossil fuels and artificially generated precursors [11]. Furthermore, the fact that agro-industrial wastes are transformed into high-priced CNMs offers an economic value to raw materials that otherwise remain untapped or discarded, which in turn creates jobs in rural areas and supports waste-to-wealth and other ecological initiatives. In addition, the various chemical compositions of agro-industrial wastes can yield CNMs with distinctive properties, which therefore expands the field of custom design and manufacture of novel materials that can be applied to different areas. Finally, the main motivation for synthesizing CNMs from agro-industrial wastes can be highlighted at the intersection of environment conservation, waste management, and maximum use of the resources, as well as for the development of new materials, which can bring both economic and social benefits.

3.2 CNMs: Types and Properties

CNMs encompass a diverse range of structures and compositions, each exhibiting unique properties and potential applications. The prominent types of CNMs, viz carbon nanotubes (CNTs), graphene, carbon nanofibers (CNFs), and carbon quantum dots (CQDs), along with their properties and characteristics are defined as follows.

3.2.1 CNTs

CNTs are cylindrical nanostructures composed of rolled-up graphene sheets, resulting in excellent mechanical, electrical, and thermal properties. The CNTs can be categorized into single-walled carbon nanotubes (SWCNTs) and multi-walled carbon nanotubes (MWCNTs) depending on the number of graphene layers forming the tube structure [12]. The SWCNTs impart unique one-dimensional electronic properties, making them ideal candidates for applications in electronics, nanoelectromechanical systems (NEMS), and field-effect transistors. The MWCNTs, with their higher mechanical strength and thermal conductivity, find applications in reinforcing materials, nanocomposites, and energy storage devices [13]. The extraordinary mechanical characteristics of CNTs, encompassing high tensile strength and Young's modulus, make them fit for applications in aerospace technology, automotive, and varied structural materials.

3.2.2 *Graphene*

Graphene, a two-dimensional (2D) honeycomb lattice of carbon atoms, represents one of the most celebrated CNMs due to its extraordinary mechanical, electrical, and thermal properties. Graphene exhibits exceptional electron mobility, high thermal conductivity, and optical transparency, rendering it promising for applications in electronics, optoelectronics, and thermal management [7]. The unique electronic band structure of graphene enables high carrier mobility and ballistic transport at room temperature, making it ideal for high-speed electronic devices and transparent conductive films. Additionally, graphene's large surface area and chemical reactivity make it suitable for applications in energy storage, catalysis, and sensors [6, 14]. The scalability and versatility of graphene synthesis techniques, such as chemical vapor deposition (CVD) and mechanical exfoliation, have spurred extensive research and development efforts toward commercializing graphene-based technologies. The various forms of graphene-based CNMs (Fig. 3.2) encompass graphite, graphene oxide, fullerene, etc.

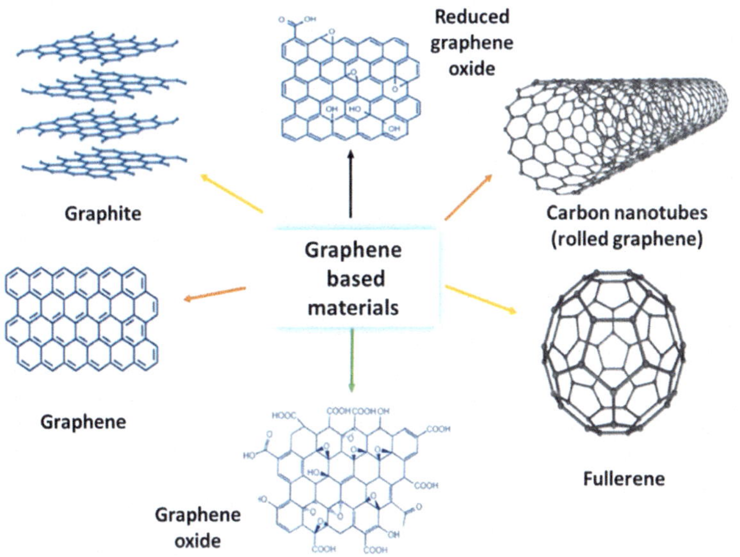

Fig. 3.2 Different graphene-based materials. The figure was reproduced with the permission [14]

3.2.3 CNFs

CNFs are nanostructured carbon materials characterized by their fibrous morphology and high aspect ratio. CNFs can be produced through various techniques, including electrospinning, catalytic synthesis, and carbonization of polymer precursors [15]. The unique structure of CNFs imparts exceptional mechanical properties, such as high tensile strength and modulus, making them suitable for reinforcement in composite materials, textiles, and structural components. Additionally, CNFs exhibit excellent electrical conductivity, thermal stability, and chemical inertness, enabling applications in conductive polymers, energy storage devices, and filtration membranes. The tunable porosity and surface functionalization of CNFs offer opportunities for tailored materials design and applications in catalysis, sensors, and biomedical devices [16].

3.2.4 Carbon Quantum Dots (CQDs)

Carbon quantum dots (CQDs) are quasi-zero-dimensional carbon nanomaterials with dimensions typically less than 10 nm [17]. CQDs exhibit unique quantum confinement effects and surface-related properties, including size-dependent optical properties, high photoluminescence quantum yield, and chemical stability [8, 18]. These distinctive characteristics make CQDs attractive for applications in bioimaging, sensing, optoelectronics, and photocatalysis. The synthesis of CQDs from carbonaceous precursors, such as biomass, carbon nanotubes, or graphene, offers a sustainable and scalable route toward functional nanomaterials with tailored optical and electronic properties [17]. Furthermore, the surface passivation and functionalization of CQDs enable precise control over their photophysical properties and biocompatibility, facilitating diverse applications in biomedical diagnostics, drug delivery, and environmental monitoring.

3.2.5 Carbon Nanohorns (CNHs)

Carbon nanohorns (CNHs) are horn-shaped carbon nanoparticles composed of graphene sheets rolled into conical structures [19]. CNHs exhibit unique properties such as high surface area, excellent adsorption capacity, and low cytotoxicity, making them promising candidates for applications in drug delivery, catalysis, and energy storage [20]. The high aspect ratio and porosity of CNHs facilitate efficient adsorption of gases, pollutants, and biomolecules, offering potential applications in environmental remediation, gas separation, and biosensing [21].

3.2.6 Carbon Nanorods (CNRs)

Carbon nanorods (CNRs) are elongated nanostructures composed of stacked graphene layers arranged in a rod-like morphology. CNRs possess exceptional mechanical properties, including high tensile strength and modulus, as well as superior thermal conductivity, making them suitable for reinforcement in composite materials and thermal management applications [22]. Additionally, the high surface-to-volume ratio of CNRs enables efficient charge transport and storage, rendering them promising for applications in energy storage devices, such as lithium-ion batteries and supercapacitors [23].

3.2.7 Carbon Nanocapsules (CNCs)

Carbon nanocapsules (CNCs) are hollow carbon nanoparticles with spherical or tubular morphology, enclosing a void space within their interior. CNCs exhibit unique encapsulation properties, enabling the confinement and controlled release of guest molecules, drugs, or nanoparticles [24]. This property makes CNCs promising candidates for targeted drug delivery, imaging agents, and nanoreactors. Additionally, CNCs can be functionalized with surface modifications to enhance biocompatibility, stability, and targeting specificity, expanding their potential applications in biomedicine and nanotechnology [25].

3.2.8 Carbon Nanobelts (CNBs)

Carbon nanobelts (CNBs) are planar carbon nanostructures with a narrow width and long length, resembling miniature belts or ribbons [26]. CNBs exhibit unique electronic properties, such as tunable bandgaps and high carrier mobility, making them attractive for applications in organic electronics, photovoltaics, and optoelectronic devices. The planar geometry of CNBs enables efficient charge transport along the longitudinal axis, facilitating high-performance electronic and photonic devices [27]. Moreover, the chemical versatility of CNBs allows for functionalization and integration into hybrid materials for tailored properties and applications.

3.2.9 Carbon Nanoscrolls (CNMSs)

Carbon nanoscrolls (CNMSs) are rolled-up graphene sheets forming tubular structures with spiral configurations. CNMSs exhibit unique mechanical properties, including flexibility, elasticity, and high aspect ratios, making them suitable for applications in nanocomposites, membranes, and nanoelectromechanical systems (NEMS) [28]. The tubular morphology of CNMSs enables the encapsulation of guest molecules or nanoparticles within their interior, offering opportunities for controlled release, sensing, and catalysis [29]. Additionally, CNMSs can be functionalized with surface modifications to tailor their properties for specific applications, such as gas storage, drug delivery, and molecular sieving.

3.3 Agro-Industrial Wastes as Precursors for Carbon Nanomaterials

Nowadays, there has been growing interest in utilizing agro-industrial wastes as precursors for the synthesis of CNMs. This innovative approach not only tackles the challenge of waste management issues but also offers a sustainable and cost-effective technique for producing valuable nanomaterials. The various selection criteria for agro-industrial wastes, the types of biomasses suitable for synthesis, and various pre-treatment methods to optimize their conversion into carbon nanomaterials are described below.

3.3.1 Selection Criteria for Agro-Industrial Wastes

The selection of appropriate agro-industrial wastes as precursors for CNMs synthesis is crucial to ensure the efficient utilization of resources and the production of high-quality nanomaterials [30, 31]. Several key criteria guide the selection process:

Abundance and Availability: The raw materials should be readily available in abundant quantities to support scalable synthesis processes. However, the wastes generated from agricultural, forestry, and food processing activities are often abundant and accessible, offering a sustainable source of raw materials for nanomaterial production.

Composition and Chemical Properties: The chemical composition of agro-industrial biomass plays a significant role in determining the properties of the resulting CNMs. Wastes rich in carbon content, such as lignocellulosic biomass, are preferred precursors due to their potential for yielding high-purity carbonaceous materials [32]. Additionally, the presence or absence of other elements or functional groups in the waste stream can influence the structure and properties of the synthesized nanomaterials.

Suitability for Conversion: The produced waste/residues should be amenable to various types of conversion into CNMs through pyrolysis, hydrothermal treatment, or other synthesis methods. Factors such as particle size, morphology, and moisture content can affect the efficiency of conversion processes and the quality of the final products.

Environmental Impact: Major consideration should be given to the environmental impact of utilizing various wastes as precursors. Wastes that are generated in large quantities but have limited recycling or valorization options represent potential targets for CNMs synthesis, giving the dual benefit of better waste management and maximized resource utilization [32].

3.3.2 Types of Agro-Industrial Wastes Suitable for Synthesis

A wide range of agro-industrial wastes have been investigated as potential precursors for CNMs synthesis, each offering unique advantages and challenges. Some of the common types include:

Biomass Residues or Farm Wastes: Crop residues, such as rice husks, wheat straw, and corn stalks, are abundant sources of lignocellulosic biomass rich in carbon content [11]. These residues can be pyrolyzed or hydrothermally treated to produce CNMs with tailored properties for various applications.

By-products of the Food Processing Sector: Waste streams generated from food processing operations, including fruit peels, vegetable trimmings, and spent coffee grounds, contain organic materials that can be converted into CNMs [33]. These by-products are often rich in sugars, cellulose, and lignin, providing potential precursors for biomass-derived carbon materials.

Forestry Outputs: Sawdust, wood chips, and forestry residues generated from timber harvesting and processing operations represent another source of carbon-rich materials. These residues can undergo pyrolysis or chemical treatment to produce carbon nanomaterials with tailored structures and properties.

Animal Manure and Wastewater Sludge: Organic wastes such as animal manure and wastewater sludge contain organic matter that can be converted into carbonaceous materials through different kinds of thermal or chemical processes. These waste streams offer potential precursors for producing CNMs while addressing environmental pollution and resource management challenges [34].

3.3.3 Pre-treatment Methods for Agro-Industrial Wastes

The pre-treatment of residue/biomass generated from agro-industrial sectors is often necessary to optimize their conversion into CNMs and improve the quality of the final products. The most common pre-treatment methods include [34–36]:

Drying: The biomass may contain high moisture content, which can hinder pyrolysis or hydrothermal treatment processes. Drying the wastes to remove excess moisture helps to improve the efficiency of subsequent conversion processes and reduce energy consumption.

Size Reduction: Agro-industrial wastes are often bulky or heterogeneous in size, requiring size reduction to facilitate uniform heating and enhance the surface area available for carbonization. Several mechanical methods such as milling or grinding can be employed to reduce the particle size of the wastes.

Chemical Activation: It involves treating the agro-industrial wastes with activating agents such as acids, bases, or metal salts to enhance their carbonization efficiency and porosity which in turn promotes the formation of micropores and mesopores in the carbonaceous materials, improving their adsorption capacity and surface area.

Demineralization: Some wastes encompass inorganic impurities or minerals that can interfere with carbonization processes or affect the properties of the resulting CNMs. Demineralization treatments, viz acid washing or leaching, can remove these impurities and improve the purity and quality of the resultant CNMs.

3.4 Synthesis Methods for Carbon Nanomaterials from Agro-Industrial Wastes

The synthesis of CNMs from varied agro-industrial wastes offers a sustainable and cost-effective approach to producing valuable nanomaterials with diverse properties and applications. Some of the important synthesis methods are underscored below (Table 3.1).

3.4.1 Pyrolysis

Pyrolysis is one of the most commonly employed methods for converting agro-industrial wastes into CNMs. In this process, the waste material is subjected to high temperatures in an inert atmosphere, leading to thermal decomposition and the formation of carbonaceous products [37]. Simplicity, scalability, and the ability to process a wide range of feedstocks are the several advantages offered by pyrolysis. Additionally, it can be tailored to produce specific types of CNMs by controlling various factors involved, viz temperature, heating rate, and residence time [38]. However, the quality and properties of the resulting nanomaterials can vary depending on the physical and chemical composition of the feedstock and environmental conditions during pyrolysis.

3.4.2 Hydrothermal Carbonization

Hydrothermal carbonization (HTC) is a wet-chemical process that involves heating biomass in the presence of water at elevated temperatures of 180–300 °C and high pressure. During HTC, biomass undergoes hydrolysis, dehydration, and polymerization reactions, resulting in the formation of carbonaceous materials [39]. HTC offers

Table 3.1 Various types of synthesis methods of carbon nanomaterials (CNMs) from agro-industrial wastes

S. No.	Synthesis method	Features	Examples	References
	Pyrolysis	Simple and scalable, wide range of applicable feedstocks, controlled heating conditions	Conversion of rice husks into biochar, production of carbon nanotubes from wood biomass, synthesis of graphene from agricultural residues	[37, 38]
	Hydrothermal carbonization	Mild reaction conditions, high carbon yield, utilization of wet and high moisture feedstocks	Alteration of sugarcane bagasse into hydrochar, production of carbon spheres from lignocellulose, synthesis of carbon nanomaterials from food waste	[39–41]
	CVD	Precise control over growth parameters, high-quality and uniform nanomaterials, scalability, and reproducibility	Growth of carbon nanotubes on catalyst substrates, synthesis of graphene on metal foils, deposition of carbon films on silicon substrates	[42]
	Solvothermal synthesis	High purity and controlled morphology, tenable size, shape, and surface properties, production of uniform nanoparticles	Production of carbon nanodots from biomass precursors, synthesis of carbon nanospheres from plant extracts, formation of carbon nanowires from cellulose	[43–45]
	Microwave-assisted synthesis	Rapid and energy-efficient, uniform heating and reaction enhancement, scalability and reproducibility	Microwave-assisted conversion of agricultural waste, synthesis of carbon quantum dots from biomass, microwave-assisted growth of carbon nanotubes	[14, 46, 47]

(continued)

Table 3.1 (continued)

S. No.	Synthesis method	Features	Examples	References
	Electrochemical synthesis	Simplicity and versatility Control over morphology and properties Tailored synthesis conditions	Electrochemical deposition of carbon nanomaterials Electrochemical synthesis of graphene from biomass Formation of carbon nanofibers using electrospinning	[48, 49]
	Plasma-enhanced chemical vapor deposition (PECVD)	Lower operating temperatures compared to conventional CVD, higher growth rates, and better control over film properties, scalability, and compatibility with various substrates	Synthesis of CNTs using plasma-enhanced CVD, production of graphene films via PECVD, deposition of carbon coatings on nanoparticles	[50, 51]
	Laser ablation	High purity and controlled size distribution, ability to produce unique structures and functionalities, versatility in processing various carbon precursors	Laser ablation of rice husks for CNMs, formation of carbon nanoparticles from biomass, laser ablation of wood biomass for graphene production	[52, 53]

(continued)

Table 3.1 (continued)

S. No.	Synthesis method	Features	Examples	References
	Template-assisted synthesis	High structural control and tunable properties, ability to produce hierarchical structures, compatibility with diverse precursor materials	Synthesis of CNTs using porous templates, formation of CNFs using NPs templates, template-assisted growth of CNMs from biomass	[54, 55]
	Flame synthesis	Simplicity and scalability, low-cost and energy-efficient production, suitable for processing a wide range of feedstocks	Flame synthesis of carbon nanotubes from agricultural waste, formation of carbon nanoparticles from bio-oils, flame synthesis of CNFs from forestry residues	[56]

advantages such as mild reaction conditions, high carbon yields, and the ability to process a wide range of feedstocks, including wet and high-moisture materials [40]. The carbonaceous materials produced via HTC exhibit unique properties, such as high surface area, porosity, and chemical stability, making them suitable for various applications, including adsorption, catalysis, and environmental remediation [41].

3.4.3 CVD

CVD is a gas-phase synthesis technique commonly used to produce high-quality carbon nanomaterials, such as CNTs and graphene, from gaseous carbon precursors. In CVD, carbonaceous gases, such as methane or ethylene, are decomposed on a catalyst surface at high temperatures, leading to the deposition of carbon atoms and the growth of nanomaterials [42]. CVD offers precise control over the growth parameters, such as temperature, pressure, and gas composition, enabling the synthesis of carbon nanomaterials with tailored structures and properties. However, CVD typically requires specialized equipment and precise process control, limiting its scalability and applicability to certain feedstocks.

3.4.4 *Solvothermal Synthesis*

Solvothermal synthesis involves the reaction of precursor materials in a solvent at elevated temperatures and pressures, resulting in the formation of CNMs. It offers several advantages such as high purity, controlled morphology, and the ability to produce uniform nanoparticles [43, 44]. By selecting appropriate solvent systems and reaction conditions, researchers can tune the size, shape, and surface properties of the synthesized CNMs. Solvothermal synthesis has been successfully employed to produce various CNMs, including carbon nanodots, carbon nanospheres, and carbon nanowires, from agro-industrial waste-derived precursors [45].

3.4.5 *Microwave-Assisted Synthesis*

Microwave-assisted synthesis is a rapid and efficient method for synthesizing CNMs from agro-industrial wastes. Here, microwave radiation is used to heat the reaction mixture, promoting rapid and uniform heating and enhancing reaction rates [46]. Microwave-assisted synthesis offers advantages such as short reaction times, high yields, and energy efficiency. Additionally, microwave irradiation can facilitate the activation of carbon precursors and promote the formation of CNMs with desired properties. Microwave-assisted synthesis has been employed to produce various CNMs, including carbon nanotubes, graphene oxide, and carbon quantum dots, from agro-industrial waste-derived precursors [14, 47].

3.4.6 *Electrochemical Synthesis*

Electrochemical synthesis involves the electrolysis of precursor materials in an electrolyte solution, leading to the deposition of CNMs on an electrode surface. Electrochemical synthesis offers advantages such as simplicity, scalability, and the ability to control the morphology and properties of the synthesized nanomaterials [48]. By adjusting parameters such as applied voltage, current density, and electrolyte composition, researchers can tailor the structure, size, and surface properties of the synthesized CNMs. Electrochemical synthesis has been employed to produce various carbon nanomaterials, including carbon nanotubes, graphene, and carbon nanofibers, from agro-industrial waste-derived precursors [49].

3.4.7 Plasma-Enhanced Chemical Vapor Deposition (PECVD)

Plasma-enhanced chemical vapor deposition (PECVD) is a variation of CVD that involves the use of plasma to enhance the decomposition of precursor gases and promote the growth of CNMs [50]. In PECVD, a plasma is generated by applying an electric field to a precursor gas, leading to the dissociation of molecular bonds and the formation of reactive species. These reactive species then interact with a substrate surface to deposit CNMs. PECVD offers advantages such as lower operating temperatures, higher growth rates, and better control over film properties compared to conventional CVD techniques [51].

3.4.8 Laser Ablation

It is a technique used to produce CNMs by irradiating a carbon-containing target with a laser beam in a controlled environment. The intense heat generated by the laser vaporizes the target material, leading to the formation of a plasma plume [57]. As the plume expands, carbon atoms condense and nucleate to form nanomaterials such as carbon nanotubes, graphene, and fullerenes. Laser ablation offers advantages such as high purity, controllable synthesis parameters, and the ability to produce nanomaterials with unique structures and properties. However, the process can be energy-intensive and requires careful control of laser parameters to achieve desired results [52, 53].

3.4.9 Template-Assisted Synthesis

Template-assisted synthesis involves using pre-existing templates or scaffolds to guide the growth of CNMs. Templates can be in the form of porous materials, nanoparticles, or biological structures, which serve as templates for nucleation and growth of CNMs. By selecting appropriate templates and controlling synthesis conditions, researchers can tailor the size, shape, and morphology of the synthesized nanomaterials [54]. Template-assisted synthesis offers advantages such as high structural control, tunable properties, and the ability to produce hierarchical structures. This method has been used to synthesize various CNMs, including carbon nanotubes, nanofibers, and nanostructures, from agro-industrial waste-derived precursors [55].

3.4.10 *Flame Synthesis*

This method involves the combustion of precursor materials in a flame environment to produce CNMs. In this technique, a fuel and an oxidizer are mixed and ignited to generate a flame, and precursor materials are introduced into the flame where they undergo pyrolysis and nucleation to form CNMs. Flame synthesis offers advantages such as simplicity, scalability, and the ability to produce nanomaterials in large quantities. Additionally, flame synthesis can be carried out at relatively low temperatures and atmospheric pressure, making it suitable for industrial-scale production [56]. This method has been used to synthesize various CNMs, including carbon nanotubes, nanofibers, and nanoparticles, from agro-industrial waste-derived precursors.

3.5 Applications of CNMs Synthesized from Agro-Industrial Wastes

CNMs synthesized from agro-industrial wastes have emerged as promising candidates for various applications due to their unique properties, sustainability, and cost-effectiveness. The diverse range of applications where these CNMs demonstrate significant potential are described in the following section (Fig. 3.3).

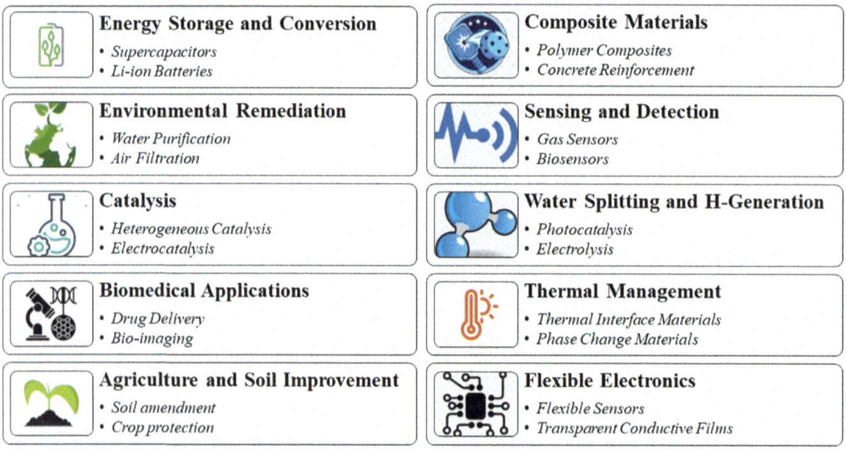

Fig. 3.3 Applications of carbon nanomaterials derived from agro-industrial wastes

3.5.1 Energy Storage and Conversion

Some CNMs generated from agro-industrial wastes exhibit exceptional properties that make them well-suited for energy storage and conversion applications.

3.5.1.1 Supercapacitors

Carbon nanotubes, graphene, and carbon nanofibers synthesized from biomass-derived precursors offer high surface areas, excellent electrical conductivity, and mechanical flexibility, making them ideal candidates for supercapacitor electrodes. These materials enable efficient charge storage and rapid charge–discharge cycles, leading to enhanced energy storage performance and longer device lifetimes [58, 59].

3.5.1.2 Li-Ion Batteries

Carbonaceous materials derived from agricultural residues and food processing by-products serve as promising electrode materials for lithium-ion batteries. The unique porous structure and high surface area of these materials facilitate lithium-ion diffusion and storage, resulting in improved battery capacity, cycling stability, and rate capability [60]. Additionally, the abundance and low cost of agro-industrial wastes make them economically viable precursors for large-scale battery production.

3.5.2 Environmental Remediation

3.5.2.1 Water Purification

Carbon nanosheets, carbon quantum dots, and carbon nanotubes derived from biomass precursors exhibit excellent adsorption properties, high surface areas, and tunable surface functionalities, making them effective adsorbents for removing pollutants and contaminants from water [58, 61]. These materials can efficiently adsorb heavy metals, organic pollutants, and dyes, leading to improved water quality and environmental sustainability.

3.5.2.2 Air Filtration

Biomass-derived carbon nanomaterials can be used as adsorbents and catalysts for air purification applications. Carbon nanofibers and graphene-based materials offer large surface areas and high porosity, enabling effective capture and degradation of volatile organic compounds (VOCs), particulate matter, and harmful gases [58].

Additionally, these materials can be functionalized to enhance their catalytic activity and selectivity for pollutant removal.

3.5.3 Catalysis

The CNMs synthesized from various agro-industrial wastes serve as efficient catalysts for various chemical reactions.

3.5.3.1 Heterogeneous Catalysis

Biomass-derived carbon nanotubes, graphene oxide, and carbon nanofibers exhibit catalytic activity and stability, making them suitable catalyst supports for heterogeneous catalysis applications [62]. These materials can be functionalized with metal nanoparticles or heteroatoms to enhance their catalytic performance in reactions such as hydrogenation, oxidation, and pollutant degradation [63].

3.5.3.2 Electrocatalysis

CNMs derived from agricultural residues and food processing by-products show promise as electrocatalysts for energy conversion and storage devices. These materials can catalyze important electrochemical reactions, such as oxygen reduction reaction (ORR) and hydrogen evolution reaction (HER), with high activity, selectivity, and durability, contributing to the development of efficient fuel cells, electrolyzers, and metal-air batteries [64].

3.5.4 Biomedical Applications

The CNMs hold great potential for biomedical applications due to their biocompatibility, low toxicity, and unique physicochemical properties [65].

3.5.4.1 Drug Delivery

Biomass-derived carbon nanoparticles, carbon quantum dots, and graphene-based materials serve as promising drug delivery vehicles for targeted and controlled release of therapeutic agents. These materials can be functionalized with targeting ligands and therapeutic payloads, allowing for site-specific drug delivery, enhanced cellular uptake, and reduced systemic toxicity [66, 67].

3.5.4.2 Bioimaging

Some of the CNMs exhibit strong optical properties and excellent biocompatibility, making them suitable contrast agents for biomedical imaging applications [65]. Carbon quantum dots and carbon nanotubes can be used for fluorescence imaging, photoacoustic imaging, and magnetic resonance imaging (MRI), enabling non-invasive visualization of biological structures and processes [68].

3.5.5 Agriculture and Soil Improvement

3.5.5.1 Soil Amendment

Biochar derived from biomass residues serves as a valuable soil amendment for enhancing soil fertility, water retention, and nutrient availability [69]. The high porosity and surface area of biochar promote microbial activity, carbon sequestration, and soil organic matter accumulation, leading to improved soil health, crop productivity, and climate resilience [70].

3.5.5.2 Crop Protection

Carbon-based nanomaterials can be used for controlled release and delivery of agrochemicals, such as pesticides and fertilizers, to enhance their efficacy and reduce environmental impact [71, 72]. Nano-encapsulation of active ingredients within carbon nanocarriers improves their stability, solubility, and targeted delivery to plants, resulting in reduced chemical usage and improved crop protection.

3.5.6 Composite Materials

Some of the CNMs also serve as reinforcing agents and functional fillers in composite materials.

3.5.6.1 Polymer Composites

Biomass-derived carbon nanofibers, carbon black, and graphene oxide can be incorporated into polymer matrices to improve mechanical strength, electrical conductivity, and thermal stability [73]. These carbon nanomaterials enhance the properties of polymer composites, leading to applications in the automotive, aerospace, construction, and electronics industries [74].

3.5.6.2 Concrete Reinforcement

They also serve as sustainable reinforcements in concrete and cementitious materials likewise the addition of CNMs improves the mechanical properties, durability, and corrosion resistance of concrete, leading to enhanced structural performance and longevity in infrastructure applications [75].

3.5.7 Sensing and Detection

3.5.7.1 Gas Sensors

Biomass-derived carbon nanotubes, graphene, and carbon quantum dots exhibit high sensitivity and selectivity toward various gases and volatile organic compounds (VOCs) [76]. These materials can be integrated into gas-sensing devices for environmental monitoring, industrial safety, and medical diagnostics.

3.5.7.2 Biosensors

CNMs can be functionalized with biomolecules or enzymes to create biosensors for detecting specific analytes, such as glucose, cholesterol, and DNA. These biosensors offer rapid, sensitive, and selective detection of biological molecules, enabling applications in healthcare, food safety, and environmental monitoring [77].

3.5.8 Water Splitting and Hydrogen Generation

3.5.8.1 Photocatalysis

Biomass-derived CNMs, such as carbon quantum dots and carbon nitride, exhibit photocatalytic activity under visible light irradiation. These materials can catalyze water-splitting reactions to generate hydrogen and oxygen, offering a renewable and sustainable approach to hydrogen production [78].

3.5.8.2 Electrolysis

CNMs serve as efficient electrocatalysts for water electrolysis, enabling the production of hydrogen gas from water using renewable energy sources such as solar or wind power. These materials enhance the kinetics of hydrogen evolution reaction (HER) and oxygen evolution reaction (OER), leading to improved efficiency and stability of electrolysis devices [79].

3.5.9 *Thermal Management*

3.5.9.1 Thermal Interface Materials

Biomass-derived CNFs, graphene, and carbon black can be incorporated into thermal interface materials (TIMs) to improve heat dissipation in electronic devices [80]. These materials enhance thermal conductivity and reduce thermal resistance at interfaces, leading to efficient heat transfer and temperature regulation in electronic components [81].

3.5.9.2 Phase Change Materials

CNMs can be integrated into phase change materials (PCMs) to enhance their thermal conductivity and stability. By encapsulating PCMs within carbon nanocarriers, such as carbon nanotubes or graphene oxide, researchers can achieve rapid heat transfer and phase change kinetics, enabling applications in thermal energy storage and management [82].

3.5.10 *Flexible Electronics*

3.5.10.1 Flexible Sensors

Biomass-derived CNMs such as graphene, and carbon nanofibers can be incorporated into flexible substrates to create wearable sensors for health monitoring, motion detection, and environmental sensing. These flexible sensors offer lightweight, conformable, and stretchable characteristics, enabling comfortable and non-invasive monitoring of physiological parameters [83, 84].

3.5.10.2 Transparent Conductive Films

Graphene and carbon nanotubes derived from agricultural residues serve as transparent conductive materials for flexible electronic displays, touchscreens, and solar cells [85]. These materials offer high electrical conductivity, optical transparency, and mechanical flexibility, enabling the development of next-generation flexible electronics and wearable devices.

3.6 Environmental and Economic Benefits of Utilizing Agro-Industrial Wastes

Agro-industrial wastes, generated from agricultural and food processing activities, represent a significant resource that can be harnessed for various beneficial purposes. By utilizing these wastes effectively in producing CNMs, not only can environmental burdens be reduced, but economic opportunities can also be created. This section of the chapter delves into the environmental and economic benefits of utilizing agro-industrial wastes mainly focusing on the reduction of waste disposal, sustainable resource management, and economic viability and cost-effectiveness.

3.6.1 Reduction of Waste Disposal

Agro-industrial wastes pose significant challenges to waste management and disposal. However, by utilizing these wastes effectively for CNMs, the burden on waste disposal systems can be mitigated.

3.6.1.1 Valorization of By-Products

Agro-industrial wastes, such as crop residues, food scraps, and processing residues, can be valorized through various conversion processes, including composting, anaerobic digestion, and bioconversion [86]. These processes transform organic wastes into valuable products, such as compost, biogas, and biofuels, reducing the volume of waste requiring disposal and minimizing environmental pollution.

3.6.1.2 Prevention of Environmental Pollution

By diverting agro-industrial wastes from landfills and incineration facilities, the release of greenhouse gases, leachate, and other pollutants into the environment can be minimized. Utilization of these wastes for productive purposes reduces the need for landfill space, mitigates soil and water contamination risks, and contributes to overall environmental preservation and sustainability [87].

3.6.2 Sustainable Resource Management

The proper utilization of agro-industrial wastes promotes sustainable resource management practices, ensuring the efficient use of natural resources and reducing environmental impacts:

3.6.2.1 Closed-Loop Systems

Incorporating agro-industrial wastes into closed-loop systems, such as circular economy models, enables the recycling and reuse of resources within the agricultural and industrial sectors. By-products and residues from one process become inputs for another, minimizing waste generation, conserving raw materials, and enhancing resource efficiency throughout the supply chain [88].

3.6.2.2 Nutrient Recycling

Agro-industrial wastes, rich in organic matter and nutrients, can be recycled back into agricultural soils as organic fertilizers and soil amendments [89]. These organic inputs improve soil health, fertility, and structure, enhancing crop productivity, reducing reliance on synthetic fertilizers, and minimizing nutrient runoff and water pollution [90, 91].

3.6.3 Economic Viability and Cost-Effectiveness

The utilization of agro-industrial wastes through the CNMs offers significant economic benefits, including cost savings, revenue generation, and the creation of new business opportunities.

3.6.3.1 Resource Use Efficiency

By converting agro-industrial wastes into value-added products, such as CNMs, revenue streams can be generated from waste streams that were previously considered liabilities [92]. These products can be sold or utilized internally, providing additional sources of income for agricultural and industrial enterprises.

3.6.3.2 Cost-Effective Solutions

Utilizing agro-industrial wastes as feedstocks for bioenergy production, such as biogas, bioethanol, and biodiesel, offers cost-effective alternatives to conventional fossil fuels. These renewable energy sources can reduce dependence on imported energy, stabilize energy costs, and mitigate greenhouse gas emissions, contributing to energy security and climate change mitigation goals [93].

3.7 Factors Influencing Carbon Nanomaterials Synthesis from Agro-Industrial Wastes

The synthesis of CNMs from agricultural wastes is influenced by various factors that affect the efficiency, yield, and properties of the final products. Understanding these factors is crucial for optimizing synthesis processes and achieving desired material characteristics (Fig. 3.4).

3.7.1 *Feedstock Selection and Availability*

The choice of feedstock plays a critical role in determining the composition, structure, and properties of the synthesized carbon nanomaterials [94]. The varied factors influencing feedstock selection and availability include:

3.7.1.1 Chemical Composition

Different agro-industrial wastes contain varying amounts of carbon, cellulose, lignin, and other organic components, which impact the yield and quality of carbon nanomaterials produced. Feedstocks with higher carbon content and lower ash content are typically preferred for carbonization and pyrolysis processes [95].

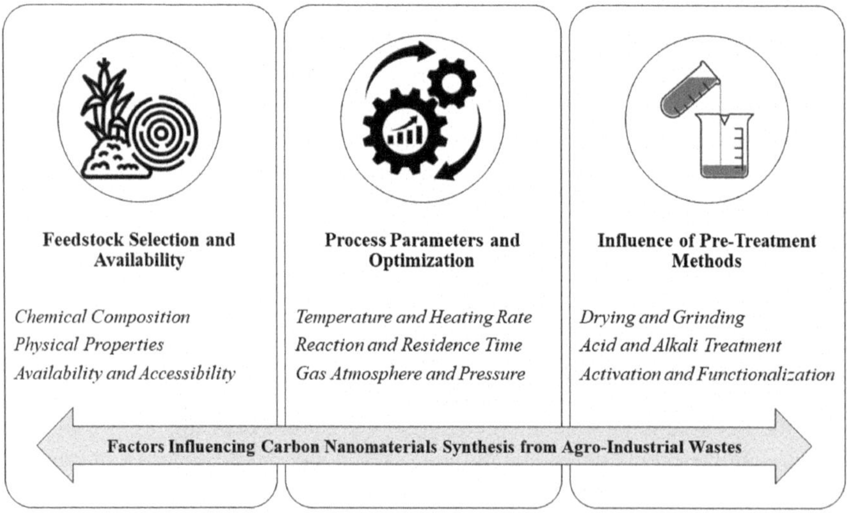

Fig. 3.4 Different factors influencing carbon nanomaterials synthesis from agro-industrial wastes

3.7.1.2 Physical Properties

Feedstock characteristics such as particle size, porosity, and moisture content influence the reaction kinetics, heat transfer, and mass transport during synthesis processes. Fine-grained feedstocks with high surface area-to-volume ratios facilitate rapid pyrolysis and carbonization, leading to higher yields and better material properties [96].

3.7.1.3 Availability and Accessibility

The availability and accessibility of agro-industrial wastes depend on factors such as geographic location, agricultural practices, seasonal variations, and industrial processing methods [95]. Access to abundant and locally sourced feedstocks reduces transportation costs and environmental impacts associated with waste disposal.

3.7.2 Process Parameters and Optimization

The optimization of process parameters is essential for controlling the synthesis conditions and tailoring the properties of CNMs. The key factors influencing process parameters and optimization include:

3.7.2.1 Temperature and Heating Rate

The temperature and heating rate during carbonization and pyrolysis processes significantly affect the formation kinetics, crystallinity, and structure of carbon nanomaterials [97]. Higher temperatures and faster heating rates promote the decomposition of organic precursors and the growth of carbonaceous structures, leading to the formation of desired CNMs [98].

3.7.2.2 Reaction Time and Residence Time

The duration of synthesis reactions and the residence time of feedstocks in reaction chambers influence the extent of carbonization, graphitization, and purification processes. Longer reaction times allow for the complete conversion of organic precursors into carbonaceous materials and the removal of impurities and volatiles [99].

3.7.2.3 Gas Atmosphere and Pressure

The choice of gas atmosphere (e.g., inert, oxidative, or reducing) and pressure conditions during synthesis processes influence the chemical reactions, phase transformations, and surface functionalization of CNMs [4]. Controlled gas environments can enhance material purity, prevent oxidation, and tailor material properties.

3.7.3 Influence of Pre-treatment Methods

Different pre-treatment methods are employed to modify the physical, chemical, and structural properties of agro-industrial wastes prior to synthesis processes. The influence of pre-treatment methods on CNMs synthesis includes:

3.7.3.1 Drying and Grinding

Pre-treatment steps such as drying and grinding reduce the moisture content and particle size of feedstocks, facilitating uniform heating, and pyrolysis reactions [34]. Drying removes excess water that can hinder carbonization processes while grinding increases surface area and enhances heat transfer efficiency.

3.7.3.2 Acid and Alkali Treatment

Chemical pre-treatment with acids or alkalis modifies the surface chemistry and functional groups of feedstocks, promoting the formation of CNMs with tailored properties. Acid treatments remove mineral impurities and enhance carbonization yields, while alkali treatments delignify biomass and increase carbon content [100].

3.7.3.3 Activation and Functionalization

Activation methods such as physical or chemical activation increase the porosity and surface area of carbon nanomaterials, improving their adsorption capacity and catalytic activity [101]. Functionalization with heteroatoms or organic groups introduces specific functionalities and enhances material compatibility for targeted applications [102].

3.8 Challenges and Future Perspectives

The CNMs synthesized from agro-industrial wastes hold immense potential for various applications, but they also pose several challenges that need to be addressed for their widespread adoption.

3.8.1 Challenges Associated with Synthesis and Characterization

Despite significant advancements in the synthesis and characterization of carbon nanomaterials from agro-industrial wastes, several challenges persist such as:

3.8.1.1 Feedstock Heterogeneity

Agro-industrial wastes exhibit inherent heterogeneity in composition, structure, and properties, posing challenges for reproducible synthesis and consistent material quality. Variations in feedstock composition can affect the yield, purity, and performance of CNMs, requiring robust characterization and process optimization strategies [99].

3.8.1.2 Controlled Synthesis Techniques

Achieving precise control over the synthesis parameters and properties of carbon nanomaterials remains a challenge. Synthesis techniques such as pyrolysis, HTC, and CVD often lack scalability, reproducibility, and control over nanomaterial morphology, size, and surface chemistry, limiting their industrial applicability [103].

3.8.1.3 Characterization Methodologies

Comprehensive characterization of carbon nanomaterials requires a multidisciplinary approach involving advanced analytical techniques. Characterization methods such as transmission electron microscopy (TEM), Raman spectroscopy, and X-ray photoelectron spectroscopy (XPS) provide valuable insights into nanomaterial structure, morphology, and composition but may lack sensitivity or specificity for certain properties [104].

3.8.2 *Potential Solutions and Innovations*

Addressing the challenges in the synthesis and characterization of carbon nanomaterials requires innovative approaches and solutions:

3.8.2.1 Feedstock Engineering

Engineering feedstocks through pre-treatment, modification, and blending can enhance their suitability for CNMs synthesis. Preprocessing techniques such as size reduction, purification, and functionalization improve feedstock homogeneity, carbon yield, and material properties, enabling tailored synthesis and enhanced performance [105, 106].

3.8.2.2 Advanced Synthesis Techniques

The development of advanced synthesis techniques with precise control over reaction conditions and nanomaterial properties is essential. Innovative approaches such as plasma-assisted synthesis, template-directed growth [54], and bottom-up assembly enable scalable, tunable, and high-throughput production of CNMs with tailored structures and functionalities [45].

3.8.2.3 Multi-modal Characterization

Integration of multi-modal characterization techniques allows for comprehensive analysis of CNMs. Combining imaging, spectroscopic, and diffraction methods with machine learning algorithms and data analytics facilitates rapid and accurate characterization of nanomaterial structure, defects, and performance under diverse conditions [107].

3.8.3 *Future Directions in Research and Development*

The future of carbon nanomaterials synthesized from agro-industrial wastes hinges on addressing current challenges and exploring new avenues for research and development:

3.8.3.1 Sustainable Synthesis Strategies

Embracing sustainable synthesis strategies that utilize renewable energy sources, green solvents, and environmentally benign catalysts is imperative. Integration of circular economy principles, waste valorization, and life cycle assessment (LCA) approaches ensures the sustainability and eco-friendliness of carbon nanomaterial production processes [108, 109].

3.8.3.2 Functional Material Design

Designing new-generation CNMs with tailored properties and functionalities for specific applications is a promising avenue. Integration of nanocomposites, hybrid materials, and hierarchical structures expands the functionality and versatility of carbon nanomaterials in fields such as energy storage, catalysis, sensing, and biomedical engineering [110, 111].

3.8.3.3 Interdisciplinary Collaboration

Fostering interdisciplinary collaboration and knowledge exchange is essential for accelerating innovation and addressing complex challenges in CNMs research [112]. Collaborative efforts between scientists, engineers, policymakers, and industry stakeholders facilitate technology transfer, standardization, and commercialization of CNMs for societal benefit.

3.9 Conclusion

Remarkably, the conversion of CNMs from agro-industrial wastes is not only the combination of the three scientific fields but also the apex of industrial production with environment protection. It acts as a gateway to green nanotechnology, which uses waste management in a fresh and novel way, and creates alternative routes to resource utilization and recovery. The main findings advocate for the use of a wide range of waste sources for material feedstocks which include materials like agricultural wastes and residues, and food processing discards to get the carbon nanomaterials with special properties and applications. Nevertheless, the issues in scaling-up the synthesis processes on an industrial level remain and that requires addressing these problems through continuous improvement of the process parameters, looking into scalability and the cost aspect. Apart from that, the most relevant obligations are compliance with given regulations and environmental safety standards which raise issues regarding the toxicity of nanomaterials, their impact on the environment, and occupational health and security. Therefore, the collaboration among the researchers, policy-makers, and the industrial entities becomes mandatory. Sustainable feedstock diversification, green technology synthesis, and the development of new characterization initiatives remain key to the future development of advanced material science. Collaboration and innovation are highly needed to foster the future of the industry. Strategies for follow-up research and development include the development of an integrated approach, environmental-friendly synthesis methods, and bringing new and more advanced characterization techniques into play to achieve the harnessing of the maximum capacity of the carbon nanomaterials generated from agro-wastes. At the end of the day, this would ensure the field delivery on its promise to be taken seriously in industries, science, and society as a result of a proper implementing of recommendations to deal with the main challenges and offer technological advancements with minimal negative effects to the environment and guarantee improved societal wellbeing through the responsible use and adoption of CNMs.

References

1. Padhan SR, Kar I, Mohanty A, Panigrahi KK (2024) Nanofertilizers: types, synthesis, methods, and mechanisms. In: Nanotechnology in the life sciences. Springer Nature Switzerland, Cham, pp 61–98
2. Anzar N, Hasan R, Tyagi M et al (2020) Carbon nanotube—a review on synthesis, properties and plethora of applications in the field of biomedical science. Sens Int 1:100003. https://doi.org/10.1016/j.sintl.2020.100003
3. Kumar S, Rani R, Dilbaghi N et al (2017) Carbon nanotubes: a novel material for multifaceted applications in human healthcare. Chem Soc Rev 46:158–196. https://doi.org/10.1039/c6cs00517a
4. Speranza G (2021) Carbon nanomaterials: synthesis, functionalization and sensing applications. Nanomaterials (Basel) 11:967. https://doi.org/10.3390/nano11040967
5. Rahmati M, Mozafari M (2019) Biological response to carbon-family nanomaterials: Interactions at the nano-bio interface. Front Bioeng Biotechnol 7. https://doi.org/10.3389/fbioe.2019.00004
6. Goenka S, Sant V, Sant S (2014) Graphene-based nanomaterials for drug delivery and tissue engineering. J Control Release 173:75–88. https://doi.org/10.1016/j.jconrel.2013.10.017
7. Urade AR, Lahiri I, Suresh KS (2023) Graphene properties, synthesis and applications: a review. JOM (1989) 75:614–630. https://doi.org/10.1007/s11837-022-05505-8
8. Kumar A, Kumar D, Saikia M et al (2023) A review on plant derived carbon quantum dots for bio-imaging. Mater Adv 4:3951–3966. https://doi.org/10.1039/d3ma00254c
9. Sadh PK, Duhan S, Duhan JS (2018) Agro-industrial wastes and their utilization using solid state fermentation: a review. Bioresour Bioprocess 5. https://doi.org/10.1186/s40643-017-0187-z
10. Rathore SS, Padhan SR, Upadhyay PK et al (2023) Biomass management in agriculture. In: Babu S, Das A, Rathore SS, Singh R, Dhar S (eds) Frontiers in agronomy for sustainable agriculture. The Indian Society of Agronomy, New Delhi, pp 217–238
11. Padhan SR, Jat SL, Mishra P et al (2023) Corn for biofuel: status, prospects and implications. In: New prospects of maize. IntechOpen
12. Patel DK, Kim H-B, Dutta SD et al (2020) Carbon nanotubes-based nanomaterials and their agricultural and biotechnological applications. Materials (Basel) 13:1679. https://doi.org/10.3390/ma13071679

13. Salvetat J-P, Bonard J-M, Thomson NH et al (1999) Mechanical properties of carbon nanotubes. Appl Phys A Mater Sci Process 69:255–260. https://doi.org/10.1007/s003390050999
14. Adeola AO, Duarte MP, Naccache R (2023) Microwave-assisted synthesis of carbon-based nanomaterials from biobased resources for water treatment applications: emerging trends and prospects. Front Carbon 2. https://doi.org/10.3389/frcrb.2023.1220021
15. Mohamed A (2019) Synthesis, characterization, and applications carbon nanofibers. In: Carbon-based nanofillers and their rubber nanocomposites. Elsevier, pp 243–257
16. Yang X, Chen Y, Zhang C et al (2023) Electrospun carbon nanofibers and their reinforced composites: preparation, modification, applications, and perspectives. Compos B Eng 249:110386. https://doi.org/10.1016/j.compositesb.2022.110386
17. Cui L, Ren X, Sun M et al (2021) Carbon dots: synthesis, properties and applications. Nanomaterials (Basel) 11:3419. https://doi.org/10.3390/nano11123419
18. Wang L, Wang Y, Hu Y et al (2019) Magnetic networks of carbon quantum dots and Ag particles. J Colloid Interface Sci 539:203–213. https://doi.org/10.1016/j.jcis.2018.12.037
19. Moreno-Lanceta A, Medrano-Bosch M, Melgar-Lesmes P (2020) Single-walled carbon nanohorns as promising nanotube-derived delivery systems to treat cancer. Pharmaceutics 12:850. https://doi.org/10.3390/pharmaceutics12090850
20. Kagkoura A, Tagmatarchis N (2020) Carbon nanohorn-based electrocatalysts for energy conversion. Nanomaterials (Basel) 10:1407. https://doi.org/10.3390/nano10071407
21. Pandit J, Alam MS, Javed MN et al (2023) Emerging roles of carbon nanohorns as sustainable nanomaterials in sensor, catalyst, and biomedical applications. Handbook of green and sustainable nanotechnology. Springer International Publishing, Cham, pp 1721–1747
22. Singh JP, Nandi T, Ghosh SK, Prasad NE (2020) Preparation and isolation of carbon nanorods and "carbon nanoflowers" through combustion of candle wax for heat transfer application. Combust Sci Technol 192:1066–1087. https://doi.org/10.1080/00102202.2019.1609951
23. Pachfule P, Shinde D, Majumder M, Xu Q (2016) Fabrication of carbon nanorods and graphene nanoribbons from a metal–organic framework. Nat Chem 8:718–724. https://doi.org/10.1038/nchem.2515
24. Maity S, Tomar MS, Wasnik K et al (2022) *Azadirachta indica* seed derived carbon nanocapsules: cell imaging, depolarization of mitochondrial membrane potential, and dose-dependent control death of breast cancer. ACS Biomater Sci Eng 8:3608–3622. https://doi.org/10.1021/acsbiomaterials.2c00463
25. Wang L, Liu F, Ning Y et al (2020) Biocompatible mesoporous hollow carbon nanocapsules for high performance supercapacitors. Sci Rep 10. https://doi.org/10.1038/s41598-020-61138-4
26. Cheung KY, Watanabe K, Segawa Y, Itami K (2021) Synthesis of a zigzag carbon nanobelt. Nat Chem 13:255–259. https://doi.org/10.1038/s41557-020-00627-5
27. Monavari SM, Marsusi F, Memarian N, Qasemnazhand M (2023) Carbon nanotubes and nanobelts as potential materials for biosensor. Sci Rep 13. https://doi.org/10.1038/s41598-023-29862-9
28. Perim E, Machado LD, Galvao DS (2014) A brief review on syntheses, structures, and applications of nanoscrolls. Front Mater 1:00031. https://doi.org/10.3389/fmats.2014.00031
29. Huang S, Wang B, Feng M et al (2015) Carbon nanoscrolls fabricated from graphene nanoribbons using Ni nanowire templates: a molecular dynamics simulation. Surf Sci 634:3–8. https://doi.org/10.1016/j.susc.2014.09.016
30. Samaddar P, Ok YS, Kim K-H et al (2018) Synthesis of nanomaterials from various wastes and their new age applications. J Clean Prod 197:1190–1209. https://doi.org/10.1016/j.jclepro.2018.06.262
31. Nath PC, Ojha A, Debnath S et al (2023) Biogeneration of valuable nanomaterials from Agro-wastes: a comprehensive review. Agronomy (Basel) 13:561. https://doi.org/10.3390/agronomy13020561
32. Soffian MS, Abdul Halim FZ, Aziz F et al (2022) Carbon-based material derived from biomass waste for wastewater treatment. Environ Adv 9:100259. https://doi.org/10.1016/j.envadv.2022.100259

33. Zhu Y, Luan Y, Zhao Y et al (2023) Current technologies and uses for fruit and vegetable wastes in a sustainable system: a review. Foods 12:1949. https://doi.org/10.3390/foods12101949
34. Judith JV, Vasudevan N (2021) Synthesis of nanomaterial from industrial waste and its application in environmental pollutant remediation. Environ Eng Res 27:200672–200670. https://doi.org/10.4491/eer.2020.672
35. Jeet K, Kaur R (2023) A novel approach for recycling agricultural wastes in the synthesis of nanomaterials and their composites. Mater Res Express 10:105006. https://doi.org/10.1088/2053-1591/acdecb
36. Flores-Contreras EA, González-González RB, Pablo Pizaña-Aranda JJ et al (2024) Agricultural waste as a sustainable source for nanoparticle synthesis and their antimicrobial properties for food preservation. Front Nanotechnol 6. https://doi.org/10.3389/fnano.2024.1346069
37. Bernd MGS, Bragança SR, Heck N, das Filho LCP (2017) Synthesis of carbon nanostructures by the pyrolysis of wood sawdust in a tubular reactor. J Mater Res Technol 6:171–177. https://doi.org/10.1016/j.jmrt.2016.11.003
38. Devi M, Rawat S, Sharma S (2020) A comprehensive review of the pyrolysis process: from carbon nanomaterial synthesis to waste treatment. Oxford Open Materials Science 1. https://doi.org/10.1093/oxfmat/itab014
39. Ischia G, Fiori L (2021) Hydrothermal carbonization of organic waste and biomass: a review on process, reactor, and plant modeling. Waste Biomass Valorization 12:2797–2824. https://doi.org/10.1007/s12649-020-01255-3
40. Hagner M, Tiilikkala K, Lindqvist I et al (2020) Performance of liquids from slow pyrolysis and hydrothermal carbonization in plant protection. Waste Biomass Valorization 11:1005–1016. https://doi.org/10.1007/s12649-018-00545-1
41. Titirici M-M, Antonietti M (2010) Chemistry and materials options of sustainable carbon materials made by hydrothermal carbonization. Chem Soc Rev 39:103–116. https://doi.org/10.1039/b819318p
42. Manawi Y, Ihsanullah SA et al (2018) A review of carbon nanomaterials' synthesis via the chemical vapor deposition (CVD) method. Materials (Basel) 11:822. https://doi.org/10.3390/ma11050822
43. Li J, Wu Q, Wu J (2015) Synthesis of nanoparticles via solvothermal and hydrothermal methods. Handbook of nanoparticles. Springer International Publishing, Cham, pp 1–28
44. Ndlwana L, Raleie N, Dimpe KM et al (2021) Sustainable hydrothermal and solvothermal synthesis of advanced carbon materials in multidimensional applications: a review. Materials (Basel) 14:5094. https://doi.org/10.3390/ma14175094
45. Sahana S, Gautam A, Singh R, Chandel S (2023) A recent update on development, synthesis methods, properties and application of natural products derived carbon dots. Nat Products Bioprospect 13. https://doi.org/10.1007/s13659-023-00415-x
46. Jha A (2021) Microwave assisted synthesis of organic compounds and nanomaterials. In: Nanofibers—synthesis, properties and applications. IntechOpen
47. Tohamy H-AS, El-Sakhawy M, Kamel S (2023) Microwave-assisted synthesis of amphoteric fluorescence carbon quantum dots and their chromium adsorption from aqueous solution. Sci Rep 13. https://doi.org/10.1038/s41598-023-37894-4
48. Yan Y, Nashath FZ, Chen S et al (2020) Synthesis of graphene: potential carbon precursors and approaches. Nanotechnol Rev 9:1284–1314. https://doi.org/10.1515/ntrev-2020-0100
49. Lebedeva O, Kultin D, Kustov L (2021) Electrochemical synthesis of unique nanomaterials in ionic liquids. Nanomaterials (Basel) 11:3270. https://doi.org/10.3390/nano11123270
50. Hamedani Y, Macha P, Bunning TJ et al (2016) Plasma-enhanced chemical vapor deposition: where we are and the outlook for the future. In: Chemical vapor deposition—recent advances and applications in optical, solar cells and solid state devices. InTech
51. Franz G (2021) Plasma enhanced chemical vapor deposition of organic polymers. Processes (Basel) 9:980. https://doi.org/10.3390/pr9060980
52. Zhang D, Wada H (2021) Laser ablation in liquids for nanomaterial synthesis and applications. Handbook of Laser Micro- and Nano-Engineering. Springer International Publishing, Cham, pp 1481–1515

53. Sportelli M, Izzi M, Volpe A et al (2018) The pros and cons of the use of laser ablation synthesis for the production of silver nano-antimicrobials. Antibiotics (Basel) 7:67. https://doi.org/10.3390/antibiotics7030067
54. Zhang X, Zhang L, Li Y et al (2019) Template-assisted, sol-gel fabrication of biocompatible, hierarchically porous hydroxyapatite scaffolds. Materials (Basel) 12:1274. https://doi.org/10.3390/ma12081274
55. Liu Y, Goebl J, Yin Y (2013) Templated synthesis of nanostructured materials. Chem Soc Rev 42:2610–2653. https://doi.org/10.1039/c2cs35369e
56. Serrano-Bayona R, Chu C, Liu P, Roberts WL (2023) Flame synthesis of carbon and metal-oxide nanoparticles: flame types, effects of combustion parameters on properties and measurement methods. Materials (Basel) 16:1192. https://doi.org/10.3390/ma16031192
57. AlMalki FA, Khashan KS, Jabir MS et al (2022) Eco-friendly synthesis of carbon nanoparticles by laser ablation in water and evaluation of their antibacterial activity. J Nanomater 2022:1–8. https://doi.org/10.1155/2022/7927447
58. Sawant SA, Patil AV, Waikar MR et al (2022) Advances in chemical and biomass-derived graphene/graphene-like nanomaterials for supercapacitors. J Energy Storage 51:104445. https://doi.org/10.1016/j.est.2022.104445
59. Han Y, Ha H, Choi C et al (2023) Review of flexible supercapacitors using carbon nanotube-based electrodes. Appl Sci (Basel) 13:3290. https://doi.org/10.3390/app13053290
60. Fu W, Wang Y, Kong K et al (2023) Materials and processing of lithium-ion battery cathodes. Nanoenergy Advances 3:138–154. https://doi.org/10.3390/nanoenergyadv3020008
61. Dutta V, Verma R, Gopalkrishnan C et al (2022) Bio-inspired synthesis of carbon-based nanomaterials and their potential environmental applications: a state-of-the-art review. Inorganics 10:169. https://doi.org/10.3390/inorganics10100169
62. Julkapli NM, Bagheri S (2015) Graphene supported heterogeneous catalysts: an overview. Int J Hydrogen Energy 40:948–979. https://doi.org/10.1016/j.ijhydene.2014.10.129
63. Shetty A, Molahalli V, Sharma A, Hegde G (2022) Biomass-derived carbon materials in heterogeneous catalysis: a step towards sustainable future. Catalysts 13:20. https://doi.org/10.3390/catal13010020
64. Meng Y, Huang X, Lin H et al (2019) Carbon-based nanomaterials as sustainable noble-metal-free electrocatalysts. Front Chem 7. https://doi.org/10.3389/fchem.2019.00759
65. Das S, Mondal S, Ghosh D (2024) Carbon quantum dots in bioimaging and biomedicines. Front Bioeng Biotechnol 11. https://doi.org/10.3389/fbioe.2023.1333752
66. Mahor A, Singh PP, Bharadwaj P et al (2021) Carbon-based nanomaterials for delivery of biologicals and therapeutics: a cutting-edge technology. C 7:19. https://doi.org/10.3390/c7010019
67. Debnath SK, Srivastava R (2021) Drug delivery with carbon-based nanomaterials as versatile nanocarriers: progress and prospects. Front Nanotechnol 3. https://doi.org/10.3389/fnano.2021.644564
68. Murjani BO, Kadu PS, Bansod M et al (2022) Carbon nanotubes in biomedical applications: current status, promises, and challenges. Carbon Lett 32:1207–1226. https://doi.org/10.1007/s42823-022-00364-4
69. Pandao MR, Lalitha GR, Kishore AJ et al (2023) A comprehensive review on biochar. Int J Environ Clim Change 13:3453–3461. https://doi.org/10.9734/ijecc/2023/v13i113520
70. Jena T, Dayal P, Padhan S et al (2023) Role of biochar in soil carbon dynamics. Innovations in nutrient management: improving efficiency and crop production. Elite Publishing House, New Delhi, India, pp 110–121
71. Sow S, Ranjan S, Seleiman MF et al (2024) Artificial intelligence for maximizing agricultural input use efficiency: exploring nutrient, water and weed management strategies. Phyton (B Aires) 1–10. https://doi.org/10.32604/phyton.2024.052241
72. Saini S, Burman RR, Padaria RN, et al (2024) Vocational opportunities for agricultural migrants in northern India: insights from grassroots. Indian J Agric Sci 94:56–62. https://doi.org/10.56093/ijas.v94i3.148637

73. Rashid AB, Haque M, Islam SMM, Uddin Labib KMR (2024) Nanotechnology-enhanced fiber-reinforced polymer composites: recent advancements on processing techniques and applications. Heliyon 10:e24692. https://doi.org/10.1016/j.heliyon.2024.e24692
74. Latif Z, Ali M, Lee E-J et al (2023) Thermal and mechanical properties of nano-carbon-reinforced polymeric nanocomposites: a review. J Compos Sci 7:441. https://doi.org/10.3390/jcs7100441
75. Yang J, He X (2023) Durability and sustainability of cement and concrete composites. Materials (Basel) 16:5693. https://doi.org/10.3390/ma16165693
76. Yin F, Yue W, Li Y et al (2021) Carbon-based nanomaterials for the detection of volatile organic compounds: a review. Carbon N Y 180:274–297. https://doi.org/10.1016/j.carbon.2021.04.080
77. Sireesha M, Jagadeesh Babu V, Kranthi Kiran AS, Ramakrishna S (2018) A review on carbon nanotubes in biosensor devices and their applications in medicine. Nanocomposites 4:36–57. https://doi.org/10.1080/20550324.2018.1478765
78. Khan MA, Zhao H, Zou W et al (2018) Recent progresses in electrocatalysts for water electrolysis. Electrochem Energy Rev 1:483–530. https://doi.org/10.1007/s41918-018-0014-z
79. Jawhari AH, Hasan N (2023) Nanocomposite electrocatalysts for hydrogen evolution reactions (HERs) for sustainable and efficient hydrogen energy—future prospects. Materials (Basel) 16:3760. https://doi.org/10.3390/ma16103760
80. Khan J, Momin SA, Mariatti M (2020) A review on advanced carbon-based thermal interface materials for electronic devices. Carbon N Y 168:65–112. https://doi.org/10.1016/j.carbon.2020.06.012
81. Jiang Y, Song S, Mi M et al (2023) Improved electrical and thermal conductivities of graphene–carbon nanotube composite film as an advanced thermal interface material. Energies 16:1378. https://doi.org/10.3390/en16031378
82. Mantilla Gilart P, Yedra Martínez Á, González Barriuso M, Manteca Martínez C (2012) Development of PCM/carbon-based composite materials. Sol Energy Mater Sol Cells 107:205–211. https://doi.org/10.1016/j.solmat.2012.06.014
83. Huang H, Su S, Wu N et al (2019) Graphene-based sensors for human health monitoring. Front Chem 7. https://doi.org/10.3389/fchem.2019.00399
84. Song Z, Zhou S, Qin Y et al (2023) Flexible and wearable biosensors for monitoring health conditions. Biosensors (Basel) 13:630. https://doi.org/10.3390/bios13060630
85. Zhou Y, Azumi R (2016) Carbon nanotube based transparent conductive films: progress, challenges, and perspectives. Sci Technol Adv Mater 17:493–516. https://doi.org/10.1080/14686996.2016.1214526
86. Padhan SR, Jat SL, Singh AK et al (2023) Nitrogen dose and placement in conservation agriculture for augmenting cropgrowth, productivity and profitability of Indian mustard (*Brassica juncea*). Indian J Agric Sci 93:1326–1332. https://doi.org/10.56093/ijas.v93i12.140934
87. Bolan NS, Thangarajan R, Seshadri B et al (2013) Landfills as a biorefinery to produce biomass and capture biogas. Bioresour Technol 135:578–587. https://doi.org/10.1016/j.biortech.2012.08.135
88. Saini S, Jirli B, Padhan SR (2023) Awareness mapping of national agriculture market (e-nam) provisions in Rajasthan. J Commun Mobiliz Sustain Dev 18(2):431–439
89. Radheshyam, Jat SL, Parihar CM et al (2023) Better crop establishment, residue recycling and diversification for sustaining non-basmati rice (Oryza sativa) in western IGP. Indian J Agric Sci 93:621–625. https://doi.org/10.56093/ijas.v93i6.134894
90. Kadam PV, Jat SL, Singh AK et al (2022) Innovative approaches to enhance maize growth and yield via residue and nitrogen management in conservation agriculture practices. Maize Journal 11(2):65–74
91. Ghuman RS, Singh H, Singh KB (2024) Tillage, crop residue management and nitrogen application impacts on soil properties under maize–wheat (*Zea mays–Triticum aestivum*) cropping system. Indian J Agron 69:137–143. https://doi.org/10.59797/ija.v69i2.5498
92. Nungula EZ, Raza MA, Nasar J, et al (2024) Cadmium in soil and plants: A review. In: Springer water. Springer Nature Switzerland, Cham, pp 21–43

93. Saleem M (2022) Possibility of utilizing agriculture biomass as a renewable and sustainable future energy source. Heliyon 8:e08905. https://doi.org/10.1016/j.heliyon.2022.e08905
94. Zaytseva O, Neumann G (2016) Carbon nanomaterials: production, impact on plant development, agricultural and environmental applications. Chem Biol Technol Agric 3. https://doi.org/10.1186/s40538-016-0070-8
95. Mujtaba M, Fernandes Fraceto L, Fazeli M et al (2023) Lignocellulosic biomass from agricultural waste to the circular economy: a review with focus on biofuels, biocomposites and bioplastics. J Clean Prod 402:136815. https://doi.org/10.1016/j.jclepro.2023.136815
96. Khot LR, Sankaran S, Maja JM et al (2012) Applications of nanomaterials in agricultural production and crop protection: a review. Crop Prot 35:64–70. https://doi.org/10.1016/j.cropro.2012.01.007
97. Amalina F, Syukor Abd Razak A, Krishnan S et al (2022) Advanced techniques in the production of biochar from lignocellulosic biomass and environmental applications. Clean Mater 6:100137. https://doi.org/10.1016/j.clema.2022.100137
98. Lee S-H, Kim J-H, Kim W-S, Roh J-S (2022) The effect of the heating rate during carbonization on the porosity, strength, and electrical resistivity of graphite blocks using phenolic resin as a binder. Materials (Basel) 15:3259. https://doi.org/10.3390/ma15093259
99. Tiwari SK, Bystrzejewski M, De Adhikari A et al (2022) Methods for the conversion of biomass waste into value-added carbon nanomaterials: Recent progress and applications. Prog Energy Combust Sci 92:101023. https://doi.org/10.1016/j.pecs.2022.101023
100. Maksimova TA, Mishakov IV, Bauman YI et al (2022) Effect of pretreatment with acids on the N-functionalization of carbon nanofibers using melamine. Materials (Basel) 15:8239. https://doi.org/10.3390/ma15228239
101. Zhou C, Wang Y (2020) Recent progress in the conversion of biomass wastes into functional materials for value-added applications. Sci Technol Adv Mater 21:787–804. https://doi.org/10.1080/14686996.2020.1848213
102. Shahcheragh SK, Bagheri Mohagheghi MM, Shirpay A (2023) Effect of physical and chemical activation methods on the structure, optical absorbance, band gap and urbach energy of porous activated carbon. SN Appl Sci 5. https://doi.org/10.1007/s42452-023-05559-6
103. Saleh HM, Hassan AI (2023) Synthesis and characterization of nanomaterials for application in cost-effective electrochemical devices. Sustainability 15:10891. https://doi.org/10.3390/su151410891
104. Jagadeesh P, Rangappa SM, Siengchin S (2023) Advanced characterization techniques for nanostructured materials in biomedical applications. Adv Ind Eng Polym Res. https://doi.org/10.1016/j.aiepr.2023.03.002
105. Hess JR, Ray AE, Rials TG (2019) Editorial: advancements in biomass feedstock preprocessing: Conversion ready feedstocks. Front Energy Res 7. https://doi.org/10.3389/fenrg.2019.00140
106. Zhang C, Ottenheim C, Weingarten M, Ji L (2022) Microbial utilization of next-generation feedstocks for the biomanufacturing of value-added chemicals and food ingredients. Front Bioeng Biotechnol 10. https://doi.org/10.3389/fbioe.2022.874612
107. Choudhary K, DeCost B, Chen C et al (2022) Recent advances and applications of deep learning methods in materials science. Npj Comput Mater 8. https://doi.org/10.1038/s41524-022-00734-6
108. Ncube A, Mtetwa S, Bukhari M et al (2023) Circular economy and green chemistry: the need for radical innovative approaches in the design for new products. Energies 16:1752. https://doi.org/10.3390/en16041752
109. Osman AI, Fang B, Zhang Y et al (2024) Life cycle assessment and techno-economic analysis of sustainable bioenergy production: a review. Environ Chem Lett. https://doi.org/10.1007/s10311-023-01694-z
110. Yadav D, Amini F, Ehrmann A (2020) Recent advances in carbon nanofibers and their applications—a review. Eur Polym J 138:109963. https://doi.org/10.1016/j.eurpolymj.2020.109963

111. Kausar A, Ahmad I, Maaza M, Bocchetta P (2023) Self-healing nanocomposites—advancements and aerospace applications. J Compos Sci 7:148. https://doi.org/10.3390/jcs7040148
112. Saini S, Jirli B, Padhan SR (2023) Analysis of factors promoting the usage of electronic National Agriculture Market in Rajasthan. India. Curr Sci 125(6):643–648

Chapter 4
Biological Waste-Derived Carbon Dots and Their Applications

Subhamoy Banerjee and Ruchira Mukherjee

Abstract Carbon dots (CDots) belong to a zero-dimensional material family which has garnered a lot of scientific attention due to their superior physico-chemical properties like high photostability, water solubility, high quantum yield, tunable optical properties, etc. CDots have gained much attention from the biomedical field due to their superior water solubility, biocompatibility, low toxicity, and ease of surface functionalization resulting in more efficient detection and varieties of target identification. CDots are synthesized using various methods from various sources. Due to the current emphasis on recycling materials, biological waste is an attractive source to synthesize CDots. Biological waste materials are abundant, and it is increasing day by day. Waste-derived products used to have lower production costs given a suitable synthesis method. Biological waste has high carbon content. As a result, large-scale synthesis of CDots at a lower price is possible. Different synthesis methods have been designed to synthesize CDots from a variety of biological waste materials. Different synthesis methods have specific applications like imaging and screening, drug delivery, cancer theranostics, tissue engineering, etc. In the current chapter, we have discussed different methods used to synthesize CDots from biological wastes and their various applications.

Keywords Carbon dots · Bio-waste · Fluorescence · Theranostics · Functionalization · Catalysis

4.1 Introduction

Biological waste is any kind of material that contains biohazardous substances. It is broadly categorized into category 1 and 2. Category 1 consists of biological waste that may cause harm to humans. Category 2 waste is non-infectious waste materials like

S. Banerjee (✉) · R. Mukherjee
Department of Basic Science and Humanities, Institute of Engineering and Management, University of Engineering and Management, Management House, 1, Sector-V, Salt Lake Electronics Complex, Kolkata, West Bengal 700091, India
e-mail: subhamoy.banerjee@iem.edu.in

N. Talreja et al. (eds.), *Waste-Derived Carbon Nanostructures*, Nanostructure Science and Technology, https://doi.org/10.1007/978-3-031-75247-6_4

tissue culture and laboratory waste which do not fall under category 1. The philosophy behind using waste for nanoparticle synthesis is valorization of waste and biomass contains many materials useful for nanoparticle synthesis. Industrial production cost may be reduced significantly if it is derived from waste materials. Significant attention has been paid to synthesized carbon dots from biomass-derived waste materials. Biomass contains a substantial amount of carbon and other heteroatoms. It is used as a green and environment-friendly precursor material facilitating large-scale production of CDots. Biological waste is a cheap and abundant source of precursor material. There are diverse sources of biological waste like household garbage, agriculture waste, perennial grass, fishery materials, animal husbandry-based waste materials, poultry waste, etc. Those waste materials have high protein content, waste materials contain cellulose, hemicellulose, lignin, etc.

Carbon-based nanomaterials (CBNMs) have garnered a lot of interest due to their unique properties. Based on their dimension, carbon nanomaterials are subdivided into four types: Three-dimensional (e.g., nanodiamonds), two-dimensional (e.g., graphene nanosheet), one-dimensional (e.g., carbon nanotubes), and zero-dimensional (e.g., (carbon dots). CDots are characterized as fluorescent nanomaterials below 10 nm in size. CDots are quasi-spherical nanoparticles with sp2/sp3 hybridization at the inner core. The high quantum yield of the CDots depends on the change of its electronic configuration while interacting with the neighboring molecules, and the presence of functional groups (e.g., carboxyl groups) on its surface. CQD-based iron ion detection study reported a 75.5% quantum yield owing to the partial transfer of excited electrons from CDots to d orbital iron ions [1]. Carbon dots are broadly categorized into three types: graphene quantum dots (GQD), carbon quantum dots (CQD), and carbonized polymerized dots (CPDs) [2].

4.1.1 Graphene Quantum Dots (GQDs)

GQDs are graphene blocks with two-dimensional (2D) nanoparticles. It has multiple layers of atoms with several functional groups attached to it. Generally, the thickness is less than 10 nm. GQDs have a graphene lattice inside the core. GQD is a zero-dimensional material that results in quantum confinement and edge effect. The reduction of crystal size to the nanometer scale results in significantly different electron distribution. The honeycomb framework of graphene core in GQDs results in spin splitting in a singlet–triplet stage which is equivalent to an electron–hole pair. This configuration results in improved electronic and optical properties. The opening of the bandgap leads to optical absorption in the broad-spectrum range. GQDs can be formed from various types of graphene-based materials like graphene oxide, carbon nanotubes, fullerenes, etc. GQDs have the following advantages over other graphene-based materials enhanced photoluminescence, better solubility, nontoxicity, higher biological half-life, wide range of surface functionalization. As a result, GQDs enjoy various applications in solar cell fabrication, light- emitting diode, bioimaging, battery and capacitor, theranostic agent, etc.

4.1.2 Carbon Quantum Dots (CQD)

CQDs are carbon nanoparticles with a size of less than 10 nm. They have many unique properties with a wide range of applications. As the nanomaterials have a high surface-to-volume ratio, surface functionalization plays an important role in the functioning of the nanoparticles. There are many approaches taken to functionalize the surface of the CQDs like covalent binding, coordination, π-π interaction, and sol–gel technology [3]. The abundance of oxygen-containing functional groups increases the chance of covalent binding. Photoluminescence is a property of CQDs having multiple applications. Doping of N, C, or P increases the optical tunability of CQDs. The introduction of Mg-citric acid chelate and retention of the carboxyl group leads to N passivation and an increase in photoluminescence.

Inorganic nanoparticle-CQD core–shell hybrid helps the nanocomposite to have fluorescence and magnetic properties which have applications as theranostic agents and catalysts. TiO_2 CQD nanohybrid was synthesized using TiO_2 as a precursor and has shown efficient dye degradation capacity [4]. A novel glucose oxide-bound silver nanoparticle and CQD (CQD/AgNP-GOx) was fabricated to create glucose-specific biosensors [5]. CQD-based metal–organic framework/Nickel foam-based electrocatalyst exhibited efficient water-splitting activity [6]. Hyperbranched polyurethane-supported Pd–Ag-based bimetallic CQD was synthesized and used as a catalyst to convert aryl boronic acid to phenol using ipso-hydroxylation [7]. We found many such applications of synthesized core–shell CQDs in the last 10 years.

4.1.3 Carbonized Polymerized Dots

Carbonized polymer dots (CPDs) represent innovative fluorescent nanomaterials that are an emerging group of CDots consisting of organic polymer chains surrounding a carbon core. CPDs are typically synthesized from monomers of polymers or polymers through methods such as condensation, cross-linking, or mild carbonization. CPDs exhibit excellent fluorescence, high aqueous solubility, biocompatibility, chemical stability, and high surface area available for functionalization. The very small structure of CPDs is comparable to or smaller than the exciton Bohr radius. As a result, quantum confinement effect rules and discrete energy bands appear. The confinement of electrons and holes in the carbon core and polymeric matrix lead to discrete energy bands that exhibit fluorescence upon excitation. The conjugated π electron system also contributes to fluorescence. The delocalized electron system in conjugated structure is excited by an external energy source like a photon, and upon returning to a ground state, it releases energy in the form of bright light, which we know as fluorescence. The surface states and defects act as trap states for charge carriers, e.g., electrons. Upon release from the traps, fluorescence is observed. Surface passivation and modifications alter the density and nature of the surface states; thus, fluorescence can be tuned. The presence of oxygen on the surface confers high aqueous solubility.

CPDs have various applications as chemical sensors for sensing Cu^{2+} and S^{-2} [8], water content 1,4-dioxane [9], and biosensing [10]. Overall various types of CDots have stable luminescent properties, and catalytic properties of varying efficiency and these CDots have found varieties of applications.

4.2 Synthesis of CDots

CDots are synthesized from various sources. In the current chapter, bio-waste-derived CDots synthesis will be discussed. There are two types of synthesis methods used for the preparation of CDots. The methods are top-down and bottom-up approaches. The top-down approach involves breaking down the bulk material into very small particles, down to the nanodimension. A bottom-up approach is building nanostructure from the molecular level using a suitable synthesis method. Top-down approaches like arc discharge, laser ablation, electrochemical oxidation, etc., are not suitable for the synthesis of CDots from biological wastes. Also, it has other limitations like long synthesis time, low quantum yield, and presence of impurities. In this chapter, we will discuss different bottom-up approaches with special emphasis on waste materials. The most common bottom-up approaches are as follows-

4.2.1 Sonochemical/Ultrasonic Fabrication of CDots

The sonochemical synthesis of CDots is achieved by using ultrasonic waves. Ultrasonic waves create very high heat, pressure, and fast thermal exchange in a very small area for a given amount of time. Generally, it measured that during sonochemical synthesis, the temperature reaches more than 5000 K, and pressure reaches 1000 atmospheres, and the heating and cooling rate is 10^{18} K/sec [11, 12]. This unique condition is briefly generated due to acoustic cavitation which facilitates one-pot synthesis of CDots. One of the common chemical precursors is polyethylene glycol (PEG). PEG-400 was ultrasonicated for 0.5–3 h at different amplitudes and temperatures to synthesize CDots with a quantum yield of emission of 16% [13]. In sonochemical synthesis, the degree of carbonization and crystallinity contributes to the photostability [11]. Different metal-doped CDots have been synthesized using ultrasonic fabrication methods like Sn@CD [14], Ga@CD [15]. As acoustic cavitation takes place in the solution, hence, physico-chemical properties of the solvent play a crucial role in efficient CDot synthesis. Different properties like viscosity, surface tension, vapor pressure, and boiling point are critical parameters to control acoustic cavitation. Solvent having lower viscosity and surface tension results in bubble growth which helps in creating highly stable CDots. Lower viscosity facilitates faster mass transfer. The addition of a solute like PEG-400 lowers the surface tension of the solution. The formation of CDots in the presence of PEG-400 is a multistep process which is as follows—condensation, dehydration, and dehydrogenation. The power of the sonication is usually kept at 20–30 W. As the frequency of the sound wave affects the formation and destruction of the bubble, the optimum frequency range of sonication is 100–40 kHz. The time of ultrasonication varies, and it depends on multiple factors like concentration of solute, solvent, dimension of the nanoparticle, etc. Hence, the timing varies from one report to another.

4.2.2 Hydrothermal/Solvothermal Synthesis of CDots

A solvothermal method is a popular synthesis technique for CDot preparation due to its comparative simple one-pot synthesis approach. Solvothermal is an umbrella term where a specific solvent is used with selected solutes in a sealed pressurized vessel for synthesis purposes. When the solvent is water, then we call it hydrothermal synthesis. In CDot synthesis, a hydrothermal method is the most common type of solvothermal synthesis method. Using this method, Fan et al. synthesized blue-emitting water-soluble CDot from urea and malonic acid and used it for rapid identification of picric acid [16]. In another study, N and S co-doped CDots were synthesized using citric acid and cystamine dihydrochloride as a sensor for dichromate ions and multicolor bioimaging [17]. Here, we shall discuss in detail on CDot synthesis from waste biomass instead of pure chemicals. Varieties of sources have been used in CDot synthesis. In many cases, authors used locally available waste biomass resources like tapioca peel and pea peel [18], rice husk [19], orange pericarps [20], wheat straw [21], tobacco stems [22], bagasse [23], fish scales [24], sweet potato [25], olive pomace [26], and many other natural sources. In general, the natural resources are collected, dried, or may be used directly in a pressurized vessel. For example, waste orange peels have been used to synthesize CDots [27] where orange waste peels were collected from a local market. It was washed and dried first in sunlight followed by heating at 150 °C for 10 h in an oven for carbonization. Then 2 g of orange peel was treated with 0.1 M H_2SO_4 solution and washed with water. These acidified orange peels are mixed with sodium hypochlorite solution to normalize the pH to 7. These oxidized orange peels are kept in Teflon-coated autoclave at 180 °C for 12 h synthesis of CDots [27]. One-step CDot synthesis was obtained from tangerine juice [28], polyethylene terephthalate (PET) waste, [29] etc.

4.2.3 Pyrolysis or Carbonization

Pyrolysis is one of the commonly used techniques for synthesizing carbon dots. Pyrolysis follows following steps:

Precursor Selection: Pyrolysis involves the decomposition of organic precursors at higher temperatures in the absence of oxygen. The choice of precursor material plays a crucial role in determining the properties of the resulting carbon dots. Common precursors include organic compounds like citric acid, glucose, urea, and various polymers.

Preparation of Precursor Solution: The precursor material is dissolved or dispersed in a suitable solvent. The solvent choice depends on the precursor's solubility and the desired properties of the carbon dots. Water, ethanol, and other organic solvents are frequently used.

Pyrolysis: The precursor solution is subjected to high temperatures typically ranging from 100 °C to 400 °C in a controlled environment. This thermal treatment induces chemical reactions such as dehydration, carbonization, and fragmentation of the precursor molecules. The absence of oxygen prevents combustion and ensures the formation of carbon-rich products instead of ash.

Nucleation and Growth: During pyrolysis, carbon nuclei form as a result of the decomposition of the precursor molecules. These nuclei act as seeds for the growth of carbon dots. The growth process involves the aggregation of carbon species and their subsequent condensation into small carbonaceous particles.

Surface Passivation: To improve the stability and biocompatibility of the carbon dots, surface passivation is often performed. Passivating agents such as polyethylene glycol (PEG), polyvinylpyrrolidone (PVP), or various amino acids are introduced during or after the synthesis process. These agents help prevent aggregation and enhance the dispersibility of carbon dots in various solvents.

Pyrolysis of sago industrial waste at 400 °C generated water-dispersible CDots [30]. Carbon black recovered from waste tire powder was used to synthesize CDots. The tires were pretreated with H_2SO_4 or KOH to obtain precursor carbon black, and the pyrolysis process was performed at 950 °C in an inert environment [31]. In another study, mangosteen peel was used as raw material for CDot synthesis. The washed and dried mangosteen peel was mixed in urea solution and heated at 200 °C to synthesize CDots [32]. Reverse osmosis (RO) membranes are carbon-based polymeric materials with limited shelf life. Waste RO membrane was subjected to pyrolysis at 600 °C followed by H_2O_2 assisted hydrothermal method to synthesize N- and S-doped water-dispersible CDots (size range 1.3 nm to 6.8 nm) [33]. CDot has been synthesized from collagen waste. Collagen was treated with NaOH solution and pyrolyzed at 350 °C for 2 h. The resultant CDots were purified with 0.22 μm membrane filter [34].

4.2.4 Electrochemical Synthesis of CDots

Electrochemical synthesis of carbon dots offers several advantages including simplicity, scalability, and controllability over the size and properties of the resulting carbon dots. Here's a general overview of the electrochemical synthesis process:

Selection of Precursor: The choice of precursor is crucial in electrochemical synthesis. Common carbon-containing precursors include organic compounds like citric acid, glucose, or other carbon-rich materials such as graphite powder or carbon nanotubes.

Choice of Electrolyte Solution: An appropriate electrolyte solution is prepared, typically consisting of a solvent and supporting electrolytes. Common solvents include water, organic solvents like ethanol, or ionic liquids. Supporting electrolytes such as sodium chloride or potassium chloride may be added to enhance the conductivity of the electrolyte solution.

Electrode Setup: The electrochemical cell is set up with two electrodes—a working electrode and a counter/reference electrode. The working electrode can be made of various materials including glassy carbon, platinum, or indium tin oxide (ITO). The choice of electrode material can influence the properties of the synthesized carbon dots.

Electrochemical Reaction: The electrochemical synthesis is initiated by applying an appropriate potential between the working and counter electrodes. This potential triggers the electrochemical oxidation or reduction of the carbon-containing precursor molecules, leading to the formation of carbon dots.

Control of Reaction Parameters: The size, morphology, and properties of the synthesized carbon dots can be controlled by adjusting various parameters including the applied potential, electrolyte composition, reaction time, and temperature. These parameters can be optimized to tailor the properties of the carbon dots for specific applications.

Collection and Purification: After the electrochemical synthesis, the carbon dots are collected from the electrolyte solution using methods such as centrifugation, filtration, or dialysis. Purification steps may also be employed to remove any impurities or unreacted precursors from the synthesized carbon dots.

Industrial waste orange peel was collected and washed with distilled water. Then, it was subjected to the hydrothermal carbonization method at 180 °C for 60 min. The collected aqueous solution rich in 5 hydroxymethyl furfural (5HMF) was electrolyzed using a platinum (Pt) electrode under constant current (150 mA/cm^2) or constant potential difference (6 V) [35]. Waste agarose gel was used as raw material to synthesize CDots. Sodium citrate and urea were used for one-pot electrolysis of urea and sodium citrate mixture to synthesize CDots. In this method, a Pt electrode was used and 5V DC current for 1 h was applied for the synthesis method [36].

4.2.5 *Microwave-Assisted Synthesis of CDots*

Microwave-assisted synthesis of CDots utilizes microwave energy to convert waste materials into CDots. Microwave technology has been proven to be an excellent technique for CDots synthesis from biological waste. Material type and the time of interaction play a crucial role in determining the outcome of the reaction. The ability of the polar molecules to absorb microwave energy and convert it to heat energy is crucial to the effectiveness of the microwave-assisted chemical reaction. This method is known as dielectric heating. Lemon and onion biomass were mixed in an ammonia solution and microwave at 1450 W for 6 min to produce a 6.15 nm average size of CDots [37]. Microwave-assisted hydrothermal synthesis of CDots from starch using aqueous phosphoric acid solution was achieved which exhibited different fluorescence in toluene and water [38]. Fluorescent CDot was synthesized from biomass waste palm kernel shell using microwave irradiation. In the presence of water and diethylene glycerol (DEGs), the palm kernel shell powder mixture was microwaved at 385 W for 1 to 5 min. The solution was filtered and centrifuged to collect CDots [39]. Oil palm fronds (OPF) are one of the major sources of biomass waste in Malaysia. OPF has been used to synthesize CDots using a microwave-based synthesis method. DEG is used as a reactant solution for OPF, and synthesis was conducted in a microwave oven with 385 W power for 30s. Thus, synthesized CDots exhibited higher photoluminescence [40]. A one-step microwave-assisted synthesis method was applied to produce CDots from papaya seeds. Sundried papaya seeds were mixed with orthophosphoric acid and irradiated in the microwave for the 90s resulting in CDots with a mean size of 1.45 nm [41]. Banana peels were ground and mixed in acetic acid, and CDot was synthesized in a microwave synthesizer machine at 15 bar pressure, 180 °C for 15 min. Then, the sample was cooled to 55 °C, and the particles were dispersed in milli-Q water followed by filtration [42]. Many such studies have been conducted using microwave-assisted synthesis method.

4.2.6 Template-Dependent Synthesis of CDots

Template-dependent synthesis of nanoparticles offers unique advantages over solvent-dependent synthesis methods. The advantages are control of particle size, structure of nanoparticles, and morphology. This method is classified into two types: soft template and hard template-dependent synthesis. Oleylamine-capped synthesis of CDots is an example of soft template synthesis. Oleylamine (OA) was mixed with octadecane in a specific ratio after OA was mixed with citric acid and diluted nitric acid. The mixture was heated at 250 °C for 2 h in an inert environment (in the presence of Ar) to synthesize CDots with high quantum yield [43]. Mesoporous silica is used as a template for CDot synthesis. The 3-Aminopropyl-triethoxysilane (APTES) is used as a precursor and mesoporous silica as a template. Mesoporous silica with an average particle size of 500 nm was chosen as the template. APTES mixed in ethanol with silica beads were heated at 250 °C for 2 h. CDots were synthesized inside the pores of the silica beads [44]. Amino-modified mesoporous SiO_2 nanoparticles were used as scaffold/template to synthesize CDots. Benzene 1, 4 diboronic acid was used as a precursor for CDots [45]. Gadolinium-encapsulated CDot has been synthesized using mesoporous silica nanoparticles as a template. Mesoporous silica nanoparticles (MSN) have been synthesized using cetyltrimethylammonium bromide (CTAB) using the co-condensation method. Thus, synthesized MSN upon calcination produces CDots (size range 3–9.7 nm in diameter) [46]. In Table 4.1, different types of sources of biomass are mentioned (Fig. 4.2).

Table 4.1 Synthesized CDots from different types of biomass waste and their respective synthesis methods and applications

Source of biomass waste	Synthesis method	Property	Ref.
Tapioca and pea peel	Hydrothermal	Fluorescence enhancement and quenching, amorphous	[18]
Rice husk	Hydrothermal	Average size 8.3 nm, identify Fe^{3+} ion	[19]
Orange pericarp	Hydrothermal	Average size 6–7 nm, Ca^{2+}, Zn^{2+}, and trypsin detection	[20]
Wheat straw	Hydrothermal	< 2 nm, biomedical imaging	[21]
Tobacco stem	Hydrothermal	Average diameter 2.8 nm, detect tetracycline antibiotic	[22]
Sugarcane industrial waste	Hydrothermal	Average size 1.7 nm, inhibition toward Gram + ve and –ve bacteria	[23]
Fish scales	Hydrothermal	Average size 4.7 nm, detect malachite green	[24]
Sweet potato	Hydrothermal	Detection of Fe^{3+}	[25]
Olive solid waste	Pyrolysis and oxidation	Detection of Fe^{3+} in water	[26]
Orange waste peels	Hydrothermal carbonization	CDot-embedded ZnO NPs perform photocatalysis of naphthol blue black azo dyes	[27]
Tangerine juice	Hydrothermal	Average size 1.92 nm, used as gilded ink	[28]
PET waste	Hydrothermal	Average size 2 nm, used as a light-blocking film when used with PVA	[29]
Sago waste	Pyrolysis carbonization	Optical sensing of metal ions	[30]
Mangosteen peel	Pyrolysis	Average size 2–15 nm based on temperature	[32]
Collagen waste	Pyrolysis	Average size 4.1 ± 2.2 nm, detection of Fe^{3+}	[34]
Lemon and onion juice	Microwave-assisted carbonization	Average size 4.23–8.22 nm, detection of riboflavin	[37]
Starch	Microwave-assisted hydrothermal synthesis	Average size > 10 nm	[38]
Palm kernel shell and palm fronds	Microwave-assisted	Average size 6–7 nm, detection of heavy metal ions	[39, 40]
Papaya seeds	Microwave-assisted	Average size 1.45 nm, detection of Fe^{3+}	[41]
Banana peels	Microwave-assisted	Average size 35 nm, detection of Cu^{2+} and tryptophan	[42]

4.3 Characterization of CDots

Synthesized CDots needed to be characterized to ascertain size, morphology, surface charge, composition, crystallinity, etc. [47]. The characterization techniques are broadly classified into the following techniques:

4.3.1 Microscopy

Various microscopic techniques are employed to characterize CDots. The techniques are scanning electron microscopy (SEM), transmission electron microscopy (TEM), and atomic force microscopy (AFM). SEM offers a few nm resolution and a 3D image of the CDot. TEM and high-resolution TEM (HRTEM) offer spatial resolution to atomic dimension and lattice arrangement of the CDot. AFM offers 2D and 3D images of the particle with a resolution at the atomic level.

4.3.2 Mass Spectrometry

Mass spectrometry (MS) is employed to characterize chemical composition based on their mass. For CDots, two MS techniques are used, namely electrospray ionization quadrupole time-of-flight tandem mass spectrometry (ESIQ-TOF–MS/MS) and matrix-assisted laser desorption/ionization time-of-flight mass spectrometry (MALDI-TOF MS), both of which uses soft ionization technique [47].

4.3.3 Spectroscopy

Spectroscopy is another very important technique to study different optical, magnetic, and surface topological characteristics of CDot. Commonly employed techniques are photoluminescence spectroscopy (PL), ultraviolet–visible spectroscopy (UV–Vis), infrared (IR) spectroscopy, Raman spectroscopy, energy-dispersive X-ray (EDX) spectroscopy, X-ray photoelectron spectroscopy (XPS), dynamic light scattering (DLS), and nuclear magnetic resonance (NMR). PL and UV–Vis are commonly used to calculate the quantum yield of CDot and IR to characterize functional groups present on CDots (especially when it is biomass-derived). Raman spectroscopy is employed to determine the carbon state of the CDot. EDX is applied to analyze the elemental composition of CDot. XPS is used to determine the electron state of the elements present in CDot. DLS measures a hydrodynamic diameter of CDot, and zeta potential measures the surface charge on it. NMR determines chemical bond formation, the presence of types of elements, etc.

4.3.4 Diffraction Technique

The X-ray diffraction (XRD) technique is used to determine the crystal lattice information, phase purity, etc., via studying diffraction patterns.

4.4 Application of CDots

CDots have various unique properties which have attracted scientists to apply them in different fields from sensors to theranostic agents. In this section, we will discuss various applications of CDots with a focus on CDots synthesized from biomass or waste products. Major applications are: 1. Electronic devices and charge storage, 2. Sensor, 3. Bioimaging, 4. Antimicrobial agent, 5. Drug delivery (Fig. 4.3).

4.4.1 Electronic Devices and Charge Storage

Carbon shows unique electronic properties like high electrical conductivity, and chemical and heat stability. CDots synthesized from rosemary leaves are used to increase the photoconversion efficiency of TiO_2-based dye-sensitized solar cells (DSSC) [48]. In another study, one-step hydrothermal synthesis of CDot/TiO_2 composite material coated was used as a photovoltaic cell for DSSC and exhibited high photovoltaic conversion efficiency [49]. CDot-ZnO nanoparticle-based hybrid photovoltaic nanostructure was formed, and higher transition conversion efficiency was identified [50]. CDots were synthesized from citric acid, and ethylenediamine and a composite was formed with NiO (NiO@CDots). This composite was used as a photocathode with higher photovoltaic conversion efficiency [51]. CDot doping in semiconductors consisting of n-TiO_2 and p-NiO heterojunction increases the charge transfer efficiency. On the contrary, CDot doping in n-TiO_2 and p-CuO heterojunction decreases charge transfer efficiency [52].

4.4.2 Sensor

The presence of different functional groups like hydroxyl, epoxy, carbonyl, and carboxyl groups on the carbon core makes CDots water-soluble, and they can be easily functionalized with different substrates. The presence of the functional groups on the CDot confers various important properties like optical property, biocompatibility, and target specificity which makes it useful as a sensor. The absorbance range of the CDot varies from UV to NIR (near-infrared) range. CDots show two absorption peaks, one at 280 nm and the other at 350 nm. The 280 nm peak is generated due to the π-π* transition of the C = C bond and 350 nm corresponds to the n-π* electronic transition of the C = O bond [53]. The mode of CDot-based sensor functionality depends on fluorescence quenching or prevention quenching in the presence of a target analyte. There are various types of quenching methods like static and dynamic quenching, fluorescence resonance energy transfer (FRET), inner field effect (IFE), and photo-induced electron transfer (PET). The detection range can reach up to picomolar or femtomolar level. Commonly used pesticides such as atrazine, chlorpyrifos, imidacloprid, lindane, and tetradifon were detected using CDot which is synthesized from gelatin. Imidacloprid was detected through enhancement of CDot fluorescence which is described as a turn-on state. Other pesticides were detected by quenching the fluorescence of CDots stated as turn-off [54]. Neurotransmitter dopamine and ascorbic acid are crucial parameters for neurodegenerative disorders. CDots synthesized from tea were used to detect dopamine and ascorbic acid via the fluorescence quenching method [55]. CDot synthesized using corn bract used to detect Hg^{2+} ion detection using ratiometric fluorescence assay. Fluorescence at 678 nm was quenched and 470 nm was slightly altered. The fluorescence intensity ratio was used to detect Hg^{2+} ions with a lower detection range of 9 nM, and the detection limit is 0–40 μM [56]. CDot functionalized polydimethylsiloxane (PDMS) coated paper strip was used to detect folic acid. The fluorescence quenching response was recorded for 1–300 μM with a limit of detection 0.28 μM [57]. Cu^{2+} plays a crucial role in plant growth and development. CDot-based sensing platform was developed for real-time imaging-based detection of Cu^{2+} distribution in plant systems [58]. Ochratoxin A (OTA) is a fungal toxin used to contaminate food and feed. Nitrogen sulfur co-doped CDot has been synthesized, and an aptamer-based FRET sensor was designed to detect ochratoxin A. OTA-based aptamer was designed, and cDNA was synthesized as a recognition and transformation element. Upon sensing OTA, c-DNA is released and causes the change in entropy and the release of Cu^{2+} which oxidizes O-phenylenediamine (oPD) to generate 2,3-diaminophenazine (DAP) which generates yellow fluorescence which in turn triggers FRET between N/S co-doped CDot and DAP and generated blue fluorescence. The ratiometric fluorescence change was used to estimate the concentration of OTA [59]. CDot synthesized from water amaranth was used to detect Cd^{2+} ions. Hence, synthesized CDot was covalently functionalized with pyrene carboxaldehyde which used ratiometric fluorescence and was used as Cd^{2+}nanosensor with a lower detection limit of 15 nM [60]. Many such applications of CDot as sensors have been explored.

4.4.3 Bioimaging Agent

Being small in size, nontoxic, and tunable fluorescent, CDots have been extensively used as bioimaging agents for the last decade. The choice of probes for in vivo imaging is crucial for getting proper results. CDot has been exploited as an in vivo probe for different studies. CDot was synthesized, and cresyl violet (CV) was functionalized on it. It was used as a red/near-IR sensing fluorescent probe for imaging purposes. The probe was used to image tumors present in nude mice [61]. N-doped CDot has been synthesized from 3-(3,4-dihydroxyphenyl)-L-alanine (L-DOPA), L-histidine, and L-arginine by heating and carbonization. Low oxidation level in doped N and N-CDots increase their signal intensity. HeLa and HepG2 cells were imaged at higher resolution with these probes [62]. CDots doped with P and Mn confer higher fluorescence and MRI properties. Functionalization with hyaluronic acid increased target specificity in vivo and targeted CD44 surface antigen on the cancer cells which could be imaged using MRI [63]. Fe-doped CDot is synthesized from dopamine hydrochloride (DA) and o-phenylenediamine (oPD), and $FeCl_3$ is used as a dopant. Thus synthesized Fe-CDots were used as fluorescent probes with emission peaks at the NIR-II region which was used in vivo imaging of gastric pH change [64]. CDots have been used as lysosome-targeting fluorescent probes. p-Phenylenediamine and thiourea were used as precursors for hydrothermal synthesis of CDots. Imaging of the lysosome-targeting ability of CDots both in in vitro and in vivo conditions was acquired [65]. 2-nitro-4-aminodiphenylamine (NAP)-derived CDots have been synthesized, and their asymmetric electronic structure made them sensitive to changes in polarity. Cancer cells have lower polarity compared to normal cells and effective imaging of cancer cells has been done [66]. This work was further extended by modifying CDots with 2-formylphenylboronic acid pinacol ester (2-FAPE), and it was highly sensitive to polarity changes. It was used to image and quantify lipid droplets within cytoplasm [67]. Fluorescent CDots have been synthesized from 2,4-dihydroxybenzaldehyde and 2,3-dimethylbenzothiazole iodide using a microwave-based synthesis method. Thus, synthesized fluorescent CDot was used for nucleolus imaging as it was able to bind RNA in the nucleolus [68].

4.4.4 Antimicrobial Agent

CDots have photosensitizing properties which can be used in photodynamic therapy. Many classes of drug-resistant bacteria can be treated by photodynamic inactivation (PDI). CDots are generally considered nontoxic. However, their photonic activation ranges from UV to NIR where CDots show a strong photodynamic effect. Excited CDots release reactive oxygen species (ROS) which can kill microorganisms. Also, ROS induces intracellular protein inactivation, mitochondrial dysfunction, membrane disintegration, etc. [69]. Garlic-derived CDots have been synthesized using one-step hydrothermal; synthesis. Thus, synthesized CDot was used to kill methicillin-resistant *Staphylococcus aureus* (MRSA) and multidrug-resistant *Salmonella typhimurium* (MRST). CDots effectively inhibit biofilm formation in a dose-dependent manner [70]. CDot synthesized from pomegranate peels and its microbicidal effect was checked on pathogenic bacteria. CDot was synthesized using low-temperature carbonization. *S. aureus* and *K. pneumonia* were found to be susceptible to these synthesized CDots, and *P. aeruginosa* showed moderate resistance [71]. In another study, CDots were synthesized from gum Arabic using microwave energy, and it was conjugated with ciprofloxacin hydrochloride, a broad-spectrum antibiotic. Besides Vero cells, it was tested on Gram-positive bacteria *Bacillus subtilis* and *Staphylococcus aureus* and Gram-negative *Escherichia coli* and *Pseudomonas aeruginosa* and exhibited differential sensitivities toward CDots [72]. 2, 2′-(ethylenedioxy)bis(ethylamine) (EDA)-CDots were synthesized from carbon nanopowder, and methylene blue (MB) or toluene blue (TB) was conjugated with it, and it was used as a photosensitizer. *Escherichia coli* cells were incubated with synthesized photosensitizer-conjugated CDots under white light (36 W), and MB or TB-conjugated CDots showed significant inhibitory activities [73]. Photodynamic antimicrobial chemotherapy (PACT) is an approach to kill microbes where CDots facilitate photodynamic therapy. Meta-phenylenediamine was used to synthesize nitrogen-doped carbon dots (N@CDots) using the hydrothermal synthesis method. It was tested on Gram-negative *Escherichia coli* and Gram-positive *Staphylococcus aureus* to check inhibitory concentration [74].

4.4.5 Drug Delivery Agent

The availability of different functional groups on the CDots opens up a unique opportunity to deliver drugs to the target site. Besides drug delivery, there are reports on gene and peptide delivery. Also, there are several reports on the multimodal application of CDots. Green fluorescent CDot was synthesized from citric acid and 3,5-aminobenzoic acid used for gene delivery and biosensing of pH change and NO_2^-. ROX-labeled DNA has been transferred to nucleolus which was monitored through imaging[75]. Folic acid-conjugated CDots were synthesized, their efficacy was checked on HeLa cells, and also it facilitated the delivery of the anticancer drug doxorubicin [76]. CDots were synthesized from ginger extract using the hydrothermal synthesis method, and it was tested on the hepatocellular carcinoma cell line (HepG2). HepG2 cell line was found to be inhibited by CDots [77]. In another approach, CDots were synthesized from chlorogenic acid (ChA), which is a natural product of coffee. ChACDot suppressed HepG2-induced tumors inside mice and recruited tumor-infiltrating immune cells. It was found that ChACDot exhibited glutathione oxidase-like activity and induced cancer cell ferroptosis [78]. Ferric oxide-CDots (Fe_3O_4-CDot) were synthesized and modified with ferrocene dicarboxylic acid (Fc) upon formation of the amide bond and crosslinked into Fe_3O_4-CDot@Fc nanocomplex. CDot facilitated the Fenton reaction of Fe_3O_4 by improving electron transfer efficiency. At 808 nm, it significantly improved thermal conversion efficiency, increased cellular oxygen level, and inhibited HIF1α. Also, the system induced immune-mediated cell death, and in vivo upon conjugation with PD-L1 antibody, it ablated the primary tumor and inhibited distant cells [79]. Red carbon dots are N-doped CDots having red fluorescence which are excellent bioimaging agents. A nanomedicine delivery system comprising Her-2 aptamer, polyethylene glycol (PEG), red CDot, and docitaxel was developed aiming at Her-2 positive breast cancer cells. The system was designed for pH-triggered drug release in low pH (pH 5.5 to 6.5) cancer microenvironment. Also, red fluorescence facilitated real-time drug delivery monitoring in situ [80]. Quercetin and p-phenylenediamine (p-PD)-derived red CDot was synthesized. It showed inhibition of amyloid β (Aβ) aggregation, rapid depolymerization of Aβ fibrils, scavenging of ROS, and red fluorescence got 'turned on' which helped in real-time imaging of Aβ plaques in situ [81]. In another study, CeO_2 nanoparticle was synthesized on the cell membrane of macrophage cells named RAW cell lines. CDots was biomimetically synthesized on cerium oxide nanoparticle (CeO_2) using it as template. RAW cell membrane protected the system from phagocytosis and inflammatory activities. Thus, synthesized CDot scavenged ROS and reduced oxidative stress. The role of CDots as photosensitizer dissociated Aβ aggregation by chelation of excess copper ion. It helped to improve cell survival rate, thus providing a promising nanosystem for Alzheimer therapeutics [82]. In another study, CDot was synthesized from 2-acrylamido-2-methylpropanesulfonic acid. The anticancer drug riluzole was delivered using CDot. Multiple sulfonated, carboxylic, and amide groups were present on the CDot. The CDot-mediated drug delivery induced the death of the glioblastoma cell line [83]. Besides multiple studies on cancer and Alzheimer's therapeutics, CDot-based therapeutics have been tested on ulcerative colitis (UC). Rhein is a bioactive compound of Chinese medicine. CDot was synthesized from rhein and doped with L-Arginine to synthesize red/NIR-I fluorescent CDots. In vitro studies showed that thus synthesized CDot scavenged ROS and protected cells from inflammation. Test on DSS-induced mouse model also showed reduction of intestinal inflammation [84].

4.5 Conclusion

CDots synthesized from biomass or waste are gaining traction for several reasons discussed in the introduction section (Sect. 4.1). Broadly, there are two types of synthesis methods namely the top top-down and bottom-up approach (Fig. 4.2). Generally, top-down approaches are not employed for biomass-derived CDots. There are various bottom-up approaches have been discussed in the synthesis section (Sect. 4.2). Synthesized CDot needs to be characterized and various characterization methods are employed which are briefly discussed in the characterization section (Sect. 4.3). Different applications of CDots have been discussed. In different fields from electronics to healthcare, various applications of CDots have been reported (Fig. 4.1). Some of the latest applications have been discussed in the application section (Sect. 4.4). Despite the large array of applications and associated promises made by the reports, some limitations of carbon dots need to be addressed and solved. One such limitation is complex surface modification when it is derived from biomass or waste. There is always a chance of undesirable interactions with neighboring molecules [85]. Some methods are plagued with low quantum yield problems. Stability inside the body is also a problem. Bioavailability and usefulness as a mainstream drug molecule is still an open challenge for the scientific community. However, given the rapid progress in this field, CDot can be called a promising molecule with application in various fields in the near future.

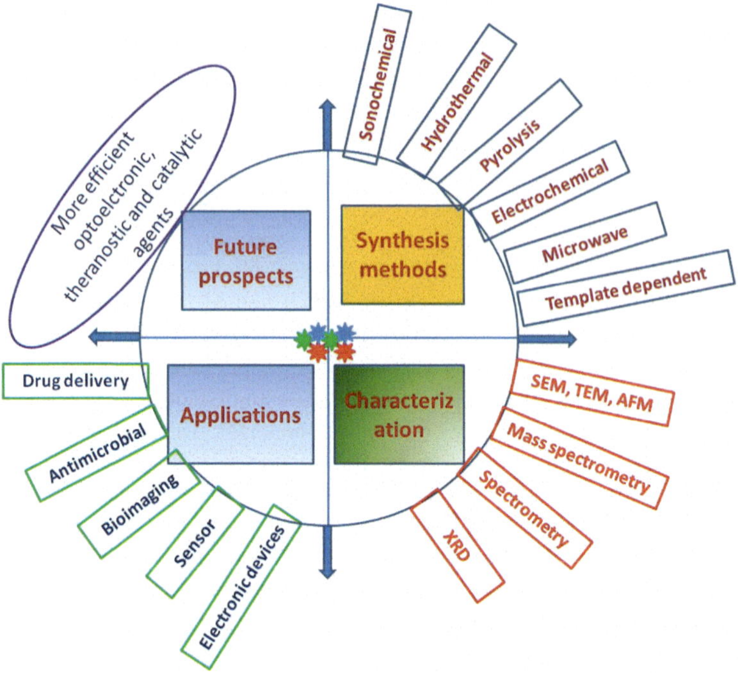

Fig. 4.1 Different types of CDots synthesis methods and characterization, application, and future aspects

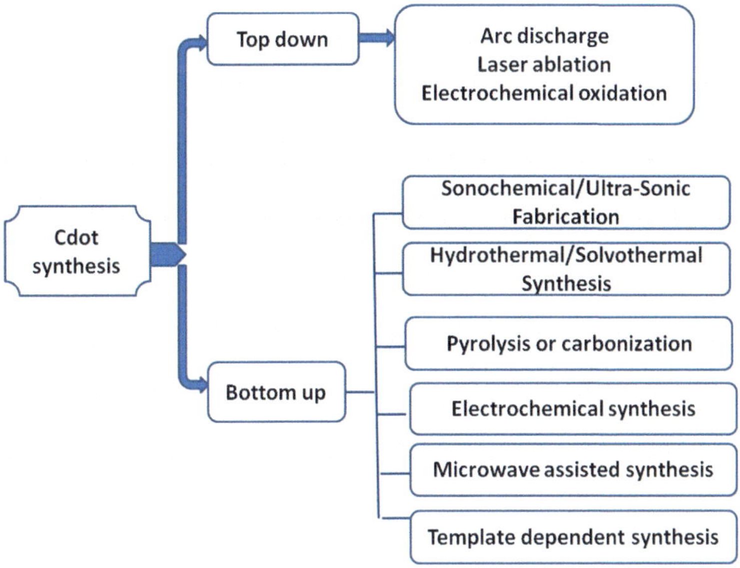

Fig. 4.2 CDot synthesis is principally subdivided into two types—top-down and bottom-up approaches. Different types of bottom-up approaches are most commonly employed to synthesize CDots

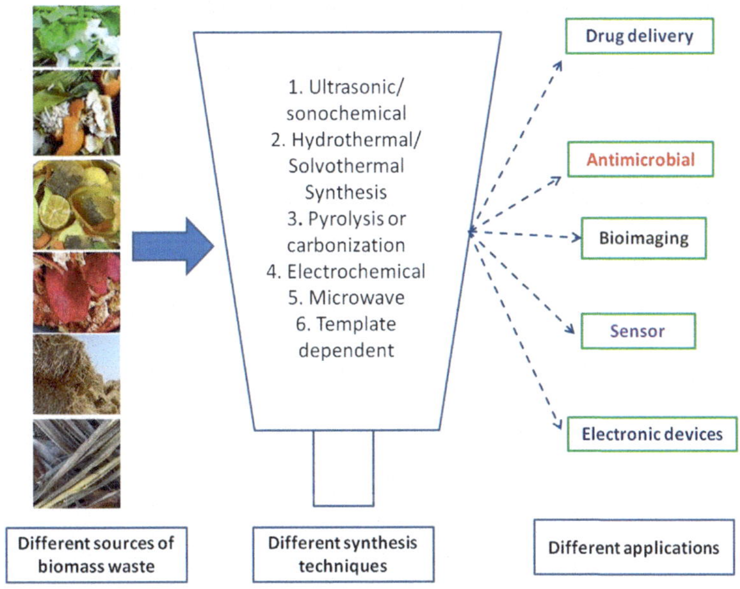

Fig. 4.3 Overview of various sources of biomass waste is converted into CDots using a different technique which has different applications based on their properties and further modifications

References

1. Siahcheshm P, Heiden P (2023) High quantum yield carbon quantum dots as selective fluorescent turn-off probes for dual detection of Fe2+/Fe3+ ions. J Photochem Photobiol A 435:114284. https://doi.org/10.1016/j.jphotochem.2022.114284
2. Cui L, Ren X, Sun M, Liu H, Xia L (2021) Carbon dots: synthesis, properties and applications. Nanomaterials 11(12). https://doi.org/10.3390/nano11123419
3. Wang Y, Hu A (2014) Carbon quantum dots: synthesis, properties and applications. J Mater Chem C 2(34):6921–6939. https://doi.org/10.1039/C4TC00988F
4. Kumar MS, Yasoda KY, Kumaresan D, Kothurkar NK, Batabyal SK (2018) TiO2-carbon quantum dots (CQD) nanohybrid: enhanced photocatalytic activity. Mater Res Exp 5(7):075502. https://doi.org/10.1088/2053-1591/aacbb9
5. Buk V, Pemble ME, Twomey K (2019) Fabrication and evaluation of a carbon quantum dot/gold nanoparticle nanohybrid material integrated onto planar micro gold electrodes for potential bioelectrochemical sensing applications. Electrochim Acta 293:307–317. https://doi.org/10.1016/j.electacta.2018.10.038
6. Hedau B, Park S-J, Kang B-C, Ha T-J (2024) Bifunctional carbon quantum dot-embedded metal-organic framework nanohybrid as a highly efficient electrocatalyst for water splitting and CO_2 reduction. Carbon 216:118527. https://doi.org/10.1016/j.carbon.2023.118527
7. Bayan R, Karak N (2018) Hyperbranched polyurethane-supported Pd-Ag@CQD nanocomposite: a high performing heterogeneous catalyst. ChemistrySelect 3(40):11210–11218. https://doi.org/10.1002/slct.201802403
8. Xu T, Zhang X, Yang H, Li H, Zhang S, Yang Z et al (2023) Promising carbonized polymer dots for visually sequential sensing of Cu2+ and S2−. Chem Eng J 467:143465. https://doi.org/10.1016/j.cej.2023.143465
9. Huo Z, Xu W, Wang Z, Xu S (2021) A carbonized polymer dot (CPD) nanosensor for trace water detection with a wide detection range. Dyes Pigm 196:109805. https://doi.org/10.1016/j.dyepig.2021.109805
10. Cheng X, Huang Y, Li D, Yuan C, Li Z-L, Sun L et al (2019) A sensitive polymer dots fluorescent sensor for determination of α-L-fucosidase activity in human serum. Sens Actuators, B Chem 288:38–43
11. Kumar R, Kumar VB, Gedanken A (2020) Sonochemical synthesis of carbon dots, mechanism, effect of parameters, and catalytic, energy, biomedical and tissue engineering applications. Ultrason Sonochem 64:105009. https://doi.org/10.1016/j.ultsonch.2020.105009
12. Xu H, Zeiger BW, Suslick KS (2013) Sonochemical synthesis of nanomaterials. Chem Soc Rev 42(7):2555–2567. https://doi.org/10.1039/c2cs35282f
13. Kumar VB, Porat Z, Gedanken A (2016) Facile one-step sonochemical synthesis of ultrafine and stable fluorescent C-dots. Ultrason Sonochem 28:367–375. https://doi.org/10.1016/j.ultsonch.2015.08.005
14. Kumar VB, Tang J, Lee KJ, Pol VG, Gedanken A (2016) In situ sonochemical synthesis of luminescent Sn@ C-dots and a hybrid Sn@ C-dots@ Sn anode for lithium-ion batteries. RSC Adv 6(70):66256–66265
15. Parveen K, Rafique U, Akhtar MJ, Ashokumar M (2018) Ultrasound-assisted synthesis of gallium hybrids for environmental remediation application. Ultrason Sonochem 49:222–232. https://doi.org/10.1016/j.ultsonch.2018.08.013
16. Fan YZ, Zhang Y, Li N, Liu SG, Liu T, Li NB et al (2017) A facile synthesis of water-soluble carbon dots as a label-free fluorescent probe for rapid, selective and sensitive detection of picric acid. Sens Actuat B Chem 240:949–955. https://doi.org/10.1016/j.snb.2016.09.063
17. Chen J, Liu J, Li J, Xu L, Qiao Y (2017) One-pot synthesis of nitrogen and sulfur co-doped carbon dots and its application for sensor and multicolor cellular imaging. J Colloid Interface Sci 485:167–174. https://doi.org/10.1016/j.jcis.2016.09.040
18. George E, Meerajanaky M, Sunny M, Shinoy D, Krishnan A, Kurian M (2023) Synthesis and characterisation of waste biomass derived carbon dot decorated zinc nanoferrites. Mater Today: Proceed. https://doi.org/10.1016/j.matpr.2023.11.074

19. Kundu A, Maity B, Basu S (2022) Rice husk-derived carbon quantum dots-based dual-mode nanoprobe for selective and sensitive detection of Fe(3+) and fluoroquinolones. ACS Biomater Sci Eng 8(11):4764–4776. https://doi.org/10.1021/acsbiomaterials.2c00798
20. Tolou-Shikhzadeh-Yazdi S, Shakibapour N, Hosseini S, Mokaberi P, Malaekeh-Nikouei B, Chamani J (2023) High-efficient synthesis of carbon quantum dots from orange pericarp as fluorescence turn-on probes for Ca(2+) and Zn(2+) ion detection and their application in trypsin activity characterization. Iran J Basic Med Sci 26(2):190–199. https://doi.org/10.22038/IJBMS.2022.67323.14758
21. Yuan M, Zhong R, Gao H, Li W, Yun X, Liu J et al (2015) One-step, green, and economic synthesis of water-soluble photoluminescent carbon dots by hydrothermal treatment of wheat straw, and their bio-applications in labeling, imaging, and sensing. Appl Surf Sci 355:1136–1144. https://doi.org/10.1016/j.apsusc.2015.07.095
22. Yang H, Wei Y, Yan X, Nie C, Sun Z, Hao L et al (2022) High-efficiency utilization of waste tobacco stems to synthesize novel biomass-based carbon dots for precise detection of tetracycline antibiotic residues. Nanomaterials 12(18). https://doi.org/10.3390/nano12183241
23. Pandiyan S, Arumugam L, Srirengan SP, Pitchan R, Sevugan P, Kannan K et al (2020) Biocompatible carbon quantum dots derived from sugarcane industrial wastes for effective nonlinear optical behavior and antimicrobial activity applications. ACS Omega 5(47):30363–30372. https://doi.org/10.1021/acsomega.0c03290
24. Du H, He L, Zhang M, Manyande A, Chen H (2023) Carbon quantum dots derived from fish scales as fluorescence sensors for detection of malachite green. J Food Measure Characteriz 17(4):3368–3376. https://doi.org/10.1007/s11694-023-01864-1
25. Shen J, Shang S, Chen X, Wang D, Cai Y (2017) Facile synthesis of fluorescence carbon dots from sweet potato for Fe(3+) sensing and cell imaging. Mater Sci Eng, C Mater Biol Appl 76:856–864. https://doi.org/10.1016/j.msec.2017.03.178
26. Sawalha S, Silvestri A, Criado A, Bettini S, Prato M, Valli L (2020) Tailoring the sensing abilities of carbon nanodots obtained from olive solid wastes. Carbon 167:696–708. https://doi.org/10.1016/j.carbon.2020.06.011
27. Prasannan A, Imae T (2013) One-pot synthesis of fluorescent carbon dots from orange waste peels. Ind Eng Chem Res 52(44):15673–15678. https://doi.org/10.1021/ie402421s
28. Aslan M, Eskalen H (2021) A study of carbon nanodots (carbon quantum dots) synthesized from tangerine juice using one-step hydrothermal method. Fulleren Nanotubes Carbon Nanostruct 29(12):1026–1033. https://doi.org/10.1080/1536383X.2021.1926452
29. Wang R, Li S, Huang H, Liu B, Gao L, Qu M et al (2023) Preparation of carbon dots from PET waste by one-step hydrothermal method and its application in light blocking films and LEDs. J Fluoresc 33(4):1305–1315. https://doi.org/10.1007/s10895-022-03132-9
30. Tan XW, Romainor ANB, Chin SF, Ng SM (2014) Carbon dots production via pyrolysis of sago waste as potential probe for metal ions sensing. J Anal Appl Pyrol 105:157–165. https://doi.org/10.1016/j.jaap.2013.11.001
31. González-González RB, González LT, Madou M, Leyva-Porras C, Martinez-Chapa SO, Mendoza A (2022) Synthesis, purification, and characterization of carbon dots from non-activated and activated pyrolytic carbon black. Nanomaterials 12(3):298
32. Aji MP, Susanto, Wiguna PA, Sulhadi (2017) Facile synthesis of luminescent carbon dots from mangosteen peel by pyrolysis method. J Theoretic Appl Phys 11(2):119–126. https://doi.org/10.1007/s40094-017-0250-3
33. Liang L, Veksha A, Mohamed Amrad MZB, Snyder SA, Lisak G (2021) Upcycling of exhausted reverse osmosis membranes into value-added pyrolysis products and carbon dots. J Hazard Mater 419:126472. https://doi.org/10.1016/j.jhazmat.2021.126472
34. Qin X, Fu C, Zhang J, Shao W, Qin X, Gui Y et al (2022) Direct preparation of solid carbon dots by pyrolysis of collagen waste and their applications in fluorescent sensing and imaging. Front Chem 10:1006389. https://doi.org/10.3389/fchem.2022.1006389
35. Michenzi C, Espro C, Bressi V, Celesti C, Vetica F, Salvitti C et al (2023) Electrochemical bottom-up synthesis of biomass-derived carbon dots for promoting Knoevenagel condensation. Molecular Catalysis 544:113182. https://doi.org/10.1016/j.mcat.2023.113182

36. Hou Y, Lu Q, Deng J, Li H, Zhang Y (2015) One-pot electrochemical synthesis of functionalized fluorescent carbon dots and their selective sensing for mercury ion. Anal Chim Acta 866:69–74. https://doi.org/10.1016/j.aca.2015.01.039
37. Monte-Filho SS, Andrade SIE, Lima MB, Araujo MCU (2019) Synthesis of highly fluorescent carbon dots from lemon and onion juices for determination of riboflavin in multivitamin/mineral supplements. J Pharmaceutic Anal 9(3):209–216. https://doi.org/10.1016/j.jpha.2019.02.003
38. Shibata H, Abe M, Sato K, Uwai K, Tokuraku K, Iimori T (2022) Microwave-assisted synthesis and formation mechanism of fluorescent carbon dots from starch. Carbohyd Polym Technol Appl 3:100218. https://doi.org/10.1016/j.carpta.2022.100218
39. Ang WL, Boon Mee CAL, Sambudi NS, Mohammad AW, Leo CP, Mahmoudi E et al (2020) Microwave-assisted conversion of palm kernel shell biomass waste to photoluminescent carbon dots. Sci Rep 10(1):21199. https://doi.org/10.1038/s41598-020-78322-1
40. Ang WL, Alqasem QA, Mohammad AW (2021) Microwave-assisted synthesis of photoluminescent carbon dots from palm fronds biomass wastes. IOP Conferen Series: Mater Sci Eng 1195(1):012008. https://doi.org/10.1088/1757-899X/1195/1/012008
41. George HS, Selvaraj H, Ilangovan A, Chandrasekaran K, Kannan VR, Parthipan P et al (2023) Green synthesis of biomass derived carbon dots via microwave-assisted method for selective detection of Fe3+ ions in an aqueous medium. Inorg Chem Commun 157:111348. https://doi.org/10.1016/j.inoche.2023.111348
42. Chopra T, Parkesh R (2024) Microwave-assisted synthesis of functionalized carbon nanospheres using banana peels: pH-dependent synthesis, characterization, and selective sensing applications. ACS Omega 9(4):4555–4571. https://doi.org/10.1021/acsomega.3c07544
43. Kwon W, Lee G, Do S, Joo T, Rhee SW (2014) Size-controlled soft-template synthesis of carbon nanodots toward versatile photoactive materials. Small 10(3):506–513. https://doi.org/10.1002/smll.201301770
44. Nelson DK, Razbirin BS, Starukhin AN, Eurov DA, Kurdyukov DA, Stovpiaga EY et al (2016) Photoluminescence of carbon dots from mesoporous silica. Opt Mater 59:28–33. https://doi.org/10.1016/j.optmat.2016.03.051
45. Liu M, Li T, Zhang C, Zheng Y, Wu C, Zhang J et al (2021) Fluorescent carbon dots embedded in mesoporous silica nanospheres: a simple platform for Cr(VI) detection in environmental water. J Hazard Mater 415:125699. https://doi.org/10.1016/j.jhazmat.2021.125699
46. Chen H, Wang GD, Sun X, Todd T, Zhang F, Xie J et al (2016) Mesoporous silica as nanoreactors to prepare gd-encapsulated carbon dots of controllable sizes and magnetic properties. Adv Func Mater 26(22):3973–3982. https://doi.org/10.1002/adfm.201504177
47. Mansuriya BD, Altintas Z (2021) Carbon dots: classification, properties, synthesis, characterization, and applications in health care-an updated review (2018–2021). Nanomaterials. 11(10). https://doi.org/10.3390/nano11102525
48. Rezaei B, Irannejad N, Ensafi AA, Kazemifard N (2019) The impressive effect of eco-friendly carbon dots on improving the performance of dye-sensitized solar cells. Sol Energy 182:412–419. https://doi.org/10.1016/j.solener.2019.02.072
49. Padmanathan S, Prakasam A (2020) Design and fabrication of hybrid carbon dots/titanium dioxide (CDs/TiO2) photoelectrodes for highly efficient dye-sensitized solar cells. J Mater Sci: Mater Electron 31(4):3492–3499. https://doi.org/10.1007/s10854-020-02897-8
50. Efa MT, Imae T (2019) Effects of carbon dots on ZnO nanoparticle-based dye-sensitized solar cells. Electrochim Acta 303:204–210. https://doi.org/10.1016/j.electacta.2019.02.012
51. Etefa HF, Imae T, Yanagida M (2020) Enhanced photosensitization by carbon dots co-adsorbing with dye on p-Type semiconductor (Nickel Oxide) solar cells. ACS Appl Mater Interfaces 12(16):18596–18608. https://doi.org/10.1021/acsami.0c02413
52. Yadeta TF, Imae T (2023) Effect of carbon dot on photovoltaic performance of n-TiO2/p-NiO and n-TiO2/p-CuO heterojunctions in dye-sensitized solar cells. Appl Surf Sci 637:157880. https://doi.org/10.1016/j.apsusc.2023.157880
53. Molaei MJ (2020) Principles, mechanisms, and application of carbon quantum dots in sensors: a review. Anal Methods 12(10):1266–1287. https://doi.org/10.1039/C9AY02696G

54. Mandal P, Sahoo D, Sarkar P, Chakraborty K, Das S (2019) Fluorescence turn-on and turn-off sensing of pesticides by carbon dot-based sensor. New J Chem 43(30):12137–12151. https://doi.org/10.1039/C9NJ03192H
55. Baruah U, Gogoi N, Konwar A, Jyoti Deka M, Chowdhury D, Majumdar G (2014) Carbon dot based sensing of dopamine and ascorbic acid. J Nanopartic 2014:178518. https://doi.org/10.1155/2014/178518
56. Zhao J, Huang M, Zhang L, Zou M, Chen D, Huang Y et al (2017) Unique approach to develop carbon dot-based nanohybrid near-infrared ratiometric fluorescent sensor for the detection of mercury ions. Anal Chem 89(15):8044–8049. https://doi.org/10.1021/acs.analchem.7b01443
57. Li W, Zhang X, Miao C, Li R, Ji Y (2020) Fluorescent paper–based sensor based on carbon dots for detection of folic acid. Anal Bioanal Chem 412(12):2805–2813. https://doi.org/10.1007/s00216-020-02507-w
58. Lin J, Huang X, Kou E, Cai W, Zhang H, Zhang X et al (2023) Carbon dot based sensing platform for real-time imaging Cu2+ distribution in plants and environment. Biosens Bioelectron 219:114848. https://doi.org/10.1016/j.bios.2022.114848
59. Du H, Li X, Xu S, Cheng G, Xue Q, Xu H (2023) N/S-Co-doped carbon dot–based FRET ratiometric fluorescence aptasensing platform modulated with entropy-driven DNA amplifier for ochratoxin A detection. Anal Bioanal Chem 415(19):4649–4660. https://doi.org/10.1007/s00216-023-04778-5
60. Keerthana P, Kumar Das A, Bharath M, Ghosh M, Varghese A (2023) A ratiometric fluorescent sensor based on dual-emissive carbon dot for the selective detection of Cd2+. J Environ Chem Eng 11(2):109325. https://doi.org/10.1016/j.jece.2023.109325
61. Yang S-T, Liu J-H, Wang P, Yang S, Ge L, Yan S et al (2018) High-performance red/near-IR carbon dots as fluorescence probes for tumor imaging in vivo. Chem Select 3(23):6374–6381. https://doi.org/10.1002/slct.201800814
62. Xu Y, Wu M, Liu Y, Feng X-Z, Yin X-B, He X-W et al (2013) Nitrogen-doped carbon dots: a facile and general preparation method, photoluminescence investigation, and imaging applications. Chem Euro J 19(7):2276–2283. https://doi.org/10.1002/chem.201203641
63. Lin J-S, Tsai Y-W, Dehvari K, Huang C-C, Chang J-Y (2019) A carbon dot based theranostic platform for dual-modal imaging and free radical scavenging. Nanoscale 11(43):20917–20931. https://doi.org/10.1039/C9NR05746C
64. Ci Q, Wang Y, Wu B, Coy E, Li Jj, Jiang D et al (2023) Fe-doped carbon dots as NIR-II fluorescence probe for in vivo gastric imaging and pH detection. Adv Sci 10(7):2206271. https://doi.org/10.1002/advs.202206271
65. Gao P, Wang J, Zheng M, Xie Z (2020) Lysosome targeting carbon dots-based fluorescent probe for monitoring pH changes in vitro and in vivo. Chem Eng J 381:122665. https://doi.org/10.1016/j.cej.2019.122665
66. Hu J, Guo Y, Geng X, Wang J, Li S, Sun Y et al (2022) Tuning asymmetric electronic structure endows carbon dots with unexpected huge stokes shift for high contrast in vivo imaging. Chem Eng J 446:136928. https://doi.org/10.1016/j.cej.2022.136928
67. Hu J, Sun Y, Geng X, Wang J, Guo Y, Qu L et al (2022) High-fidelity carbon dots polarity probes: revealing the heterogeneity of lipids in oncology. Light: Sci Appl 11(1):185. https://doi.org/10.1038/s41377-022-00873-x
68. Yin X, Sun Y, Yang R, Qu L, Li Z (2020) RNA-responsive fluorescent carbon dots for fast and wash-free nucleolus imaging. Spectrochim Acta Part A Mol Biomol Spectrosc 237:118381. https://doi.org/10.1016/j.saa.2020.118381
69. Dong X, Liang W, Meziani MJ, Sun YP, Yang L (2020) Carbon dots as potent antimicrobial agents. Theranostics. 10(2):671–686. https://doi.org/10.7150/thno.39863
70. Wang Z, Sheng L, Yang X, Sun J, Ye Y, Geng S et al (2023) Natural biomass-derived carbon dots as potent antimicrobial agents against multidrug-resistant bacteria and their biofilms. Sustain Mater Technol 36:e00584. https://doi.org/10.1016/j.susmat.2023.e00584
71. Qureshi WA, Vivekanandan B, Jayaprasath JA, Ali D, Alarifi S, Deshmukh K (2021) Antimicrobial activity and characterization of pomegranate peel-based carbon dots. J Nanomater 2021:9096838. https://doi.org/10.1155/2021/9096838

72. Thakur M, Pandey S, Mewada A, Patil V, Khade M, Goshi E et al (2014) Antibiotic conjugated fluorescent carbon dots as a theranostic agent for controlled drug release, bioimaging, and enhanced antimicrobial activity. J Drug Deliv 2014:282193. https://doi.org/10.1155/2014/282193
73. Dong X, Bond AE, Pan N, Coleman M, Tang Y, Sun YP et al (2018) Synergistic photoactivated antimicrobial effects of carbon dots combined with dye photosensitizers. Int J Nanomed 13:8025–8035. https://doi.org/10.2147/IJN.S183086
74. Saravanan A, Maruthapandi M, Das P, Ganguly S, Margel S, Luong JHT et al (2020) Applications of N-doped carbon dots as antimicrobial agents, antibiotic carriers, and selective fluorescent probes for nitro explosives. ACS Appl Bio Mater 3(11):8023–8031. https://doi.org/10.1021/acsabm.0c01104
75. Wang Q, Shi L, Zhao J, Shuang S (2022) Fluorescent carbon dots with real-time nucleolus-monitoring capability for gene delivery and biosensing of NO2– and pH. Appl Surf Sci 599:153902. https://doi.org/10.1016/j.apsusc.2022.153902
76. Fahmi MZ, Sholihah NF, Wibrianto A, Sakti SCW, Firdaus F, Chang JY (2021). Simple and fast design of folic acid-based carbon dots as theranostic agent and its drug release aspect. Mater Chem Phys 267:124596. https://doi.org/10.1016/j.matchemphys.2021.124596
77. Li CL, Ou CM, Huang CC, Wu WC, Chen YP, Lin TE et al (2014) Carbon dots prepared from ginger exhibiting efficient inhibition of human hepatocellular carcinoma cells. J Mater Chemistr B 2(28):4564–4571. https://doi.org/10.1039/c4tb00216d
78. Yao L, Zhao MM, Luo QW, Zhang YC, Liu TT, Yang Z et al (2022) Carbon quantum dots-based nanozyme from coffee induces cancer cell ferroptosis to activate antitumor immunity. ACS Nano 16(6):9228–9239. https://doi.org/10.1021/acsnano.2c01619
79. Li G, Bao Y, Zhang H, Wang J, Wu X, Yan R et al (2024) Enhanced catalytic activity of Fe(3)O(4)-carbon dots complex in the Fenton reaction for enhanced immunotherapeutic and oxygenation effects. J Colloid Interface Sci 668:618–633. https://doi.org/10.1016/j.jcis.2024.04.151
80. Xu D, Guo D, Zhang J, Tan X, Deng Z, Hou X et al (2024) Innovative tumor interstitial fluid-triggered carbon dot-docetaxel nanoassemblies for targeted drug delivery and imaging of HER2-positive breast cancer. Int J Pharm 657:124145. https://doi.org/10.1016/j.ijpharm.2024.124145
81. Wei Z, Dong X, Sun Y (2024) Quercetin-derived red emission carbon dots: a multifunctional theranostic nano-agent against Alzheimer's beta-amyloid fibrillogenesis. Colloids Surf, B 238:113907. https://doi.org/10.1016/j.colsurfb.2024.113907
82. Chi M, Liu J, Li L, Zhang Y, Xie M (2024) In-situ growth of CeO(2) on biofilms: Innovative nanoparticles for photothermal therapy & multi-pronged attack on Alzheimer's disease. Colloids Surf B 238:113887. https://doi.org/10.1016/j.colsurfb.2024.113887
83. Algarra M, Soto J, Pino-Gonzalez MS, Gonzalez-Munoz E, Ducic T (2024) Multifunctionalized carbon dots as an active nanocarrier for drug delivery to the glioblastoma cell line. ACS Omega 9(12):13818–13830. https://doi.org/10.1021/acsomega.3c08459
84. Xia J, Wang J, Liu F, Chen Z, Chen C, Cheng X et al (2024) Red/NIR-I-fluorescence carbon dots based on rhein with active oxygen scavenging and colitis targeting for UC therapeutics. Adv Healthcare Mater e2304674. https://doi.org/10.1002/adhm.202304674
85. Chary K, Sharma A, Singh A (2023) Carbon quantum dots in healthcare: a promising solution for sustainable healthcare and biomedical practices. E3S Web Conferen 453. https://doi.org/10.1051/e3sconf/202345301017

Chapter 5
Waste-Derived Cellulose Nanomaterials–Based Membranes for Water Filtration Applications

Mohd Jahir Khan, Abrar Ahmad, and Chularat Sakdaronnarong

Abstract Employing waste materials as a source of cellulose not only offers a sustainable method for producing nanomaterials (NMs) but also gives an environmentally conscious option for managing garbage. The development of low-cost, versatile nanostructures from agricultural waste has garnered significant attention in the past few years. Although many methods have been developed to produce high-value carbon nanomaterials, the most popular method nowadays seems to be the use of green technology due to its eco-friendly nature, non-toxicity, low fabrication cost, easy modification, and abundant availability of the starting precursor. It is therefore impressive that low-value biomass waste can be used for the processing of environmentally friendly materials to produce high-value products. Cellulose materials are naturally occurring and have unique structural, mechanical, and optical properties. It offers a lot of untapped potential for improving water treatment technology. The primary goal of this chapter is to demonstrate how cellulose and carbon-based nanomaterials are made from agricultural waste and their range of uses in water filtration. The production of nanocellulose-based membranes has been discussed in this chapter. The challenges and future directions of nanocellulose in water treatment are also covered in this chapter, along with a summary of their recent advancements and applications.

Keywords Nanocellulose · Membranes · Wastewater · Water filtration · Ecofriendly

M. J. Khan (✉) · C. Sakdaronnarong
Faculty of Engineering, Department of Chemical Engineering, Mahidol University, Nakhon Pathom 73170, Thailand
e-mail: khanmohdjahir@gmail.com

A. Ahmad
Faculty of Sciences, Department of Biochemistry, King Abdulaziz University, Jeddah 21589, Saudi Arabia

N. Talreja et al. (eds.), *Waste-Derived Carbon Nanostructures*, Nanostructure Science and Technology, https://doi.org/10.1007/978-3-031-75247-6_5

5.1 Introduction

Industrialization has resulted in substantial enhancements in multiple facets of human existence, encompassing economic progress, quality of life, and technological innovations. However, it also poses environmental and human sustainability challenges, including resource depletion, deforestation, and water shortages, which affect the overall quality of life due to human consumption and agriculture [1, 2]. Industry wastewater contributes to freshwater pollution, requiring various water purification methods like UV disinfection, ozonation, membrane filtration, and conventional methods like sedimentation, filtration, and coagulation [3]. Among these methods, membrane technology is important for the production of high-quality water because of its high ability to remove pollutants, sustainability, and efficiency. Different fields, including food concentration, gas separation, water and wastewater treatment, recycling, and saltwater desalination, are widely exploiting advanced membrane technology. The development of membrane technology has increased in recent decades due to the industrial shift toward environmentally friendly and sustainable methods. Various materials, including composite, metallic, ceramic, organic–inorganic hybrid, and polymeric, make up water treatment membranes [4–6]. It is possible for synthetic polymers like polyvinyl alcohol, polyamide, and polypropylene, or those with inorganic nanoparticles like SiO_2 and Al_2O_3, to make membranes better than natural materials. Conversely, the non-renewable nature, lack of biodegradability, and high production costs limit the practical use of synthetic polymers [7]. Industry prefers polymeric membranes like cellulose acetate, polyamide, polysulfone, and polyethersulfone because of their affordability, accessibility, and versatility [8, 9].

Nanocellulose membranes are suitable for a variety of applications because of their exceptional mechanical flexibility, chemical adaptability, and selectivity. A wide variety of natural materials, including wood, rice straw, sugarcane, and bacteria, contain cellulose, a common polysaccharide with exceptional mechanical strength and stiffness, which serves as the main structural element of plants and is frequently associated with hemicellulose, pectin, lignin, and other materials [10, 11]. Cellulose materials can be divided into two groups: the first group includes nanostructured materials, which include cellulose microcrystals and cellulose microfibrils. The second category is nanofibers, which consist of cellulose nanofibrils and cellulose nanocrystals (CNCs) [12]. During the selection of methods and raw materials for the production of the required and selective membranes, cost is always considered a major factor, and raw materials play a significant role in influencing the overall process. To reduce the cost of cellulose-based membranes, researchers are utilizing naturally available waste materials from different sources, like agricultural and industrial waste, through different extraction processes, including chemical and mechanical. This not only reduces the production cost of the membranes but also enhances sustainability goals. For example, corn stover is a single source of agricultural waste, and its annual production is 1 billion tons. It contains 35% cellulose, 20% hemicellulose, and 12% lignin, making it a valuable resource for cellulose production [13]. Estimates place the global production of agricultural waste, which includes corn stalks, wheat straw, and rice straw, at 140 billion metric tons. The majority of agricultural wastes include cellulose, hemicellulose, and lignin. Thailand, in particular, is a major agricultural producer, with 46% of its land used for agricultural practices. The country produces a variety of crops, including rice, corn, palm oil, and sugarcane [14, 15]. In 2023 alone, Thailand's rice production reached 22.2 billion tons. The residues from these crops, especially rice, contain large amounts of cellulose. By utilizing these agricultural wastes properly, they become an inexpensive source of cellulose [16]. Palm oil meets one-third of the global demand for vegetable oil, and its empty fruit bunches (EFBs) are a rich source of cellulose. Each hectare of palm oil cultivation produces approximately 4 tons of biomass annually. Paper and membrane manufacturing widely utilize EFBs due to their high cellulose content, which includes 14.1–30.5% lignin, 23.7–65.0% cellulose, and 20.6–33.5% hemicellulose [17–19]. Additionally, sugar and alcohol mills produce around 30% of bagasse during the processing of sugarcane. Cellulosic components found in this bagasse include 14–23% lignin, 26–47% cellulose, and 19–33% hemicellulose. Worldwide, including China and Thailand, produce around 100,000 metric tons of bagasse annually [20]. Besides these, the percentage of cellulose, hemicellulose, and lignin are varied in different agriculture waste materials as shown in Table 5.1. Cellulose-based materials have a wide range of applications, including medical applications, and tissue engineering [21], CO_2 adsorbents [22], desalination, sensors, and electrical and energy-related applications [12]. These waste materials save costs and encourage environmental sustainability standards, in addition to producing membranes for wastewater filtering applications.

5.2 Extraction of Cellulose

The common sources of cellulose are agricultural, industrial, and textile wastes. These waste materials contain substantial amounts of cellulose. The common steps applied in cellulose extractions are pretreatment, pulping, and bleaching. Pretreatment is the removal of non-cellulosic constituents such as lignin, hemicellulose, and extractives using mechanical, chemical, or enzymatic methods, while pulping is the disintegrating of the unprocessed material into a fibrous substance, usually using chemical (such as the kraft process) or mechanical techniques. Bleaching is the improvement of the quality and brightness of the cellulose by removing any remaining lignin and other contaminants.

5.2.1 Extraction of Cellulosic NMs from Agricultural Waste

Agricultural waste is classified into four different categories such as crop residues; animal waste; processing waste; and hazardous waste [34]. Agricultural wastes are the primary source of the cellulose. The common agricultural wastes are shown in

Table 5.1 Cellulose, hemicellulose, and lignin in different agriculture waste materials

Agriculture waste	Hemicellulose	Cellulose	Lignin	Ref
Bamboo leaves	25.55	34.14	35.03	[23]
Oil palm fiber	40.40	30.40	21.70	[24]
Corn stover	28.94	43.97	21.82	[23]
Palm kernel shell	21.60	27.70	44.00	[24]
Pine wood sawdust	10.50	48.60	25.30	[25]
Rice husk	23.90	37.20	12.80	[26]
Wheat straw	36.00	38.00	22.00	[27]
Larch wood block	12.00	40.00	31.00	[28]
Sugarcane bagasse	27.40	46.55	20.61	[23]
Sugarcane peel	26.40	41.11	24.31	[23]
Coconut shell	56.29	17.89	25.82	[29]
Rice straw	32.53	54.67	12.83	[30]
Poplar	22.00	47.00	20.00	[31]
Wheat straw	37.80	32.90	8.50	[32]
Corn stover	28.94	43.97	21.82	[23]
Switchgrass	32.40	37.50	16.30	[33]

Fig. 5.1, and various methods (mechanical, chemical, and biological) applied for the cellulose extraction and cellulose-based NMs synthesis are shown in Fig. 5.2.

Fig. 5.1 Different sources of cellulose-based agricultural waste materials: **a** rice husk, **b** rice straw, **c** baggage, **d** coconut, **e** banana, and **f** pineapple leaf

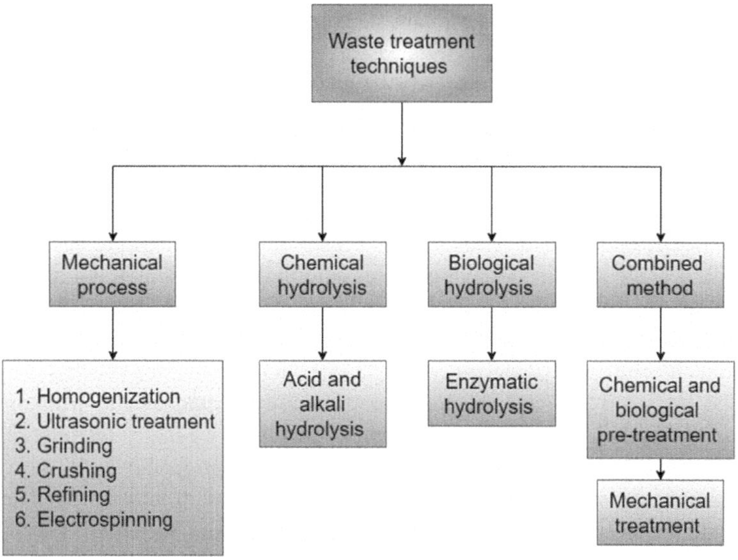

Fig. 5.2 Different treatment techniques used for the extraction of cellulose from waste materials

5.2.1.1 Mechanical Methods

Mechanical extraction techniques such as homogenization, grinding, crushing, ultrasonic treatment, refining, and electrospinning break down and extract cellulose fibers from agricultural waste. Hammer mills, ball mills, and disc mills typically perform crushing and grinding [35]. Homogenization employs high-pressure forces, while ultrasonic treatment uses high-frequency sound waves. Electrospinning, which applies an external electric force to a polymer solution, is particularly common in microfibrillated cellulose processing [36]. Despite facilitating the chemical and biological treatment of waste materials, the industry does not prefer these mechanical processes due to their high energy consumption and potential damage to the morphology and structure of cellulose materials, leading to products that may not meet the desired specifications. Additionally, mechanical methods cannot effectively remove the non-cellulosic parts of biomass [37]. Chemical and biological methods often take precedence to address these limitations. However, a careful approach is required when dealing with chemical treatment processes to ensure proper treatment of effluents before disposal, in accordance with environmental protocol.

5.2.1.2 Chemical Methods

Agricultural waste requires chemical treatment methods to eliminate non-cellulosic materials. These methods focus on dissolving the amorphous parts while keeping the crystalline cellulose. A pretreatment stage initiates the process, treating the agricultural waste with solvents like toluene and ethanol [38]. This step is crucial for eliminating impurities like lipids and waxes, which can interfere with subsequent processing stages. Following pretreatment, the waste undergoes an alkaline treatment. Sodium hydroxide and an organic solvent facilitate hydrolysis at this stage, breaking down complex molecules and enhancing the performance of the cellulose material. The alkaline treatment not only improves cellulose accessibility, but also removes hemicellulose and lignin, making the cellulose fibers more accessible for further processing. The next stage involves acid hydrolysis, which is important for refining the cellulose and ensuring that it is free from non-cellulosic components [36]. After acid hydrolysis, the material undergoes a bleaching process. Agents such as sodium chlorite and hydrogen peroxide carry out the bleaching process. This step is critical for removing any residual non-cellulosic substances and whitening the cellulose fibers, thereby improving their quality and purity. Finally, the treated material undergoes a thorough washing process [15, 39]. Washing removes any remaining chemical residues and ensures that the cellulose fibers are of high purity. This final step is vital for producing cellulose that meets the stringent requirements for various industrial applications, including textiles, paper production, and membrane fabrication. The extraction of cellulose from different agricultural waste materials and synthesis of nanocrystals using different concentrations of sulfuric acid are shown in Table 5.2.

5.2.2 Synthesis of Cellulosic NMs from Industrial Waste

Industrial activities generate significant amounts of biomass waste due to human activities. Various types of firms within the industry produce large quantities of waste containing cellulosic materials, with the wood and paper industries being the primary contributors. The wood industry, which includes sectors such as furniture, crafting, wood panels, and apartment components, generates varying amounts of waste depending on the specific application [40]. According to the 2014 European Wood Waste Statistics report, the EU-27 produced 60 million tons of wood waste annually [41], while the USA alone produces approximately 300 million tons each year, with at least 12 million tons ending up in landfills. Only about 15% of wood waste globally undergoes recycling. The Wood Recyclers Association reports that the UK generates around 4.5 million tons of waste wood annually. Similarly, the pulp and paper industry is a significant contributor to biomass waste. In 2022, the global production of paper reached 414 million metric tons. In 2021, European paper production was 90.5 million tons. According to a monitoring report, the recycling rate of paper in 2022 was 70.5%, a decrease from 72.5% in 2021.

The wood chips undergo pretreatment with steam to soften the chip material. The extraction and drying processes yield pulp. The traditional Kraft process further refines this pulp by chemically breaking down lignin and hemicellulose. This process mostly uses sodium hydroxide and sodium sulfide at high temperatures and pressures,

Table 5.2 Extraction and production of CNCs from different waste materials

Sources	Method	Yield (%)	Crystallinity index (%)	Ref
Brewery spent grain	The experiment involved 51 wt% H_2SO_4 at 50 °C, followed by 41 min of treatment. The process involved pretreatment, bleaching, and hydrolysis	43.24	76.8	[42]
Rice husk	The experiment involved 10.0 mol/L of H_2SO_4 at 50 °C, followed by 40 min of continuous stirring. The process involved pretreatment, bleaching, and hydrolysis	–	59	[43, 47]
Cotton linter	The sample was initially treated with NaOH at 80 °C for 5 h, followed by treatment with different concentrations of sulfuric acid at varying rpm and time intervals	59–72	82	[44, 48]
Sisal fibers	Extracted cellulose was initially pretreated, bleached, and then treated with HNO_3 (70% + HAc) at 120 °C, followed by hydrolysis with 60 wt% H_2SO_4 for 30 min of continuous stirring at 45 °C	–	75	[45]
Potato peel	Initially treated with NaOH at 80 °C for 2.5 h, followed by bleaching with sodium chlorite. The pretreated material is hydrolyzed with 64 wt% H_2SO_4 at 45 °C for 90 min	41–42	85	[46]

(continued)

Table 5.2 (continued)

Sources	Method	Yield (%)	Crystallinity index (%)	Ref
Empty fruit bunch	The experiment involved a 65–85%-wt $H_3PW_{12}O_{40}$ at 80–100 °C, followed by 25–35 h of treatment under mechanical stirring. The process involved pretreatment, bleaching, and hydrolysis	44.8	73.3	[47]
Mango peel	Material first undergoes a 4-h treatment with NaOH at 100 °C, after which it undergoes bleaching. The pretreated material is hydrolyzed with 11.21 M H_2SO_4 at 40 °C, followed by 10 min of constant stirring	22.8	90.6	[48]
Cocoa pod husk	Initially treated with NaOH at 80 °C for 2 h, then bleached with $NaOCl_2$. The pretreated material undergoes hydrolysis with 64 wt% H_2SO_4 at 45 °C, followed by 60 min of vigorous stirring	25	67.6	[49]
Waste paper	Initially treated with NaOH, followed by bleaching with NaClO. The pretreated material undergoes hydrolysis with 60 wt% H_2SO_4 at 45 °C, followed by 60 min of constant stirring	19	75.9	[50]

(continued)

Table 5.2 (continued)

Sources	Method	Yield (%)	Crystallinity index (%)	Ref
Khat waste	Initially treated with a mixture of acetic/formic acid (40%/40%), then NaOH, followed by bleaching with H_2O_2. The pretreated material undergoes hydrolysis with 64 wt% H_2SO_4 at 45 °C, followed by 60 min of magnetic stirring	49	81.60	[51]
Corn cob	Initially treated with NaOH, followed by bleaching with $NaClO_2$. The pretreated material undergoes hydrolysis with 64 wt% H_2SO_4 at 45 °C, followed by 60 min of vigorous stirring	34.5	55.9	[52]
Sugarcane straw	Pretreated with ethanol, followed by bleaching with $NaClO_2$. The pretreated material is hydrolyzed with 64 wt% H_2SO_4 at 45 °C for 60 min	21.8	62.7	[53]
Eucalyptus Kraft pulp	Milled fiber obtained through mechanical treatment is further treated by hydrolysis with 60 wt% H_2SO_4 at 45 °C for 30 min	65	76	[54]

leaving primarily cellulose fibers [55]. Researchers also study the morphology of extracted cellulose after treating wood chips with nitric acid and ethanol. Ethanol enhances the solubility of oxidized lignin through the reduction of oxidation and hydrolysis of cellulose [56].

5.2.3 *Synthesis of Cellulosic NMs from Textile Waste*

After industrial and agricultural residues, the textile industry produces the largest amount of cellulose-enriched waste. This industry uses a variety of raw materials, including synthetic and natural materials, depending on product demands and consumer preferences, with cotton being the most significant. The textile industry used 23% cotton [57]. For centuries, cotton has been a staple in numerous products, and its consumption has steadily increased, leading to a rise in both pre-consumer and post-consumer waste. Much of this textile waste ends up in landfills, which is a major concern. It is estimated that global cotton production will reach 25 million tons in 2024, with China leading the way with over 6.5 million tons harvested in 2021. The textile industry manufactures approximately 100 billion garments annually. According to a United Nations report, people now buy 60% more clothes than they did in the past, but they use them for a shorter time. This leads to the incineration of 25% of clothing waste and the landfilling of 57%. There is an urgent need to utilize these wastes effectively. The best way to do this is to extract cellulose materials from textile waste. Cotton, which contains 93% cellulose, can have its cellulose content increased to 99% after bleaching and scoring. Cellulose is extracted from cotton using either chemical or mechanical methods. By dissolving the amorphous parts of cellulose through acid hydrolysis, the chemical extraction method gets rid of most of the unwanted materials and makes CNCs [58]. The mechanical extraction method could deform the cellulosic fibers, resulting in lower-quality cellulose materials compared to virgin fibers. Mechanical methods cannot effectively treat blended fabrics. Therefore, chemical treatment methods are highly preferred for extracting cellulose from waste cotton material [59, 60]. The various chemicals used in the dyeing process complicate nanocellulose processing and extraction from most post-consumer fibers. Dyed cotton for cellulose extraction is always discouraged, as it necessitates bleaching to eliminate the color from the waste cotton. It also contains many unknown additives, requiring multiple processes to achieve pure cellulose.

5.3 Synthesis of Nanocellulose

Nanocellulose is a cellulose-driven renewable material that is classified based on synthesis process, particle size, and morphology. Different kinds of renewable sources, such as agricultural, industrial, and textile waste, produce different kinds of nanocellulose materials. These include CNCs, cellulose nanofibers (CNFs), and bacterial nanocellulose (BNC) (Fig. 5.3). Nanocellulose demonstrates remarkable mechanical characteristics, such as elevated tensile strength and stiffness.

5.3.1 Properties of Nanocellulose

Nanocellulose has extensive applications due to its unique intrinsic properties including biocompatibility, biodegradability, mechanical, and thermal stability as well as hydrophilic and optical properties. The mechanical properties of nanocellulose, including high tensile strength and stiffness, make it a highly effective agent of reinforcement in composite materials. Deterioration of nanocellulose typically

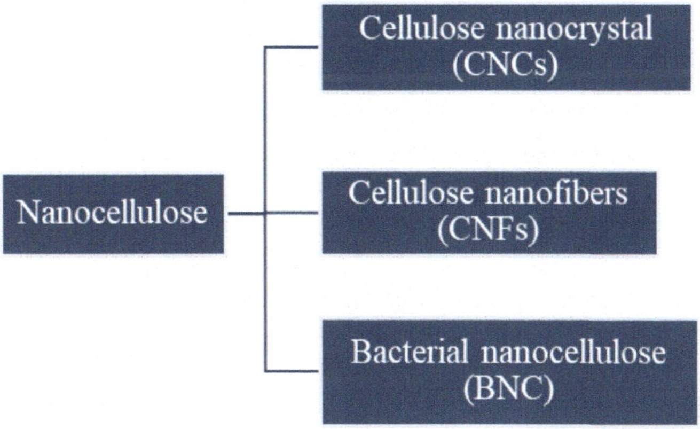

Fig. 5.3 Classification of nanocellulose

begins around 200–250 °C which can be increased further by chemical modifications, making it acceptable for high-temperature applications. The biocompatibility of nanocellulose ensures its safety for utilization in biomedical applications, including drug delivery systems, tissue engineering scaffolds, and wound dressings. Nanocellulose is a biodegradable material that can be decomposed by microbes. Nanocellulose is naturally hydrophilic due to the presence of plenty of hydroxyl groups on its surface. This feature enables it to establish strong hydrogen bonds with water molecules, which contributes to its high water retention capacity and utility in applications such as hydrogels and films. Nanocellulose can make transparent films because of its small size, which stops visible light from dispersing. Figure 5.4 illustrates the detailed properties of nanocellulose, which make it a demanding material.

5.3.2 CNCs

The chemical treatment method yields rod-shaped CNCs by hydrolyzing cellulose-based sources. Sulfuric acid is the preferred acid for CNCs synthesis because it produces a stable and evenly mixed solution and enhances the CNC's crystallinity index. The acids used during the hydrolysis process and the materials used for nanocellulose extraction greatly vary the CNC's dimensions [61–63]. A number of characterization tools, including X-ray diffraction (XRD), atomic force microscopy (AMF), and scanning electron microscopy (SEM), are used to find out about the crystalline properties, size, and shape of the CNCs [64].

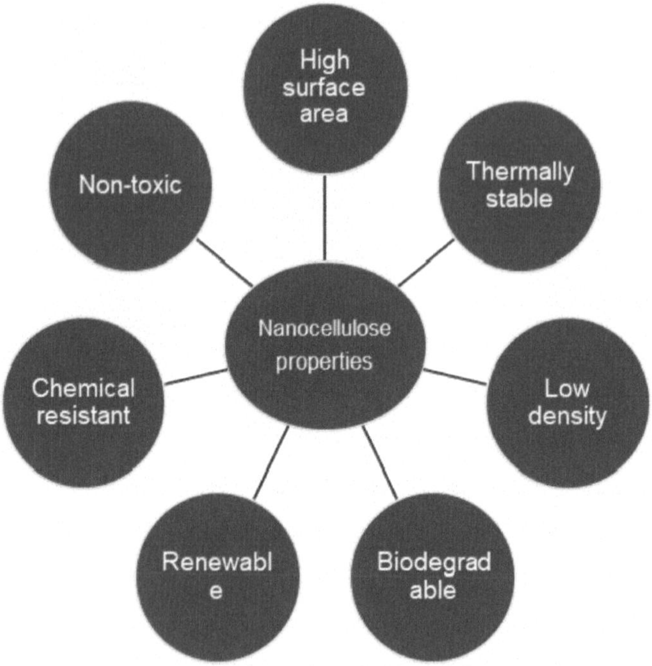

Fig. 5.4 Characteristics of nanocellulose material

5.3.2.1 CNFs

CNFs are ultra-fine fibers obtained by mechanically processing cellulose-containing materials. This process does not separate the amorphous parts of the cellulose, so CNF has less crystallinity than CNCs. A wide range of applications utilize CNF due to its unique properties, including biodegradability, high barrier properties, renewability, low density, and high mechanical strength, making it strong and ideal component in composite materials [65]. Compared to cellulose nanocrystal synthesis methodology, cellulose nanofiber synthesis is simpler. However, several factors prevent the widespread commercial use of CNF as is has a relatively larger diameter and length compared to CNCs, resulting in a higher surface-to-volume ratio. This means CNF contains more hydroxyl groups that form strong hydrogen bonds, causing the fibers to agglomerate [66]. To separate these fibers, high shear energy is required. Therefore, the synthesis of CNF employs mechanical methods like homogenizers, ball milling, and disc milling, resulting in an energy-intensive process. Mechanical methods cause the cellulose fiber's structure to break down and alter the gel's behavior, as the synthesis process generates high temperatures [67, 68].

5.3.2.2 Bacterial Nanocellulose (BNC)

As the world strives for green engineering and environmental sustainability, bacterial nanocellulose (BNC) is gaining significant interest. This is because of its environmentally friendly nature, excellent biocompatibility, and biodegradability [69]. Bacteria synthesized NC, the purest form of cellulose, distinguishing it from CNCs and cellulose nanofibers (CNF). Unlike plant-derived cellulose, BNC is free from other polysaccharides such as hemicellulose, lignin, and pectin, which contribute to its high purity and crystallinity, resulting in superior mechanical strength [70]. However, BNC shares chemical similarities with plant cellulose, despite being approximately 100 times thinner [71]. The synthesis of cellulose from bacteria is a complex, multistage process that involves the fermentation of gram-negative bacteria, particularly *Acetobacter* genus (e.g., *Komagataeibacter europaeus*). Despite its advantages, large-scale production is limited due to high costs and low productivity [72]. Bacterial cellulose boasts several advantageous properties: it is a porous material with a high degree of polymerization, excellent water-holding capacity, a high specific surface area, superior shape retention, and significant liquid absorption capacity. These characteristics make it suitable for a wide range of applications [73].

5.4 Nanocellulose Membrane in Water Filtration

Conventional water treatment methods like coagulation and flocculation have limitations due to high operating costs and low efficiency in eliminating contaminants [74]. These drawbacks can be solved by pressure-driven membrane systems that are capable of eliminating impurities as small as 5 nm with a typical filtration efficiency of more than 95% [75]. Membrane filtration techniques, such as reverse osmosis, ultrafiltration, microfiltration, and nanofiltration, offer an environmentally friendly alternative and reduce chemical dependency. These pressure-driven membranes use high pressure to separate contaminants from the water stream. Membrane separation is the process of partially separating a feed that contains a mixture of two or more components by utilizing a membrane [76]. Polymeric membranes, as the predominant membrane material for water filtration, greatly value nanocellulose for its high surface area and abundant free hydroxyl groups, which facilitate surface modification [77]. The filtration effectiveness of nanocellulose-based membranes is essentially determined by the rate of water permeation and the individual pollutant's removal capacity. Membrane rejection is achieved through two independent mechanisms: size exclusion or the membrane's affinity to the target pollutant. The size exclusion technique selectively rejects contaminants based on their size relative to membrane pores. Nanocellulose membranes primarily use the size exclusion technique, with pore size set by a randomly aligned CNF or CNC network, known as "micro/nano-pores." While, affinity membranes reject contaminants through electrostatic interactions. Affinity of the cellulose membranes can be altered by the surface modification of the nanocellulose [77, 78]. Oxidation, esterification, silyation, polymer grafting, and

Table 5.3 Summary of cellulose derivatives, synthesis, and membrane types with synthesis procedure

Cellulose derivative	Source material	Type of membrane	Membrane synthesis method	Additives	Ref
Cellulose acetate (CA)	Cellulose + Acetic Anhydride	(CA-NPs)	Phase inversion process	Cellulose nanoparticles	[79]
Carboxymethyl cellulose (CMC)	Cellulose + Monochloroacetic acid	PVA/CMC	Radical polymerization	Cellulose acetate	[80]
Bacterial cellulose	*Acetobacter xylinum*	BC/ hydroxyapatite	Layer-by-layer deposition	Na_2HPO_4 and $CaCl_2$	[81]
Cellulose nanocrystal	Cellulose acid hydrolysis	PVA/Am-CNC nanocomposite	Solution casting	Potassium peroxydisulphate	[82]

amidation are used for the surface modification of nanocellulose materials [74]. For instance, researchers prepared cellulose acetate membranes incorporating cellulose nanoparticles using the phase inversion method to enhance their function-related properties and selectivity [79]. Similarly, different methods use the cellulose derivative to create unique separation membranes with specific targets. Table 5.3 illustrates various derivatives of cellulose materials, their synthesis procedures, and the type of membrane used, along with their respective synthesis processes.

5.5 Conclusion

The application of waste-derived cellulose NMs for water filtration membranes marks a significant step forward in both sustainable development and water purification technologies. This technology addresses the management of waste challenges while also providing an environmentally acceptable water treatment option by converting waste from agriculture and industry into valuable nanocellulose. Cellulose-based NMs possess remarkable characteristics, including excellent mechanical strength, extensive surface area, and customizable surface chemistry. These characteristics make it possible to create membranes that exhibit better filtration capabilities. These membranes efficiently clear a broad spectrum of pollutants, such as heavy metals, dyes, and microbes, guaranteeing the delivery of purified and secure water. In a nutshell, membranes made from cellulose NMs produced from trash have the potential to greatly transform water filtration applications. Their offerings combine environmental sustainability, affordability, and high performance, enabling new solutions to worldwide water scarcity and pollution concerns.

References

1. Diop S (2002) Vital water graphics: an overview of the state of the world's fresh and marine waters. UNEP, pp 1–88
2. Shannon MA, Bohn PW, Elimelech M, Georgiadis JG, Mariñas BJ, Mayes AM (2008) Science and technology for water purification in the coming decades. Nature 452(7185):301–310
3. Popescu RC, Fufă MOM, Grumezescu AM, Holban AM (2017) Nanostructurated membranes for the microbiological purification of drinking water. In: Water purification. Elsevier, pp 421–446
4. Yu S, Pang H, Huang S, Tang H, Wang S, Qiu M, Chen Z, Yang H, Song G, Fu D (2021) Recent advances in metal-organic framework membranes for water treatment: a review. Sci Total Environ 800:149662
5. Kayvani Fard A, McKay G, Buekenhoudt A, Al Sulaiti H, Motmans F, Khraisheh M, Atieh M (2018) Inorganic membranes: preparation and application for water treatment and desalination. Materials 11(1):74
6. Zahid M, Rashid A, Akram S, Rehan ZA, Razzaq W (2018) A comprehensive review on polymeric nano-composite membranes for water treatment. J Membr Sci Technol 8(1):1–20
7. Ghosal K, Ghosh S (2023) Biodegradable polymers from lignocellulosic biomass and synthetic plastic waste: an emerging alternative for biomedical applications. Mater Sci Eng R Rep 156:100761
8. Nunes SP, Culfaz-Emecen PZ, Ramon GZ, Visser T, Koops GH, Jin W, Ulbricht M (2020) Thinking the future of membranes: Perspectives for advanced and new membrane materials and manufacturing processes. J Membr Sci 598:117761
9. Mansoori S, Davarnejad R, Matsuura T, Ismail AF (2020) Membranes based on non-synthetic (natural) polymers for wastewater treatment. Polym Testing 84:106381
10. Tashiro K, Kobayashi M (1991) Theoretical evaluation of three-dimensional elastic constants of native and regenerated celluloses: role of hydrogen bonds. Polymer 32(8):1516–1526
11. Khui PLN, Rahman MR, Bakri MKB (2023) A review on the extraction of cellulose and nanocellulose as a filler through solid waste management. J Thermoplast Compos Mater 36(3):1306–1327
12. Trache D, Tarchoun AF, Derradji M, Hamidon TS, Masruchin N, Brosse N et al (2020) Nanocellulose: from fundamentals to advanced applications. Front Chem 8:1–33. https://doi.org/10.3389/fchem.2020.00392
13. Ruan Z, Wang X, Liu Y, Liao W (2019) Chapter 3—Corn. In: Pan Z, Zhang R, Zicari S (eds) Integrated processing technologies for food and agricultural by-products. Academic Press, pp 59–72. https://doi.org/10.1016/B978-0-12-814138-0.00003-4.
14. Sundarraj AA, Ranganathan TV (2018) A review on cellulose and its utilization from agro-industrial waste. Drug Invent Today 10(1):89–94
15. Romruen O, Karbowiak T, Tongdeesoontorn W, Shiekh KA, Rawdkuen S (2022) Extraction and characterization of cellulose from agricultural by-products of Chiang Rai Province, Thailand. Polymers 14(9):1830
16. Rizwan M, Gilani SR, Durani AI, Naseem S (2021) Materials diversity of hydrogel: Synthesis, polymerization process and soil conditioning properties in agricultural field. J Adv Res 33:15–40
17. Yiin CL, Ho S, Yusup S, Quitain AT, Chan YH, Loy ACM, Gwee YL (2019) Recovery of cellulose fibers from oil palm empty fruit bunch for pulp and paper using green delignification approach. Biores Technol 290:121797
18. Obada DO, Kekung MO, Levonyan T, Norval GW (2023) Palm oil mill derived empty palm fruit bunches as a feed stock for renewable energy applications in Nigeria: a review. Bioresour Technol Rep, p 101666
19. Heniegal AM, Ramadan MA, Naguib A, Agwa IS (2020) Study on properties of clay brick incorporating sludge of water treatment plant and agriculture waste. Case Stud Constr Mater 13:e00397

20. Mahmud MA, Anannya FR (2021) Sugarcane bagasse-a source of cellulosic fiber for diverse applications. Heliyon 7(8):e07771
21. Eo MY, Fan H, Cho YJ, Kim SM, Lee SK (2016) Cellulose membrane as a biomaterial: from hydrolysis to depolymerization with electron beam. Biomater Res 20(1):16
22. Hastuti N, Khaerudini DS, Aini EN, Lukmandaru G, Hisankusuma J, Hertada MQ, Christiani MF, Razi MA, Abrori SA (2024) The potential application of cellulose acetate membrane for CO2 adsorbent. Environ Technol 5818–5826
23. Lo S-L, Huang Y-F, Chiueh P-T, Kuan W-H (2017) Microwave pyrolysis of lignocellulosic biomass. Energy Procedia 105:41–46. https://doi.org/10.1016/j.egypro.2017.03.277
24. Salema AA, Ani FN (2011) Microwave induced pyrolysis of oil palm biomass. Bioresour Technol 102(3):3388–3395. https://doi.org/10.1016/j.biortech.2010.09.115
25. Shi X, Wang J (2014) A comparative investigation into the formation behaviors of char, liquids and gases during pyrolysis of pinewood and lignocellulosic components. Bioresour Technol 170:262–269. https://doi.org/10.1016/j.biortech.2014.07.110
26. Zhang S, Dong Q, Zhang L, Xiong Y (2015) High quality syngas production from microwave pyrolysis of rice husk with char-supported metallic catalysts. Bioresour Technol 191:17–23. https://doi.org/10.1016/j.biortech.2015.04.114
27. Zhao X, Wang M, Liu H, Li L, Ma C, Song Z (2012) A microwave reactor for characterization of pyrolyzed biomass. Bioresour Technol 104:673–678. https://doi.org/10.1016/j.biortech.2011.09.137
28. Miura M, Kaga H, Sakurai A, Kakuchi T, Takahashi K (2004) Rapid pyrolysis of wood block by microwave heating. J Anal Appl Pyrolysis 71(1):187–199. https://doi.org/10.1016/S0165-2370(03)00087-1
29. Rout T, Pradhan D, Singh RK, Kumari N (2016) Exhaustive study of products obtained from coconut shell pyrolysis. J Environ Chem Eng 4(3):3696–3705. https://doi.org/10.1016/j.jece.2016.02.024
30. Park J, Lee Y, Ryu C, Park Y-K (2014) Slow pyrolysis of rice straw: analysis of products properties, carbon and energy yields. Bioresour Technol 155:63–70. https://doi.org/10.1016/j.biortech.2013.12.084
31. Shemfe MB, Gu S, Ranganathan P (2015) Techno-economic performance analysis of biofuel production and miniature electric power generation from biomass fast pyrolysis and bio-oil upgrading. Fuel 143:361–372. https://doi.org/10.1016/j.fuel.2014.11.078
32. Ding T, Li S, Xie J, Song W, Yao J, Lin W (2012) Rapid pyrolysis of wheat straw in a bench-scale circulating fluidized-bed downer reactor. Chem Eng Technol 35(12):2170–2176. https://doi.org/10.1002/ceat.201200140
33. Chen T, Liu R, Scott NR (2016) Characterization of energy carriers obtained from the pyrolysis of white ash, switchgrass and corn stover—Biochar, syngas and bio-oil. Fuel Process Technol 142:124–134. https://doi.org/10.1016/j.fuproc.2015.09.034
34. Nagendran R (2011) Agricultural waste and pollution. In: Waste. Elsevier, pp 341–355
35. Kaur P, Sharma N, Munagala M, Rajkhowa R, Aallardyce B, Shastri Y, Agrawal R (2021) Nanocellulose: ressources, physio-chemical properties, current uses and future applications. Front Nanotechnol 3:747329
36. Norrrahim MNF, Ariffin H, Yasim-Anuar TAT, Ghaemi F, Hassan MA, Ibrahim NA, Ngee JLH, Yunus WMZW (2018) Superheated steam pretreatment of cellulose affects its electrospinnability for microfibrillated cellulose production. Cellulose 25:3853–3859
37. Cao J, Xia X, Wang L, Zhang Z, Liu X (2021) A novel CNC milling energy consumption prediction method based on program parsing and parallel neural network. Sustainability 13(24):13918
38. Coelho CC, Michelin M, Cerqueira MA, Gonçalves C, Tonon RV, Pastrana LM, Freitas-Silva O, Vicente AA, Cabral LM, Teixeira JA (2018) Cellulose nanocrystals from grape pomace: production, properties and cytotoxicity assessment. Carbohyd Polym 192:327–336
39. Wang W, Liu CS, Huang FH, Li WL, Zheng C (2019) Preparation and characterization of nanocellulose from rapeseed hull. Oil Crop Sci 4:55–64

40. Glińska K, Lerigoleur C, Giralt J, Torrens E, Bengoa C (2020) Valorization of cellulose recovered from WWTP sludge to added value levulinic acid with a brønsted acidic ionic liquid. Catalysts 10(9):1004
41. Zbieć M, Franc-Dąbrowska J, Drejerska N (2022) Wood waste management in Europe through the lens of the circular bioeconomy. Energies 15:4352. https://doi.org/10.3390/en15124352
42. Matebie BY, Tizazu BZ, Kadhem AA, Venkatesa Prabhu S (2021) Synthesis of cellulose nanocrystals (CNCs) from brewer's spent grain using acid hydrolysis: characterization and optimization. J Nanomater 2021(1):7133154
43. Johar N, Ahmad I, Dufresne A (2012) Extraction, preparation and characterization of cellulose fibres and nanocrystals from rice husk. Ind Crops Prod 37(1):93–99
44. Hemmati F, Jafari SM, Taheri RA (2019) Optimization of homogenization-sonication technique for the production of cellulose nanocrystals from cotton linter. Int J Biol Macromol 137:374–381. https://doi.org/10.1016/j.ijbiomac.2019.06.241
45. Morán JI, Alvarez VA, Cyras VP, Vázquez A (2008) Extraction of cellulose and preparation of nanocellulose from sisal fibers. Cellulose 15(1):149–159. https://doi.org/10.1007/s10570-007-9145-9
46. Chen D, Lawton D, Thompson MR, Liu Q (2012) Biocomposites reinforced with cellulose nanocrystals derived from potato peel waste. Carbohydr Polym 90(1):709–716. https://doi.org/10.1016/j.carbpol.2012.06.002
47. Budhi YW, Fakhrudin M, Culsum NTU, Suendo V, Iskandar F (2018) Preparation of cellulose nanocrystals from empty fruit bunch of palm oil by using phosphotungstic acid. IOP Conf Ser Earth Environ Sci 105(1):012063. https://doi.org/10.1088/1755-1315/105/1/012063
48. Henrique MA, Silvério HA, Flauzino Neto WP, Pasquini D (2013) Valorization of an agro-industrial waste, mango seed, by the extraction and characterization of its cellulose nanocrystals. J Environ Manag 121:202–209. https://doi.org/10.1016/j.jenvman.2013.02.054
49. Akinjokun AI, Petrik LF, Ogunfowokan AO, Ajao J, Ojumu TV (2021) Isolation and characterization of nanocrystalline cellulose from cocoa pod husk (CPH) biomass wastes. Heliyon 7(4):e06680. https://doi.org/10.1016/j.heliyon.2021.e06680
50. Danial WH, Abdul Majid Z, Mohd Muhid MN, Triwahyono S, Bakar MB, Ramli Z (2015) The reuse of wastepaper for the extraction of cellulose nanocrystals. Carbohydr Polym 118:165–169. https://doi.org/10.1016/j.carbpol.2014.10.072
51. Gabriel T, Wondu K, Dilebo J (2021) Valorization of khat (Catha edulis) waste for the production of cellulose fibers and nanocrystals. PLoS ONE 16(2):e0246794. https://doi.org/10.1371/journal.pone.0246794
52. Liu C, Li B, Du H, Lv D, Zhang Y, Yu G, Mu X, Peng H (2016) Properties of nanocellulose isolated from corncob residue using sulfuric acid, formic acid, oxidative and mechanical methods. Carbohydr Polym 151:716–724. https://doi.org/10.1016/j.carbpol.2016.06.025
53. Lu S, Ma T, Hu X, Zhao J, Liao X, Song Y, Hu X (2022) Facile extraction and characterization of cellulose nanocrystals from agricultural waste sugarcane straw. J Sci Food Agric 102(1):312–321. https://doi.org/10.1002/jsfa.11360
54. Tonoli GHD, Teixeira EM, Corrêa AC, Marconcini JM, Caixeta LA, Pereira-da-Silva MA, Mattoso LHC (2012) Cellulose micro/nanofibres from Eucalyptus kraft pulp: preparation and properties. Carbohydr Polym 89(1):80–88. https://doi.org/10.1016/j.carbpol.2012.02.052
55. Wang X, Li H, Cao Y, Tang Q (2011) Cellulose extraction from wood chip in an ionic liquid 1-allyl-3-methylimidazolium chloride (AmimCl). Bioresour Technol 102(17):7959–7965. https://doi.org/10.1016/j.biortech.2011.05.064
56. Agbor VB, Cicek N, Sparling R, Berlin A, Levin DB (2011) Biomass pretreatment: fundamentals toward application. Biotechnol Adv 29(6):675–685. https://doi.org/10.1016/j.biotechadv.2011.05.005
57. Felgueiras HP, Padrão J, Antunes JC (2022) Extraction of cellulose-based polymers from textile wastes. Polymers (Basel) 14(10):2063. https://doi.org/10.3390/polym14102063
58. Foster EJ, Moon RJ, Agarwal UP, Bortner MJ, Bras J, Camarero-Espinosa S, Chan KJ, Clift MJD, Cranston ED, Eichhorn SJ, Fox DM, Hamad WY, Heux L, Jean B, Korey M, Nieh W, Ong KJ, Reid MS, Renneckar S, Roberts R, Shatkin JA, Simonsen J, Stinson-Bagby K, Wanasekara

N, Youngblood J (2018) Current characterization methods for cellulose nanomaterials. Chem Soc Rev 47(8):2609–2679. https://doi.org/10.1039/C6CS00895J
59. Haule LV, Carr CM, Rigout M (2016) Preparation and physical properties of regenerated cellulose fibres from cotton waste garments. J Clean Prod 112:4445–4451. https://doi.org/10.1016/j.jclepro.2015.08.086
60. Shuhua W, Xiaoying Y, Xiaogang C, Wensheng H, Mei N (2020) Recycling of cotton fibers separated from the waste blend fabric. J Nat Fibers 17(4):520–531. https://doi.org/10.1080/15440478.2018.1503130
61. Kurniawan TW, Sulistyarti H, Rumhayati B, Sabarudin A (2023) Cellulose nanocrystals (CNCs) and cellulose nanofibers (CNFs) as adsorbents of heavy metal ions. J Chem 2023(1):5037027
62. Yang H, Tejado A, Alam N, Antal M, van de Ven TG (2012) Films prepared from electrosterically stabilized nanocrystalline cellulose. Langmuir 28(20):7834–7842
63. Onkarappa H, Prakash G, Pujar G, Rajith Kumar C, Betageri RVVS (2020) Facile synthesis and characterization of nanocellulose from Zea mays husk. Polym Compos 41(8):3153–3159
64. Moon RJ, Martini A, Nairn J, Simonsen J, Youngblood J (2011) Cellulose nanomaterials review: structure, properties and nanocomposites. Chem Soc Rev 40(7):3941–3994
65. Abraham E, Deepa B, Pothan LA, Jacob M, Thomas S, Cvelbar U, Anandjiwala R (2011) Extraction of nanocellulose fibrils from lignocellulosic fibres: A novel approach. Carbohyd Polym 86(4):1468–1475
66. Lundahl MJ, Cunha AG, Rojo E, Papageorgiou AC, Rautkari L, Arboleda JC, Rojas OJ (2016) Strength and water interactions of cellulose I filaments wet-spun from cellulose nanofibril hydrogels. Sci Rep 6(1):30695
67. Kafy A, Kim HC, Zhai L, Kim JW, Hai LV, Kang TJ, Kim J (2017) Cellulose long fibers fabricated from cellulose nanofibers and its strong and tough characteristics. Sci Rep 7(1):17683
68. Yi T, Zhao H, Mo Q, Pan D, Liu Y, Huang L, Xu H, Hu B, Song H (2020) From cellulose to cellulose nanofibrils—a comprehensive review of the preparation and modification of cellulose nanofibrils. Materials 13(22):5062
69. Abol-Fotouh D, Hassan MA, Shokry H, Roig A, Azab MS, Kashyout AE-HB (2020) Bacterial nanocellulose from agro-industrial wastes: low-cost and enhanced production by Komagataeibacter saccharivorans MD1. Sci Rep 10(1):3491
70. El-Naggar NE-A, Mohammed AA, El-Malkey SE (2023) Bacterial nanocellulose production using Cantaloupe juice, statistical optimization and characterization. Sci Rep 13(1):51
71. Ross P, Mayer R, Benziman M (1991) Cellulose biosynthesis and function in bacteria. Microbiol Rev 55(1):35–58
72. Reshmy R, Philip E, Thomas D, Madhavan A, Sindhu R, Binod P, Varjani S, Awasthi MK, Pandey A (2021) Bacterial nanocellulose: engineering, production, and applications. Bioengineered 12(2):11463
73. Barja F (2021) Bacterial nanocellulose production and biomedical applications. J Biomed Res 35(4):310
74. Jaffar SS, Saallah S, Misson M, Siddiquee S, Roslan J, Saalah S, Lenggoro W (2022) Recent development and environmental applications of nanocellulose-based membranes. Membranes 12(3):287
75. Tan H-F, Ooi BS, Leo CP (2020) Future perspectives of nanocellulose-based membrane for water treatment. J Water Process Eng 37:101502. https://doi.org/10.1016/j.jwpe.2020.101502
76. Liu Y, Liu H, Shen Z (2021) Nanocellulose based filtration membrane in industrial waste water treatment: a review. Materials 14(18):5398
77. Sharma PR, Sharma SK, Lindström T, Hsiao BS (2020) Nanocellulose-enabled membranes for water purification: perspectives. Adv Sustain Syst 4(5):1900114. https://doi.org/10.1002/adsu.201900114
78. Mautner A (2020) Nanocellulose water treatment membranes and filters: a review. Polym Int 69(9):741–751. https://doi.org/10.1002/pi.5993
79. Al-Rawi AS, Ibrahim Al-Khateeb IK, Zaidan TA (2021) Nanocellulose acetate membranes: Preparation and application. Environ Nanotechnol Monit Manag 16:100529. https://doi.org/10.1016/j.enmm.2021.100529

80. Ibrahim MM, Koschella A, Kadry G, Heinze T (2013) Evaluation of cellulose and carboxymethyl cellulose/poly(vinyl alcohol) membranes. Carbohydr Polym 95(1):414–420. https://doi.org/10.1016/j.carbpol.2013.03.012
81. Coelho F, Cavicchioli M, Specian SS, Scarel-Caminaga RM, Penteado LA, Medeiros AI, Ribeiro SJL, Capote TSO (2019) Bacterial cellulose membrane functionalized with hydroxiapatite and anti-bone morphogenetic protein 2: a promising material for bone regeneration. PLoS ONE 14(8):e0221286. https://doi.org/10.1371/journal.pone.0221286
82. Ahmad S, Jahan Z, Sher F, Niazi MBK, Noor T, Hou H, Azhar O, Sher EK (2022) Polyvinyl alcohol and aminated cellulose nanocrystal membranes with improved interfacial compatibility for environmental applications. Environ Res 214:113793. https://doi.org/10.1016/j.envres.2022.113793

Chapter 6
Waste-Derived Graphene for the Removal of Heavy Metals: A Sustainable Approach Toward Environmental Remediation

Fozia Batool, Sobia Noreen, Humaira Yasmeen Gonadal, Anila Arshad, Tunzeel Iqbal, and Shahid Iqbal

Abstract Green and low-cost sources for remediation of environmental pollution are the focus of research nowadays. In this regard, nanoparticles derived from different waste materials are a sizzling area of research particularly graphene-based carbon dots are employed by many researchers for waste management. Graphene has a single layer of carbon atoms arranged in a hexagonal pattern which gives it a large surface area, significant conduction power, and surface properties suitable for the attachment of different waste materials from aqueous media. Agricultural waste materials are mostly employed to extract graphene and then different nanoparticles are synthesized by employing new and eco-friendly techniques to boost the potential of graphene for Environmental applications. Heavy metals are persistent pollutants as they cannot be degraded or destroyed. The only remedy to eliminate them is to remove them from aqueous media. Graphene can effectively remove heavy metals like lead, cadmium, mercury, and chromium from contaminated water. The adsorption capacity of graphene for heavy metals is high, making it an efficient remediation tool. Functionalized graphene can enhance its heavy metal adsorption capabilities. In this chapter, waste waste-derived graphene and its applications for the removal of heavy metals are thoroughly described in focus of current research.

Keywords Graphene · Waste-derived materials · Heavy metals · Wastewater · Environment

---

F. Batool (✉) · S. Noreen · H. Y. Gonadal
Institute of Chemistry, University of Sargodha, Sargodha 40100, Pakistan
e-mail: fozia.batool@uos.edu.pk

A. Arshad
School of Agricultural Engineering, Jiangsu University, Zhenjiang 212013, PR China

T. Iqbal
The Rawalpindi Women's University, Rawalpindi 43600, Pakistan

S. Iqbal
Faisalabad Campus, University of Education, University of Education, Faisalabad 37000, Pakistan

N. Talreja et al. (eds.), *Waste-Derived Carbon Nanostructures*, Nanostructure Science and Technology, https://doi.org/10.1007/978-3-031-75247-6_6

6.1 Introduction

Recycling waste material to prepare graphene-based products has attracted scientists for various applications, however, it is very challenging to make an effective synthesis of graphene [1]. Waste materials from homes to the industrial sector harm environmental and cleanliness factors. In recent developments and factors of population, total waste production in the world has been estimated to emerge at 2.2 billion tons in the year 2025 [2]. It is a fact that waste conversion depends on the type of waste precursors. For example, biomass inversion such as rice husk, and chitosan depends on the nature of the source. Graphene is an allotropic form of carbon having a single layer of atoms arranged in a hexagonal structure. It is derived from "graphite" and tells the fact that there are various double bonds in graphite allotrope. Andre Geim and Konstantin Novoselov discovered graphene and were rewarded with a Nobel Prize in 2010 [3]. Graphene, an enchanting material of the twenty-first century discovered from a graphite single sheet consists of sp^2 hybridized carbon arranged in a honeycomb appearance with miracle characteristics such as powerful mechanical strength, high conductor of electricity, chemical stability, and better surface area (Fig. 6.1). Multiple techniques have been applied for the formation of graphene [4].

Since its discovery, graphite has been used as the primary raw material in the synthesis of graphene. Because of its highly anisotropic structure, graphite exhibits very distinct in-plane and out-of-plane surface properties [5]. Graphene is often combined with other materials to create tunable optical and other electronic devices because of its remarkable physical properties, including huge nonlinearity, electrical tunability, and ultra-thin properties. Pure graphite is used as the starting material

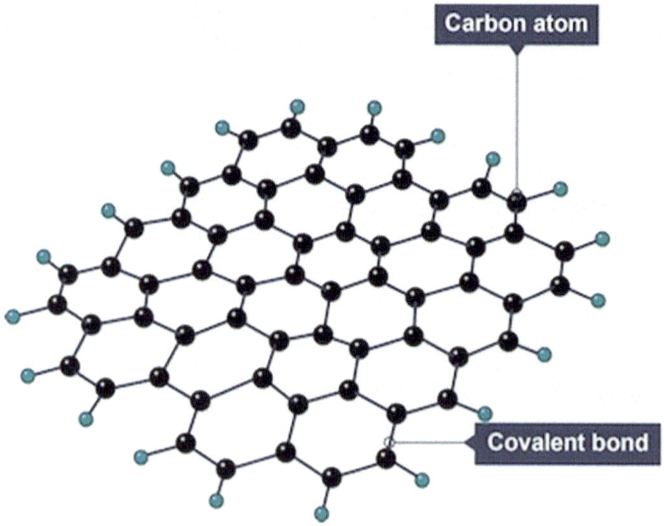

Fig. 6.1 Structure of graphene

for producing graphene and graphene derivatives by hammers process. However, it is expensive to produce graphene from the hammer process. Therefore, a basic method for the manufacture of high-quality graphene from numerous naturally occurring and industrial carbonaceous wastes such as vegetation wastes (cow dung, other animal waste, and fruit waste, semi-industrial waste (newspaper), and industrial waste including textile, machine (soot powders produced in the exhaust of diesel vehicles) was developed [6].

Graphene reduces the environmental burden of waste material dually. Graphene is synthesized from waste materials which are itself a threat to the environment and it is also utilized for the removal of several waste materials from aqueous media. It is a key priority of the present research to develop materials that are safe for our environment and pay less cost for their production. In this regard, Graphene is an emerging area of research which have many applications in different fields [7]. Figure 6.2 shows the schematic illustration of the synthesis of graphene using agricultural waste and its application in environmental remediation.

Fig. 6.2 A schematic illustration of the synthesis of graphene using agricultural waste and its application in environmental remediation

6.2 Waste-Derived Graphene: A Green Approach of Synthesis

Due to the large application of graphene in different fields, its synthesis by employing cost-effective and recent techniques is attractive for researchers. In this regard, conversion of different waste carbonaceous products generated from graphene is becoming more and more important to meet the material's low cost of production and increasing demand. The kind and makeup of carbonaceous wastes will determine the properties of the graphene material [8].

6.2.1 Agriculture Waste

Agriculture waste is a big source of graphene as it contains high carbon content as its backbone structure. In this regard, different waste materials contain variable amounts of graphene (Fig. 6.3). Every year, a vast array of biowastes from large-scale livestock or agricultural sources are released into the environment. Even though these wastes can be created in huge quantities and are biodegradable, their value-added consumption and future applications are accounted for by their use as biowaste precursors into graphene [9]. The utilization of organic matter and scarps as carbon sources is an encouraging way that results in a cost-effective, bulk production of graphene. However, it is strenuous to synthesize a huge quantity of graphene using a catalyst. Moreover, in the absence of a catalyst, when waste material and plants are used as catalysts, problems such as low quality and high flaws might exist. We show that using biomass waste as the precursor, a green technique of producing huge volumes of few-layer graphene may be achieved. We chose wheat straw as our research's beginning material because of its massive global output (about 350 million tons annually) [10]. Due to its plentiful supply, wheat straw is now receiving enough attention to be considered as a raw material to produce high-value nanomaterials such as activated carbon and nano-silica. In this work, we verified that a combination of hydrothermal and graphitization procedures could successfully transform the worthless wheat straw into value-added graphene sheets. This technique can be used to create high-quality, few-layer pure graphene sheets with mild mesoporosity, high graphitization, and interlayer spacing resembling graphite [11].

6.2.2 Rice Husk

Rice is the third most widely grown crop in the world. As a result, a significant amount of rice husk, also known as hull, the outer layer of the rice grain, is produced annually as agricultural waste. Chemically, rice husk is composed of 50% cellulose, 25–30% lignin, and 15–20% silica [12]. Researchers are paying close attention to the

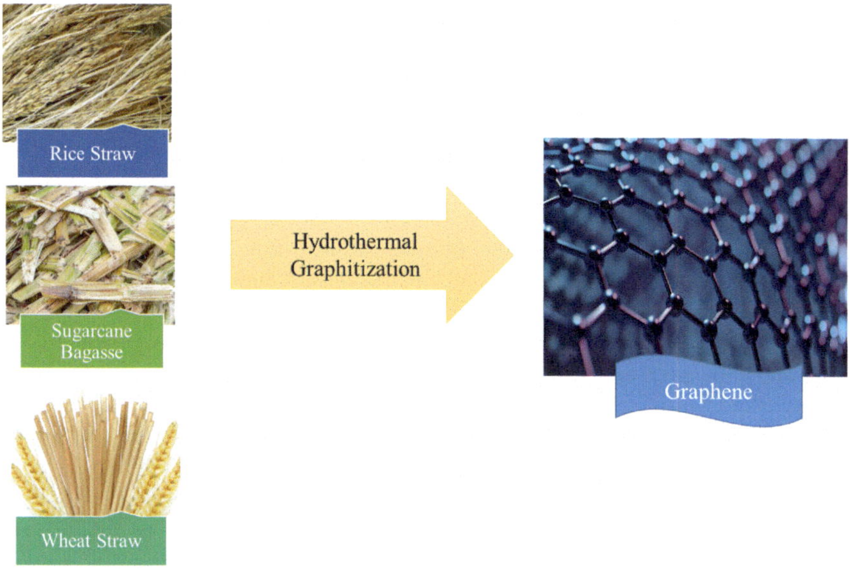

Fig. 6.3 Synthesis of graphene from agriculture waste

use of rice husk as a raw material for the synthesis of activated carbon, meso, nano, and microporous carbon, as well as graphene derivatives. Rice husk activation with KOH helped to create high porosity in the activated carbon, which in turn produced pure graphene with edges that were stable and smooth. Furthermore, KOH activation aided in the leaching of Si impurities because it is a strong base, the process of creating graphene carbon nanotubes (G-CNTs) by irradiating rice husks in a single step using microwave plasma [13]. According to Scientists, there are four steps involved in making GO from rice husk. These are carbonizing the husk to create carbonized rice husk (CRH), desiccating it using a KOH solution, activating it for two hours at 850 °C, and then exfoliating and oxidizing the resulting material to produce GO. Because of low commercial value and toil to get rid of this material, large no of countries that produce rice will either burn or waste the rice husk. This process can cause pollution by polluting the air which will in turn affect human health. Moreover, the usage of rice husk is critical to minimizing waste by using a protected and everlasting approach [13]. Lignin and cellulose components are present in bulk quantities in rice husks, and activated carbon is produced as a raw material by rice husks. Activated carbon is an effective adsorbent due to its complex microporous structure [14]. Based on multiple types of research, the rice husk is utilized as the starting material to yield the best quality graphene from biowaste because of its epic accessibility and powerful constitution.

6.2.2.1 Hemp Fiber

The hemp or cannabis is an annual herbaceous plant, which can grow in different climatic conditions. The hemp bast has been used to derive hemp fibers which have been used in the textile industry for ages. The hemp plants are also used to make bioplastic, biofuels, animal feed, various phytochemicals, and paper [15]. Hemp fibers are cultivated extensively as a result of the rising demand. However, following cultivation, hemp stems are frequently burned or thrown as waste. The hemp stem is rich in lignin, crystalline cellulose, and hemicellulosic material. As a result, these precursors have been used to reduce biowastes and produce carbon materials [16]. Similarly, hemp materials can be used to produce graphene at a reasonable cost. Wang and colleagues synthesized graphene-like linked carbon nanosheets for KOH activation and hydrothermal procedures to assess capacitance. The scientists reported that the hemp was constructed of three composite layers. The central layer was constituted of crystalline cellulose, while the outer (S1) and inner (S3) layers were made of semi-cellulose and lignin. Fibers were hydrothermally treated for 24 h at 180 °C in the presence of dilute H_2SO_4, dissolving the cellulose and lignin layers to create biochar [17].

6.2.2.2 Sugarcane Bagasse

Sugarcane bagasse is a subset of Agro waste or agricultural waste. The residual fiber is taken out after the juice is removed. To make powder, the fiber is finely processed and crushed. To get fine powder, this crushing and separation procedure is carried out multiple times. The black solid that had been generated is then gathered at room temperature [18]. It is carbon obtained from the waste of sugarcane bagasse. It is used for the synthesis of Graphene.

6.2.3 Household Waste

A huge amount of household waste is produced every year from every house. Management of this waste is itself an issue. Conversion of this waste to graphene which can be applied in a variety of fields is an excellent way to cope with this issue [19].

6.2.3.1 Disposable Paper Cups

These days, disposable paper cups are widely utilized everywhere. These cups make up a significant amount of the daily solid waste generated because they are thrown away shortly after being used. The lignin and hemicellulose from paper pulp are chemically indicated by these cups. Despite being a waste product that decomposes naturally, it contains a high amount of carbon in the form of carbs [20]. It can also be thought of as a raw ingredient used to make carbon materials. Paper cup wastes are treated with a KOH solution as part of the synthesis process to create pulp [21]. This pulp is a rich source of carbon to synthesize Graphene.

6.2.3.2 Plastic Bottles

Every aspect of life uses plastic, and the enormous amount of plastic generated globally eventually results in the production of enormous amounts of solid garbage. Researchers now face a difficult task in managing plastic solid waste to protect the environment. Given that plastic trash is a rich source of carbon, it has been suggested that it be recycled sustainably and converted into materials derived from graphene for a range of uses [22]. Figure 6.4 gives the modification of waste bottles into graphene.

6.2.4 Industrial Waste

Industrial waste has a large variety due to variations in different industrial products. Graphene is synthesized from industrial waste materials that have a sufficient amount of carbon present and easy to convert in graphene as demonstrated in Fig. 6.5. In this

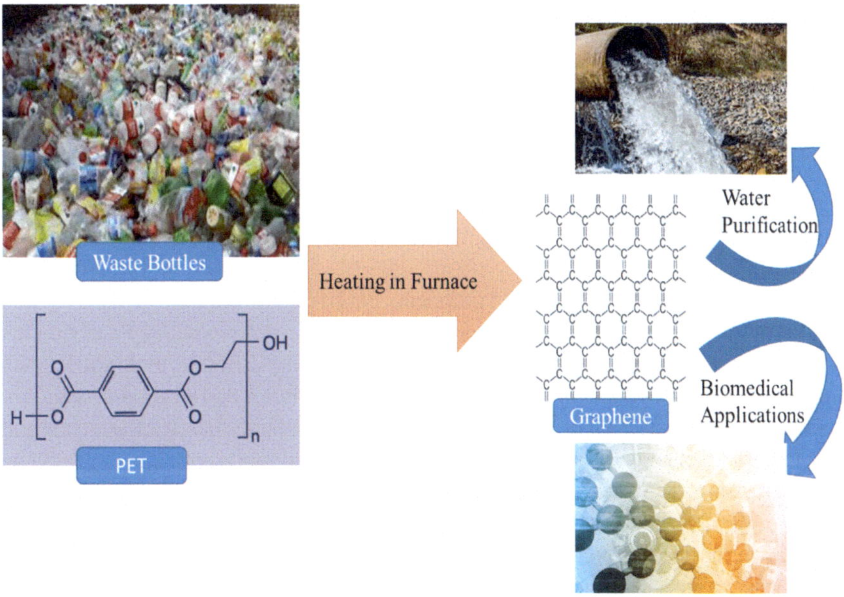

Fig. 6.4 Preparation of graphene from waste PET bottles and its applications

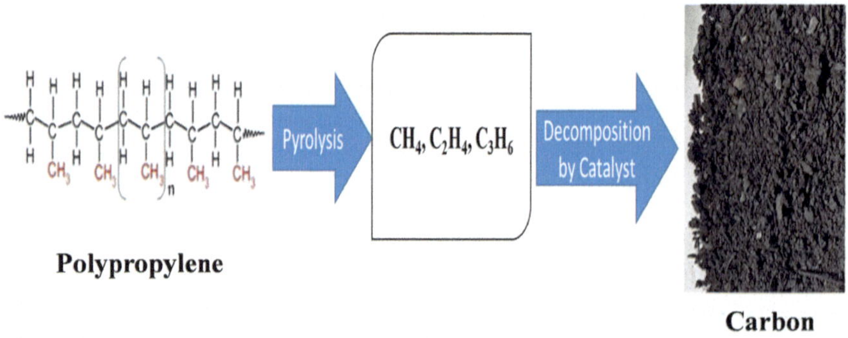

Fig. 6.5 Synthesis of graphene by industrial plastic waste

regard, plastic-based materials from different industries are very common as they are rich sources of carbon [23].

6.2.4.1 Plastics

Plastic materials are widely used in packing, medical devices, furniture construction, and transportation. This huge amount of plastic used in different industries is a big threat to pollution as well. Recycling and reforming these materials to graphene for useful purposes is an excellent way to cope with these materials. Most commonly used plastic materials like polyvinyl chloride (PVC), Polyethylene (PE), Polypropylene (PP), and polyacrylonitrile (PAN) are converted into graphene, nanotubes, and carbon dots [24].

6.2.5 Graphene Synthesis Process

Any method used to create or extract graphene, based on the required size, purity, and efflorescence of the product is referred to as graphene synthesis. Any carbon-rich compound can be a source of graphene synthesis using different techniques such as mechanical and chemical exfoliation, thermal, pyrolysis, and plasma irradiation. There are two approaches used in the synthesis of graphene that are top-down and bottom-up approaches as given in Fig. 6.6. The top-down approach is graphene synthesis by breakdown of graphite by using chemical and mechanical exfoliation. In the bottom-up approach where carbon-rich sources unite together by pyrolysis, chemical vapor deposition, and plasma source [25].

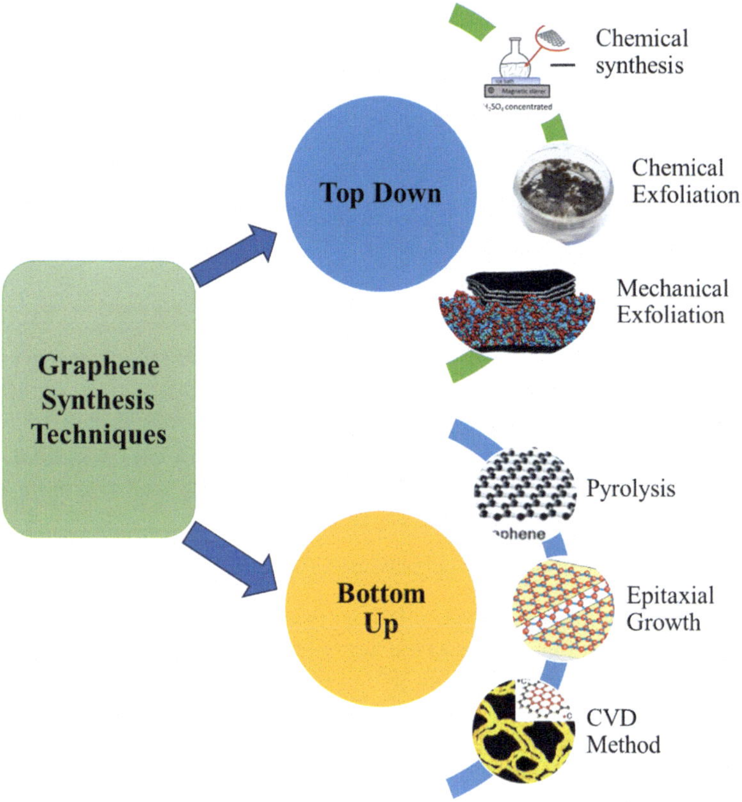

Fig. 6.6 Techniques for synthesis of graphene

6.2.5.1 Mechanical Exfoliation

In this concept, graphene is synthesized by the breakdown of graphite which is a more acceptable is layer by layer breakdown. Two mechanical routes are breaking graphite into its flakes. Normal force overcomes Vanderwaal interactions because of two graphite layers apart while exerting lateral force promotes relative motion between two layers. Another auxiliary route is the fragmentation of graphite into flakes [26]. Micromechanical exfoliation which is also called scotch tape exfoliation is given in Fig. 6.7.

Micromechanical Cleavage

This concept includes cleavage of highly ordered pyrolytic graphite HOPG with the help of Scotch tape. The method includes the application of scotch tape on graphite layers and just by applying normal force, it cleaved out. This way, the graphitic

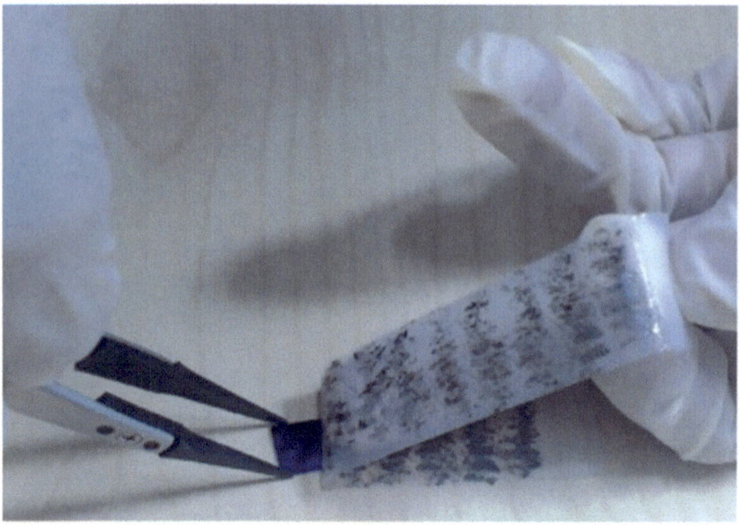

Fig. 6.7 Micromechanical exfoliation for synthesis of graphene [25]

layer becomes thin and thin and gets the desired single graphene layer. The method is time-consuming and includes labor [27]. This method is modified as a lathe-like experimental setup to generate graphene flakes to improve efficiency and save labor. The graphitic sample is trimmed into pyramidal shape and embedded into epoxy. The tool used for this setup is an ultrasharp single-crystal diamond. The diamond wage is aligned carefully concerning the amount of HOPG and is attached to an ultrasonic oscillating system. As the sample goes down toward the ultrasharp wedge, it cleaves off as graphene flakes. The resulting flakes are of thickness tens of nanometers. Further precision is required to be scaled up [28]. Another technique includes the Scotch tape method using a polymer as adhesive using three roll machines. The adhesive used is polyvinylchloride dissolved in dioctyl phthalate (DOP), which acts as scotch tape in the process. The process includes continuous exfoliation by driving graphite from the feed roll to the apron roll and then back to the feed roll resulting in graphene flakes. The process is famous in the rubber industry [29].

Sonication

It is the type of liquid-phase exfoliation in which graphite powder is dispersed into organic solvents such as N, N-dimethylformamide, and other solvents then followed by sonication and centrifugation. This process just includes the graphite powder dissolved in a solvent followed by sonication and then centrifuged to a fine solid product, that is washed with water and the resultant is solid graphene powder [30].

Ball-Milling

It is the common type used in the powder production industry. Two main factors are in this process for cleavage out the graphite flakes. One is the shear force and the other is collisions by the balls in the rolling process. Graphite is turned into graphene by using a ball mill that may be wet-milling or dry-milling [31].

Fluid Dynamics

The process includes the moving of liquid through a narrow space in the valve; cavi-ation and velocity gradient are generated due to abrupt changes in shape resulting in shear and normal force for exfoliation. By changing the width of the valve, the process can be controlled. The graphitic powder is dissolved in liquid injected by a pump into the valve. After many cycles, the dispersion can be collected, and centrifuged, resulting resulted into graphene powder [32].

Chemical Exfoliation

This process includes the change of graphite into graphene by using chemicals such as alkali metals that reduce the vanderwaal interactions between graphitic layers and as a result, graphene-intercalated compounds are formed. The base of the intercalated graphene compound is the difference in ionization energy of graphene and alkalis. The reaction requires great control as alkalis react vigorously with alcohol and water so an ice bath to dissipate heat [5].

6.2.5.2 Arc-Discharge

The process includes two electrodes and a chamber of steel. The two electrodes are of pure graphite rods. A steady current of 100–150 A is maintained during the discharge process. The discharge environment is varied. H_2, (NH_3 + He), and air are the most used environments used for the production of graphene [33].

6.2.5.3 Chemical Vapor Deposition (CVD)

The process includes the breakdown of precursor carbonaceous material on some substrate surface using high temperature and the resultant is graphene layers. Here is the pyrolytic decomposition of the attached precursor and the formation of uniform thin layers of graphene. To decrease the pyrolytic temperature, transition metals are used as they are effective catalysts in transforming the hydrocarbon into graphene [34]. For the growth of Graphene film in CVD, different substrates are employed

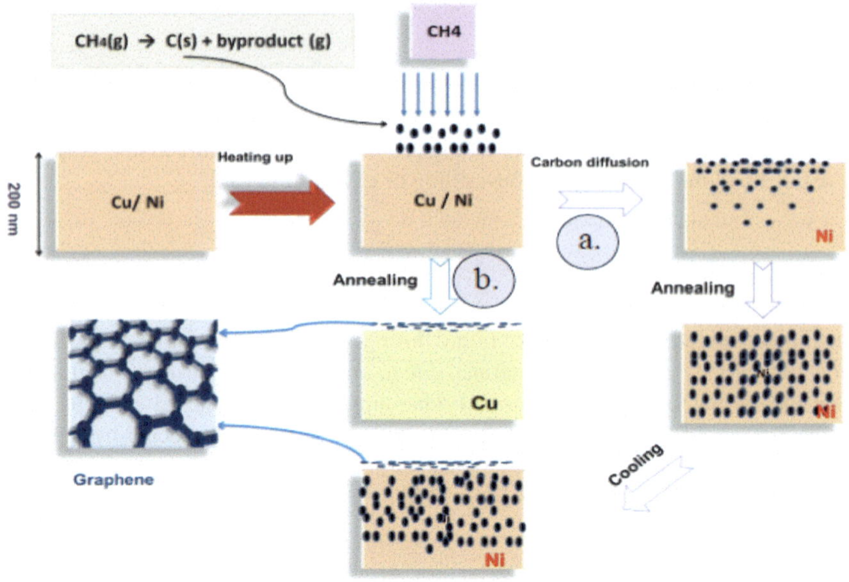

Fig. 6.8 Scheme of synthesis for CVD graphene [25]

including Copper, Nickel, and Iron as given in Fig. 6.8. As a source of carbon, mostly acetylene or methane is employed for reaction.

6.2.5.4 Thermal Decomposition

Silicon is thermally decomposed into graphene on the surface of 6H–SiC. When 6H–SiC is heated with Hydrogen as etching gas at a temperature range of 1250 °C for 20 min. As a result, graphene sheets are produced epitaxially on the hydrogen-etched surface and the number of layers is directly dependent on the decomposition temperature [35].

6.2.5.5 Plasma Decomposition

The process includes the precursor breakdown into graphene using a plasma source. During the process, the source gas is activated with high-energy electrons. The ionization, dissociation, and excitation all occur at low temperatures. Then the highly energetic gas ion as radicals fall on the precursor substrate resulting in graphene [36]. So many techniques in basic and modified forms can be used in the formation of graphene from natural precursors and graphite and used in various applications further. For the industrial production of graphene, graphene oxide reduction by chemical treatment method is the most prominent. Pyrolysis on the surface of the metal for the nanostructured graphene has been one of the researcher's choices. However, due to its low yield, this method was later replaced by the chemical treatment method. The chemical treatment method requires some hazardous and expensive chemicals [37].

6.3 Heavy Metals

Metals with a density above 5 g/cm^3 are considered heavy metals. Metal toxicity is a serious threat to humans and other animals as metals are persistent and cannot be destroyed. Their safe removal from the environment is needed for the day. They cause several diseases like cancer, kidney damage, lung issues, and respiratory problems [38]. Several methods are used for the removal of heavy metals from polluted water such as chemical precipitation, adsorption, coagulation and flocculation, solvent extraction, ultra-filtration, or ion exchange. Heavy metals exert bad impacts on human health. So it is necessary to remove the heavy metals from water [39]. There are two traditional methods implemented for the removal of heavy metals chemical and biological. A biological process is considered superior to a chemical method due to fewer amounts of chemicals and sewage produced.

6.3.1 Biological Methods

The following important processes are involved in this method (USEPA 2004).

6.3.1.1 Microbial Remediation

Microorganisms are used to remove heavy metals from the water in the microbial remediation process. Genetically engineered microorganisms are added to boost the breakdown process. This type of microbial remediation is called engineered microbial remediation. Bio-augmentation, biosparging, and biosorption techniques are used for the extraction of heavy metals from polluted water [40].

6.3.2 Advantages of Microbial Remediation Process

- Economically favorable in cost and technology.
- Environmental friendly process.
- Wastage of water is less.
- The residual products are less harmful such as carbon dioxide, water, and cell biomass.
- This process is functional along with other techniques also [41].

6.3.3 Disadvantages of the Microbial Remediation Process

- All the pollutants in water are not separated by this process.
- This process is suitable only for those compounds which are biodegradable.
- Products produced by this process are more harmful than other processes.
- This process is specific [42].

6.3.4 Phytoremediation

In this process, plants are used to remove harmful elements and to clean the environment. It is called green bioengineering technology. Plants break down in water and gather harmful metals from water in this process [30].

6.3.5 Chemical Precipitation Method

This method is used for the removal of heavy metals by converting soluble heavy metals into unsolvable solids then these solids are removed by filtration [43]. The amount of waste is high due to the chemical reaction and it may be exceeding up to 50% so this method becomes costly. It may work with harmful metals that are corrosive as a result safety problems arise (EPA 2000). The problem of precipitation of one metal in the presence of other metals may arise from small changes in pH [44].

6.3.6 Adsorption as a Green Technology

The binding of gas, liquid, or solid particles on the surface of inert material is called adsorption [45]. Inert material is called adsorbent while the particles which adsorb on the surface of inert material are called adsorbate. In this process, layer of adsorbate is formed on the adsorbent surface. The absorption process is the reverse of the adsorption process. Adsorption takes place on the surface only while absorption requires the whole volume of the substance [46].

6.3.6.1 Agro Waste as Adsorbent

Agro waste consists of lignin and cellulose as the basic components. Further constituents are hemicellulose, lipids, proteins, simple sugars, starches, water, hydrocarbons, ash, and numerous compounds that consist of different functional parts involved in the binding procedure. Metals are adsorbed on agro-waste adsorbents in different ways including chemisorption, complexation, ion exchange, microprecipitation, surface adsorption, etc. [47].

6.4 Graphene as an Adsorbent

Adsorption is currently thought to be a practical and affordable way to remove organic materials and heavy metal ions from wastewater. This procedure is simple to use and its design effectively removes harmful contaminants from wastewater, purifying the water in the process.

Several materials are utilized by researchers for the active remediation of metal ions from wastewater [48]. However, graphene has prime importance in this regard due to its several properties which make it a suitable choice for the purpose. Two properties that contribute to the good adsorption of heavy metals on graphene. One is the large surface area and the second is the pore size distribution. Large surface area results in increased exposure of active site toward the adsorbate. Thus, for this purpose, high porosity and small grain size distribution are needed. Research shows that graphene has zero porosity with a large surface area [49]. Graphene and its related structures have emerged as a novel material with promising efficaciousness in wastewater treatment applications very quickly [50]. Graphene-based materials are being used as sorbents for the decontamination process in wastewater because of their special qualities (such as a high specific surface area, a large number of functional groups, and excellent charge carrier mobility). Various characterization probes have unraveled that graphene acts as the potential adsorbent for dyes and toxic heavy metals. It is an efficient source that can remove significant amounts of this pollutant by employing a very small quantity of Graphene [51].

6.4.1 Adsorption of Heavy Metals

Several materials are utilized by researchers for the active remediation of metal ions from wastewater. However, graphene has prime importance in this regard due to its several properties which make it a suitable choice for the purpose [50]. Two properties that contribute to the good adsorption of heavy metals on graphene. One is the large surface area and the second is the pore size distribution. Large surface area results in increased exposure of active site toward the adsorbate. Thus, for this purpose, high porosity and small grain size distribution are needed. Research shows that graphene has zero porosity with a large surface area [49].

6.4.1.1 Lead (Pb II) and Cadmium (Cd II) Adsorption

Lead has been utilized by human beings from old age for several purposes. However, it causes severe toxicity as well when it is exposed up to a certain limit. Lead is naturally present in the earth's crust. Over time, its concentration has been incorporated into the environment. Pb (II) is very toxic. It is a poison that can damage the tissues of plants, animals, and humans. Lead is also present in paints, battery, and pottery industries which contaminate food and water. It has been observed that the amount of lead in food and tape water has increased about 100 fold and 760 fold respectively from the previous quantity [52]. Lead has many dangerous effects on human health. Accumulation of high levels of lead in the body causes numerous diseases. Lead is very corrosive; it inhibits enzymatic activity when incorporated into the biological system. Reduced concentration of lead in the body causes disturbance of nerve conduction while its higher concentration leads to nerve degeneration. Its macro concentrations in the body of kids pose a serious threat to kids every year, especially in metropolitan areas [53]. Similarly, Cadmium is one of the most hazardous metals reported by WHO. It is poisonous heavy metal which is harmful even in small quantities. Cadmium is present in less quantity in rocks, petroleum, and coal. It is not an essential element for humans and its presence in water is unfit for drinking purposes. Mostly cadmium deposits in kidneys and spreads cancer in humans [54]. People exposed to cadmium suffer from bone and renal disorders, lung disease, and hypertension. Cd survives for a longer time in the body. Tobacco burning produces Cd which is inhaled by humans and is more harmful as compared to ingested Cd. Cadmium is similar in size and has a similar charge to zinc so it can replace Zn in the body. Some household food items made by Cd are also a source of Cd pollution. Cd enters the environment through different synthetic activities such as electroplating, mining, smelting, and incineration processes used in plastic companies, paints, fossil fuel burning, and fertilizers. Cd is also introduced into the water by the deterioration of galvanized pipe [55]. The impacts of Cd on human health are different depending on its concentration level. Cd in lower concentrations has less bad impacts on health but in higher concentrations, it deposits in the liver and kidney and causes problems in the bone and renal system. Particularly, kidney is the major target of Cd [56]. Several methods are reported for the removal of lead from wastewater by employing physical and chemical techniques. Waste-derived graphene has got prime importance in this regard due to its effectiveness up to a small amount of metal removed and a small amount utilized for the purpose. It is considered environmentally friendly and cost-effective way as well. L. Hao et al. in 2012 reported that the optimum adsorption of lead onto SiO_2-graphene is 113.5 mg g^{-1} at ambient room temperature. At the same time, the maximum removal efficiency takes place in half an hour at pH 6. The percentage of the removal efficiency by the SiO_2-graphene was found to be 98.8 percent [57]. This research was further modified by the Deng et al. They prepared the functionalized graphene sheets with the assistance of water and ionic liquid. Ionic liquids include potassium hexafluorophosphate (PF6) and 1-octyl-3-methylimidazolium hexafluorophosphate (CP8) acted as the ionic liquids. The resulting graphene sheets were known as GNS^{PF6} and GNS^{CP8}. The sheets were used for studying the adsorption of lead and cadmium. Reports revealed the adsorption capacities as: the adsorption capacity of Pb on GNS^{CP8} was 74.18 at pH 5.1, whereas the adsorption of Pb on GNS^{PF6} was 406.4 at pH 5,1. Moreover, the adsorption capacity of cadmium on GNS^{CP8} was 30.05 at pH 6.2, and the adsorption capacity of GNS^{CP8} was 73.42 at pH 6.2, respectively. Due to the higher adsorption capacity of GNS^{PF6}, they can be reused for five consecutive cycles. Furthermore, researchers also considered the desorption experiment for GNS^{PF6} [58].

The removal of toxic Pb and Cd by the PS-GO gel was introduced by Guiyin Zhou et al. They prepared the sponge-like adsorbent polysiloxane-graphene oxide (PS-GO) gel by the sol–gel method. The GO was derived naturally from graphite. The PS-GO gel synthesis requires 0.40 ml 3-aminopropyltriethoxysilane (APTS), 0.3 ml dimethyldiethoxylsilane (DDS), and the solution of 4 ml of 5 mg per ml GO, and o.2 g of sodium dodecylsulfonate, and 0.8 g of urea. The resulting mixture underwent gelation and aging in a closed vessel at 90 °C for 6 h, with detailed procedures in the Supporting Information (SI). Low cost, excellent adsorption, and reusability make PS-GO gels a remarkable industrial method [59]. The removal efficiency of 99% is achieved when this method is used at an industrial scale for the removal of lead. In 2018, Alireza Shams et al. synthesized a novel adsorbent for Pb(II) and Cd(II). The adsorbent is known as the magnetic ethylene diamine-functionalized graphene oxide. MDFGO exhibited high removal efficiency in treating battery and electroplating wastewater. With MDFGO, Pb(II) and Cd(II) removal percentages are 99.6 ± 0.5 and 99.4 ± 0.6, respectively. The optimal conditions for this experiment are pH 6.2, shaking rate 500 rpm, sorbent dosage 33.0 mg, and adsorption time 11 mints [60]. In 2020, Samuel et al. synthesized β-cyclodextrin/graphene oxide (β-CD-GO) for the removal of cd(II) from water. After the characterization (FTIR, X-RAY, XRD, SEM, TEM), the high adsorption capacity of 196 mg per gram of Cd(II) at pH 7.0 was revealed. The study also checked if a material called β-CD-GO harms cervical cancer cells. As the amount of β-CD-GO increased from 50 to 200 µg, the cell viability dropped from 74 to 25%. This suggests that β-CD-GO could be useful for removing Cd(II) from water [61].

6.4.1.2 Arsenic As (III) and As (V)

Arsenic present in water can affect millions of people all over the world due to its extreme toxicity. A higher concentration of As in the body may lead to the death of humans. As it clots the protein, a complex is formed which stops the ATP creation during respiration. In recent years, graphene nanosheets have drawn the attention of scientists due to their unique properties, graphene nanosheets have various applications in supercapacitors, environmental remediation, sensors, etc. [62]. In groundwater, the amount of arsenic is greater than 0.01 mg per liter which makes the water toxic. The results are carcinogenic along with lacerations, coma, and liver and kidney diseases. To address this serious issue, Debabrata Nandi et al. have prepared the graphene nanosheets in 2021. They used human nails and iron as carbon as well as catalyst sources, in the synthesis technique. During synthesis, the carbonization process results in the formation of a black product. The black product was treated with dilute hydrochloric acid to remove the iron species which results in the graphene. GN exhibited a sorption efficiency of 99% for As (III) [63]. Mishra et al. Introduced HEG *(hydrogen-exfoliated graphene sheets)* by the reduction of GO with hydrogen. Further treatment with concentrated nitric acid introduces the different functional groups on the surface of HEG. These f-HEG have the maximum adsorption efficiency of 142 and 139 mg/g for As (V) and As (III) [64]. Magnetic graphene nano-platelet composites (MGNCs) decorated with Fe–Fe_2O_3 nanoparticles (NPs) showed the efficient adsorption of As (III) compared to conventional ones. Zhu et al. reported an adsorption capacity for As (III) of 11.34 mg/g by MGNCs [65]. Gollavelli et al. Yielded the SMG (smart magnetic graphene). The method of SMG production involves the reduction of GO by microwave irradiation. It facilitates the magnetic separation of As (V), Pb(II) and, Cr (VI) [66].

6.4.1.3 Copper (Cu II) and Mercury (Hg II)

Cu is another heavy metal that is important for humans but its higher concentration in water pollutes the water and then enters the body producing harmful effects on the liver and kidney [67]. The deterioration of water pipes prepared from copper and factory wastes pollutes the water. Copper also spreads diarrhea disease in babies and higher concentrations of copper cause metabolic problems [68]. Mercury is too harmful to human health. Mercury exists in inorganic and organic chemical forms. Both forms of mercury have different effects on health. The inorganic form of mercury causes spontaneous abortion, congenital malformation, and gastrointestinal dysfunction, and the organic form causes stomatitis, neurological disorders, brain and central nervous system damage, congenital malformation, and other diseases.

The researchers have looked for ways of transforming the carbon in biomass such as wood, rice husk, and, coconut shells, into the graphene and its derivatives. Ali et al. Synthesized an effective adsorbent for Hg(II) removal which is known as PANI–rGO(polyaniline and r-GO). The maximum adsorption capacity of PANI-rGO is 1000 mg^{-1}. Moreover, low adsorption capacity was recorded due to the protonation reaction occurring at low pH [69]. Sreeprasad et al. performed a batch experiment in which they used rGO-MnO_2 and rGO-Ag. After analyzing the results, they found out that the maximum adsorption capacities were almost identical i.e., 9.50 and 9.53 mg g^{-1} [70] (Table 6.1).

6.4.1.4 Chromium (Cr II and Cr VI)

Anthropogenic activities are mostly responsible for the addition of Chromium in our atmosphere. This metal has found many uses in metallurgy, pigments, leather goods synthesis, and fertilizers. Chromium III is not as hazardous as Chromium VI. Chromium VI causes cell mutations and has severe carcinogenic activities. Hence

Table 6.1 Different forms of graphene as an efficient adsorbent for metals

S. No.	Adsorbent	Adsorbate	Adsorption capacity (mg/g)	References
1	GO	Co(II)	371.9	[71]
2	Amino siloxane oligomer-modified GO	U(VI)	310.63	[72]
3	GO	Cu(II)	294	[50]
4	GO-TiO_2	(a) Pb(II) (b) Cd(II) (c) Zn(II)	(a) 65.6 (b) 72.8 (c) 88.9	[73]
5	GO modified with 2,2′-dipyridyl amine	Cd(II)	257	[74]
		Pb(II)	369	
6	Chitosan–gelatin/GO	Pb(II)	120	[75]
7	Triethylenetetramine-GO/ chitosan composite (TGOCS)	Cr(VI)	219	[76]
8	GO/silk fibroin aerogel (GO-SF)	Zn(II)	72.1	[77]
		Ni(II)	13.9	
9	Sulfonated-GO	Pb(II)	415	[78]
10	Magnetic chitosan-GO	Hg(II)	397	[79]
11	Zeolite-GO	As(V)	50	[80]
12	GO	Zn(II)	345	[50]
13	Ag-$CoFe_2O_4$-GO	Pb (II)	46.8 μg/mL	[81]
14	Chitosan/GO	Cr (VI)	310	[57]

removal of this metal from wastewater is quite important. Graphene and its derivatives have been utilized by researchers for the removal of this harmful metal from water. Removal of Cr VI was reported by Carpio [82] by employing graphene oxide functionalized material with significant removal efficiency. Similarly, cross-linked graphene oxide was employed for the removal of Cr VI by Xu et al. [83]. Synthesized material was found to contain excellent adsorption properties for metals.

6.4.2 *Removal of Air Pollutants*

Because of their large surface area, graphene and its derivatives have been considered appealing candidates for air pollution adsorption. However, due to the strong stacking of graphene sheets, their adsorption capacities are always restricted [84]. There are two reasons why 3D GBM performs better than other methods for gaseous pollutant adsorption. The first consists of the fundamental characteristics of 3D GBM, such as its large specific surface area that exposes many sites to adsorption and its porous structure that offers flexible and quick transport routes. The second is its functional groups that are decorated polar and contain oxygen, which strengthens the affinity for polar gas molecules and provides chemical adsorption sites [85].

6.4.2.1 Pollution Transformation on 3D GBM

Graphene's surface functional groups allow it to offer many active sites for chemical reactions. Because of its high conductivity, graphene can accept electrons and guarantee quick charge transportation. Even in ambient conditions, the charge carrier mobility rate of graphene can surpass 150,000 $cm^2 (V\ s)^{-1}$. Therefore, 3D GBM has an encouraging potential for changing pollutants [86]. Certain redox reactions can happen when electrons are directly transferred from an exogenous reductant to the surface functional groups of 3D GBM. Additionally, in a catalytic degradation process, the stable 3D network can function as a rapid electron transmitter and provide a good framework for catalyst dispersion, both of which can accelerate the chemical reaction [87].

6.4.2.2 Transformation of Gas Pollutants

Whether or not the gases can be changed by reaction on 3D GBM depends on their own characteristics. For example, GO aerogels can oxidize H_2S, SO_2, and HI to S, SO_3, and I_2, respectively. On the other hand, because of its comparatively low reducing capability, CO and H_2 do not undergo chemical reactions on GO aerogel. These gas reactions on 3D GBM occur at room temperature and do not require a noble metal catalyst, making it an affordable and environmentally friendly material for treating toxic gases from coal combustion and industrial production. It also helps to prevent the formation of sulfuric acid smog, acid rain, and other air pollutants [88].

6.4.2.3 Transformation of Water Pollutants

Gas detoxification benefits from 3D GBM's oxidizability. Its catalytic potential is, nevertheless, more frequently used in the purification of water pollutants, where 3D GBM loaded with metal (oxides/sulfides) NPs is more common. Graphene Nanosheets serve as superb support for the uniform catalyst anchoring and copious adsorption of reactants, while the metal NPs within 3D GBM function as a good catalyst for transformation. The additives' and graphene's combined effects produce high catalytic activity: A high preconcentration is produced close to the catalytic NPs by the high adsorption ability of 3D GBM toward water pollutants, particularly via π–π stacking interactions for aromatic compounds the uniform dispersion of metal NPs, which would otherwise aggregate easily due to their high surface energy, is made possible by 3D GBM, preventing a noticeable decrease in their original catalytic activity; the uptake of electrons by pollutant molecules is facilitated by the electron

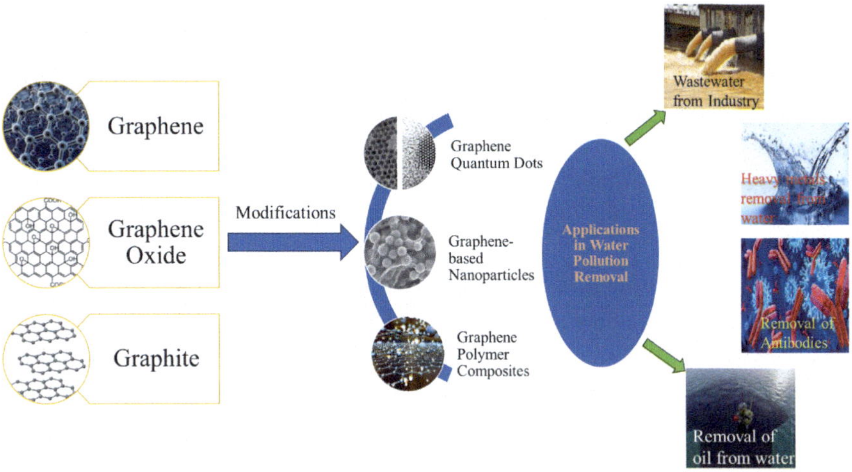

Fig. 6.9 Use of graphene-based products for wastewater treatment

transfer that occurs without obstruction from the 3D GBM to catalytic NPs [89]. Figure 6.9 shows the uses of graphene-based Products for Wastewater Treatment.

6.4.3 Dyes Pollution

Dyes are important constituents of the textile industry, as food colored-agents and anything that has to impart color consists of dyes. Dyes are organic compounds that contain three basic groups: chromophore, auxochrome, and matrix. Dyes containing different types of classes; azo-dyes, cationic, anionic, and neutral dyes. As the textile industry developed, more than 100,000 dyes were produced a year and so contamination added to water. Now, the materials that are required are durable fast, and not chemically degraded, which is an emerging issue nowadays. The wastewater from the textile industry is not satisfactorily cleaned out because 2% of dyes are directly discharged and 10% is lost during the coloring procedure [90]. Wastewater is treated by primary and secondary treatment to remove most of the harmful constituents in it. The treatment of effluents has become a top priority. During the past years many physical, chemical, and biological techniques such as flotation, precipitation, oxidation, solvent extraction, evaporation, carbon adsorption, ion exchange, membrane filtration, electrochemistry, biodegradation, and phytoremediation. Each treatment has its merits and demerits. So the best, economical, effective, and without any damage method is adsorption by different substrates [91]. Khan reported the mitigation of crystal violet using graphene quantum dots. Graphite changed into graphene oxide using a chemical exfoliation method. Characterization can be done using a UV–visible spectrophotometer, FTIR, SEM, TEM, and XRD. Different conditions are recorded to get the results. The adsorption efficiency could get 99% using graphene quantum dots. The kinetic study is reported using different isotherms [92]. Hossein Khojasteh reported mitigation of organic dye using graphene quantum dots using precursor urea. Graphene oxide from hydrazine requires worse conditions so using urea is the simple way to oxidation of graphene. So, excess urea can be removed. So, N-GQDs are synthesized and have a large surface area used in water purification [93].

6.4.4 Applications of Graphene in Other Fields

Graphene derived from waste material has found its applications not only in wastewater treatment but also in bulk of other properties making it a miracle material. Some of these applications are tissue engineering, drug delivery, antimicrobial activity, biosensors, energy conservation, and many more [94]. The reason behind its multiple applications is its unique properties which include, its mechanical strength, electronic properties, surface functionality, amphiphilicity, and broad chemical and light scattering properties. These features make graphene a material for future generations to work with and unveil its hidden applications which still need consideration [95–97]. Figure 6.10 presents some applications of waste-derived graphene in different fields.

6.5 Conclusion

Due to the unique physiological capabilities of graphene, it plays a vital role as an eco-friendly material. The adsorption capacity of graphene and its derivatives for hazardous metals removal is remarkable. However, there are still some research gaps which are yet unfilled. Thus, the implementation of these nanocomposites on a large scale still needs further research. Despite these hurdles, the study of waste-derived graphene nanocomposites is the emerging field of material science.

Fig. 6.10 Applications of waste-derived graphene

References

1. Berktas I et al (2020) Recent developments in the synthesis of graphene and graphene-like structures from waste sources by recycling and upcycling technologies: a review. Graph Technol 5:59–73
2. Sharma KD, Jain S (2020) Municipal solid waste generation, composition, and management: the global scenario. Soc Respons J 16(6):917–948
3. Novoselov K (2010) Konstantin Novoselov—Biographical. Nobel Prize Outreach AB. Retrieved September 18, p 2021
4. Ikram R et al (2020) Advances in synthesis of graphene derivatives using industrial wastes precursors; prospects and challenges. J Mater Res Technol 9(6):15924–15951
5. Adetayo A, Runsewe D (2019) Synthesis and fabrication of graphene and graphene oxide: a review. Open J Compos Mater 9(2):207
6. Zhu Y et al (2018) Mass production and industrial applications of graphene materials. Natl Sci Rev 5(1):90–101

7. Pryce D, Khalil AME, Memon FA (2022) Investigating the environmental costs of utilizing graphene-based adsorbents and pulsed power oxidation for the removal of emerging contaminants from urban wastewater. Sci Total Environ 817:152985
8. Mohan VB et al (2018) Graphene-based materials and their composites: A review on production, applications and product limitations. Compos B Eng 142:200–220
9. Nanaji K et al (2018) Robust, environmentally benign synthesis of nanoporous graphene sheets from biowaste for ultrafast supercapacitor application. ACS Sustain Chem Eng 7(2):2516–2529
10. Jindal S et al (2022) Sustainable approach for developing graphene-based materials from natural resources and biowastes for electronic applications. ACS Appl Electron Mater 4(5):2146–2174
11. Yan Y et al (2020) Synthesis of graphene: potential carbon precursors and approaches. Nanotechnol Rev 9(1):1284–1314
12. Seck PA et al (2012) Crops that feed the world 7: rice. Food Secur 4:7–24
13. Ikram R, Jan BM, Ahmad W (2020) Advances in synthesis of graphene derivatives using industrial wastes precursors; prospects and challenges. J Market Res 9(6):15924–15951
14. Menya E et al (2018) Production and performance of activated carbon from rice husks for removal of natural organic matter from water: a review. Chem Eng Res Des 129:271–296
15. Wimalasiri EM et al (2021) A framework for the development of hemp (*Cannabis sativa* L.) as a crop for the future in tropical environments. Ind Crops Prod 172:113999
16. Żuk-Gołaszewska K, Gołaszewski J (2020) Hemp production. Sustainable agriculture reviews 42: hemp production and applications, p 1–36
17. Sedan D et al (2007) Effect of calcium rich and alkaline solutions on the chemical behaviour of hemp fibres. J Mater Sci 42:9336–9342
18. Somanathan T et al (2015) Graphene oxide synthesis from agro waste. 5(2):826–834
19. Ayalon O, Brody S, Shechter M (2013) Household waste generation, recycling and prevention, vol 219. OECD Publishing Paris, France
20. Shulga O et al (2023) Consumer properties of biodegradable edible cups for hot drinks. Ukrainian Food J 12(3)
21. El Essawy NA et al (2017) Green synthesis of graphene from recycled PET bottle wastes for use in the adsorption of dyes in aqueous solution. 145:57–68
22. Sharma S et al (2014) Synthesis of graphene crystals from solid waste plastic by chemical vapor deposition. 72:66–73
23. Ravindran R, Jaiswal AK (2016) Exploitation of food industry waste for high-value products. Trends Biotechnol 34(1):58–69
24. He M, Sun Y, Han B (2022) Green carbon science: efficient carbon resource processing, utilization, and recycling towards carbon neutrality. Angew Chem 134(15):e202112835
25. Mbayachi VB et al (2021) Graphene synthesis, characterization and its applications: a review. Results Chem 3:100163
26. Yi M, Shen Z (2015) A review on mechanical exfoliation for the scalable production of graphene. 3(22):11700–11715
27. Jayasena B, Subbiah S (2011) A novel mechanical cleavage method for synthesizing few-layer graphenes. Nanoscale Res Lett 6:1–7
28. Chamoli P et al (2020) Characteristics of graphene/reduced graphene oxide. In: Handbook of nanocomposite supercapacitor materials I: characteristics, pp 155–177
29. Scott SM, Ali Z (2021) Fabrication methods for microfluidic devices: An overview. Micromachines 12(3):319
30. Choucair M, Thordarson P, Stride J (2009) Gram-scale production of graphene based on solvothermal synthesis and sonication. 4(1):30–33
31. Calderón-Ayala G et al (2017) Green synthesis of reduced graphene oxide using ball milling. 21:93–97
32. Zhang Z et al (2018) Efficient production of high-quality few-layer graphene using a simple hydrodynamic-assisted exfoliation method. Nanoscale Res Lett 13(1):416
33. Agudosi ES et al (2020) A review of the graphene synthesis routes and its applications in electrochemical energy storage. Crit Rev Solid State Mater Sci 45(5):339–377

34. Li X, Colombo L, Ruoff RS (2016) Synthesis of graphene films on copper foils by chemical vapor deposition. Adv Mater 28(29):6247–6252
35. Li N et al (2012) A facile one-step method to produce Ni/graphene nanocomposites and their application to the thermal decomposition of ammonium perchlorate. CrystEngComm 14(2):428–434
36. Li M et al (2016) Controllable synthesis of graphene by plasma-enhanced chemical vapor deposition and its related applications. 3(11):1600003
37. Dehghanzad B et al (2016) Synthesis and characterization of graphene and functionalized graphene via chemical and thermal treatment methods. RSC Adv 6(5):3578–3585
38. Kumar M, Chung JS, Hur SH (2019) Graphene composites for lead ions removal from aqueous solutions. Appl Sci 9. https://doi.org/10.3390/app9142925
39. Baysal A, Ozbek N, Akman S (2013) Determination of trace metals in waste water and their removal processes. Waste Wat-Treat Technol Recent Anal Dev 1:145–171
40. Kumar L, Bharadvaja N (2020) Microbial remediation of heavy metals. In: Microbial bioremediation & biodegradation, pp 49–72
41. Akpor OB, Muchie M (2010) Remediation of heavy metals in drinking water and wastewater treatment systems: processes and applications. Int J Phys Sci 5(12):1807–1817
42. Yap CK, Peng SHT (2019) Cleaning contaminated soils by using microbial remediation: a review and challenges to the weaknesses. Am J Biomed Sci Res 2(3)
43. Kikuchi T, Tanaka S (2012) Biological removal and recovery of toxic heavy metals in water environment. Crit Rev Environ Sci Technol 42(10):1007–1057
44. Oncel MS et al (2013) A comparative study of chemical precipitation and electrocoagulation for treatment of coal acid drainage wastewater. J Environ Chem Eng 1(4):989–995
45. Tilak A et al., Related items
46. Pourhakkak P et al (2021) Fundamentals of adsorption technology. Interface science and technology. Elsevier, pp 1–70
47. Sud D, Mahajan G, Kaur M (2008) Agricultural waste material as potential adsorbent for sequestering heavy metal ions from aqueous solutions—a review. Biores Technol 99(14):6017–6027
48. Wang J, Chen B (2015) Adsorption and coadsorption of organic pollutants and a heavy metal by graphene oxide and reduced graphene materials. Chem Eng J 281:379–388
49. Ali I et al (2019) Graphene based adsorbents for remediation of noxious pollutants from wastewater. Environ Int 127:160–180
50. Sitko R et al (2013) Adsorption of divalent metal ions from aqueous solutions using graphene oxide. Dalton Trans 42(16):5682–5689
51. Cao Y, Li X (2014) Adsorption of graphene for the removal of inorganic pollutants in water purification: a review. Adsorption 20:713–727
52. Rahman Z, Singh VP (2019) The relative impact of toxic heavy metals (THMs)(arsenic (As), cadmium (Cd), chromium (Cr)(VI), mercury (Hg), and lead (Pb)) on the total environment: an overview. Environ Monit Assess 191:1–21
53. Oladoye PO (2022) Natural, low-cost adsorbents for toxic Pb (II) ion sequestration from (waste) water: a state-of-the-art review. Chemosphere 287:132130
54. Genchi G et al (2020) The effects of cadmium toxicity. Int J Environ Res Public Health 17(11):3782
55. Hayat MT et al (2019) Environmental hazards of cadmium: past, present, and future. Cadmium toxicity and tolerance in plants. Elsevier, pp 163–183
56. Issod A (2003) WR Grace & (and) Co. v. EPA: an arbitrary outcome. Ecol LQ 30:409
57. Hao L et al (2012) SiO2/graphene composite for highly selective adsorption of Pb (II) ion. J Colloid Interface Sci 369(1):381–387
58. Deng Y et al (2011) An efficient way to functionalize graphene sheets with presynthesized polymer via ATNRC chemistry. J Polym Sci, Part A: Polym Chem 49(7):1582–1590
59. Zhou GZG et al (2015) Sponge-like polysiloxane-graphene oxide gel as a highly efficient and renewable adsorbent for lead and cadmium metals removal from wastewater

60. Ghorbani M et al (2018) Magnetic ethylene diamine-functionalized graphene oxide as novel sorbent for removal of lead and cadmium ions from wastewater samples. Environ Sci Pollut Res 25:5655–5667
61. Samuel MS et al (2020) Synthesized β-cyclodextrin modified graphene oxide (β-CD-GO) composite for adsorption of cadmium and their toxicity profile in cervical cancer (HeLa) cell lines. Process Biochem 93:28–35
62. Biswas D et al (2008) Mechanism of erythrocyte death in human population exposed to arsenic through drinking water. Toxicol Appl Pharmacol 230(1):57–66
63. Nandi D et al (2024) Hexagonal nanoplatelets of Ni–Mn oxide implanted reduced graphene oxide for high response in humidity sensing. Chem Pap 1–12
64. Mishra AK, Ramaprabhu S (2011) Functionalized graphene-based nanocomposites for supercapacitor application. J Phys Chem C 115(29):14006–14013
65. Zhu J et al (2012) Magnetic graphene nanoplatelet composites toward arsenic removal. ECS J Solid State Sci Technol 1(1):M1
66. Gollavelli G, Chang C-C, Ling Y-C (2013) Facile synthesis of smart magnetic graphene for safe drinking water: heavy metal removal and disinfection control. ACS Sustain Chem Eng 1(5):462–472
67. Madsen H, Poulsen L, Grandjean P (1990) Risk of high copper content in drinking water. Ugeskr Laeger 152(25):1806–1809
68. Jennings G, Sneed R, Clair M (1996) Metals in drinking water. Published by: North Carolina Cooperative Extension service Publication no. 1996, AG-473–1. Electronic version 3
69. Ali I et al (2022) Polyaniline modified CNTs and graphene nanocomposite for removal of lead and zinc metal ions: kinetics, thermodynamics and desorption studies. Molecules 27(17):5623
70. Sreeprasad TS et al (2011) Reduced graphene oxide–metal/metal oxide composites: facile synthesis and application in water purification. J Hazard Mater 186(1):921–931
71. Liu X et al (2016) Graphene oxides with different oxidation degrees for Co (II) ion pollution management. Chem Eng J 302:763–772
72. Zhao D et al (2017) Amino siloxane oligomer modified graphene oxide composite for the efficient capture of U (VI) and Eu (III) from aqueous solution. ACS Sustain Chem Eng 5(11):10290–10297
73. Lee Y-C, Yang J-W (2012) Self-assembled flower-like TiO_2 on exfoliated graphite oxide for heavy metal removal. J Ind Eng Chem 18(3):1178–1185
74. Zare-Dorabei R et al (2016) Highly efficient simultaneous ultrasonic-assisted adsorption of Pb (II), Cd (II), Ni (II) and Cu (II) ions from aqueous solutions by graphene oxide modified with 2, 2′-dipyridylamine: central composite design optimization. Ultrason Sonochem 32:265–276
75. Zhang N et al (2011) Fabrication of highly porous biodegradable monoliths strengthened by graphene oxide and their adsorption of metal ions. Carbon 49(3):827–837
76. Ge H, Ma Z (2015) Microwave preparation of triethylenetetramine modified graphene oxide/chitosan composite for adsorption of Cr (VI). Carbohyd Polym 131:280–287
77. Wang S et al (2019) Preparation and characterization of graphene oxide/silk fibroin hybrid aerogel for dye and heavy metal adsorption. Compos B Eng 163:716–722
78. Wei M-P et al (2018) Sulfonated graphene oxide as an adsorbent for removal of Pb^{2+} and methylene blue. J Colloid Interface Sci 524:297–305
79. Kyzas GZ, Travlou NA, Deliyanni EA (2014) The role of chitosan as nanofiller of graphite oxide for the removal of toxic mercury ions. Colloids Surf, B 113:467–476
80. Khatamian M et al (2015) Synthesis and characterization of RGO/zeolite composites for the removal of arsenic from contaminated water. RSC Adv 5(45):35352–35360
81. Dąbrowski A (2001) Adsorption—from theory to practice. Adv Coll Interface Sci 93(1–3):135–224
82. Carpio IEM et al (2014) Graphene oxide functionalized with ethylenediamine triacetic acid for heavy metal adsorption and anti-microbial applications. Carbon 77:289–301
83. Liu Y et al (2016) Graphene oxides cross-linked with hyperbranched polyethylenimines: preparation, characterization and their potential as recyclable and highly efficient adsorption materials for lead (II) ions. Chem Eng J 285:698–708

84. Zou W et al (2019) Preparation of a graphene oxide membrane for air purification. Mater Res Exp 6(10):105624
85. Carolin CF et al (2017) Efficient techniques for the removal of toxic heavy metals from aquatic environment: a review. J Environ Chem Eng 5(3):2782–2799
86. Zhang F et al (2019) 3D graphene-based gel photocatalysts for environmental pollutants degradation. Environ Pollut 253:365–376
87. Navalon S et al (2017) Active sites on graphene-based materials as metal-free catalysts. Chem Soc Rev 46(15):4501–4529
88. Nehra S, Sharma R, Kumar D (2021) Removal of air pollutants using graphene nanocomposite. In: Environmental remediation through carbon based nano composites, pp 275–291
89. Ding X et al (2022) Environmental and health effects of graphene-family nanomaterials: potential release pathways, transformation, environmental fate and health risks. Nano Today 42:101379
90. Singh K, Arora S (2011) Removal of synthetic textile dyes from wastewaters: a critical review on present treatment technologies. Crit Rev Environ Sci Technol 41(9):807–878
91. Crini G, Lichtfouse E (2019) Advantages and disadvantages of techniques used for wastewater treatment. Environ Chem Lett 17(1):145–155
92. Khan A et al (2023) Removal of crystal violet from wastewater using synthesized graphene quantum dots as adsorbents: kinetic approach. Int J Environ Sci Technol 20(12):13219–13232
93. Khojasteh H et al (2017) Facile reduction of graphene using urea in solid phase and surface modification by N-doped graphene quantum dots for adsorption of organic dyes. Diam Relat Mater 79:133–144
94. Feng L, Wu L, Qu X (2013) New horizons for diagnostics and therapeutic applications of graphene and graphene oxide. Adv Mater 25(2):168–186
95. Chung C et al (2013) Biomedical Applications of Graphene and Graphene Oxide. Acc Chem Res 46(10):2211–2224
96. Nguyen BH, Nguyen VH (2016) Promising applications of graphene and graphene-based nanostructures. Adv Nat Sci Nanosci Nanotechnol 7(2):023002
97. Olabi AG et al (2021) Application of graphene in energy storage device—a review. Renew Sustain Energy Rev 135:110026

Chapter 7
Rice Waste-Derived Carbon Nanomaterials for Environmental Applications

Raheel Akram, Naveed Anwar, Anila Arshad, and Mohammad Mansoob Khan

Abstract The potential of reusing low-value biomass waste for the processing of renewable materials to fabricate high-value products is remarkable. In recent decades, a major impact has been witnessed in the economy and environment based on the conversion of biomass waste into valuable carbon nanomaterials. Various chemical methods have been developed to synthesize high-value carbon nanomaterials; however, the use of green technology for the synthesis of carbon nanomaterials is the most common nowadays because of the abundant availability of the starting precursors, their non-toxicity, low fabrication cost, ease of modification, and eco-friendly nature. The large-scale production of rice is associated with some valuable by-products (rice husk, rice straw, and rice bran) attributed to the presence of important components like lignin, cellulose, and hemicellulose. Carbon nanostructures made from rice waste have unique physicochemical properties like high surface area, porosity, and unique functionalities that have excellent potential to be applied to solve a wide range of environmental issues. The primary goal of this chapter is to draw attention to the recent developments in the synthesis of environmentally friendly methods of producing carbon nanomaterials, such as graphene, carbon nanotubes, and carbon dots, from rice biomass waste, their complete characterization, and their applications to solve the environment-related problems.

R. Akram (✉)
Research Laboratory for Analytical Instrument and Electrochemistry Innovation, Department of Chemistry, Faculty of Science, Chiang Mai University, Chiang Mai 50200, Thailand
e-mail: raheel.akram19@yahoo.com

N. Anwar
Biofuels Institute, School of Environmental and Safety Engineering, Jiangsu University, Zhenjiang 212013, PR China

A. Arshad
School of Agriculture Engineering, Jiangsu University, Zhenjiang 212013, PR China

M. M. Khan
Chemical Sciences, Faculty of Science, Universiti Brunei Darussalam, JalanTungku Link, Gadong BE1410, Brunei Darussalam

N. Talreja et al. (eds.), *Waste-Derived Carbon Nanostructures*, Nanostructure Science and Technology, https://doi.org/10.1007/978-3-031-75247-6_7

Keywords Rice waste · Carbon nanomaterials · Carbon nanotubes · Graphene · Environmental applications

7.1 Introduction

Over the years, the rapid spread of urbanization and industrialization has led to a massive increase in the use of non-renewable energy sources worldwide, which has posed a serious threat to the environment and human health. To combat these issues, the researchers are looking forward to investigating eco-friendly and clean renewable energy sources to meet the energy-related challenges [1]. The biggest renewable energy sources include solar, wind, thermal, hydropower, and biomass energy. Among these energy sources, biomass-derived energy has attracted particular interest based on its abundant availability and low cost. Biomass is organic waste derived from renewable sources like sewage, agriculture waste, and forests by-products [2]. A huge amount of biomass, approximately 200 billion metric tons, is produced annually, so there is a huge potential to utilize the idea of 'waste to wealth' by converting biomass into promising nanomaterials [3]. Biomass utilization can be a substantial solution to environmental problems and meanwhile adding value to the economy. The potential to exploit the organic nature of waste to valuable nanomaterials is remarkable. Concerted efforts have been seen by researchers and environmentalists to utilize biomass for the production of carbon nanomaterials based on their superior performance and high demands. In contrast to chemical methods utilized to produce carbon nanomaterials, the production of carbon nanomaterials from biomass often employs green technology. These biomasses are carbon-rich materials containing valuable components like cellulose, hemicellulose, and lignin which are the basic precursors to fabricate high-valued carbon nanomaterials, e.g., graphene, fullerenes, carbon nanotubes, and carbon dots [4]. Various synthetic approaches are applied to convert biomass into carbon-based materials; however, green synthesis is highly valued as renewable energy aims to address environmental issues.

The most applied agri-waste includes wheat stems, sugarcane bagasse, corn straw, and rice residues to convert into carbon nanomaterials. Rice is a major crop to feed the masses and is cultivated across the world. As a result of the exponential rise in human population, the rice industry has had to grow constantly to meet the demand for rice, which has resulted in the release of large amounts of solid waste (mostly rice husk, rice bran, rice straw, rice dust, and broken-rice which need a huge amount of men power and cost to be disposed of [5]). According to earlier research, 1 kg of harvested rice grain is accompanied by 1–1.5 kg of rice straw, and 20% of rice husk can be produced during the optimal milling process. This indicates that around 150 million tons of rice husk and 750–1125 million tons of rice straw are produced annually as rice trash worldwide [6]. In most cases, instead of converting rice waste into valuable products, it was straightforwardly put into horticulture uses, the construction industry, and conventional energy production [7, 8]. Furthermore, rice-based biomass has been exploited for different chemical products including carbon, biofuel, and syngas which are used as adsorbents and in energy production. In recent times, more concerted efforts have been made to utilize the chemistry of its components to change valuable carbon nanomaterials. For example, rice hulk consists of 70% cellulose, hemicellulose, and lignin which are organic in nature and a promising source of carbon-derived nanomaterials [9]. These carbon nanomaterials have excellent properties and different groups have exploited their chemistry to address the environmental challenges. At this stage, we do realize that it is demanding to compile these contributions in a single chapter to present sustainable solutions to environmental issues. Herein, we will emphasize utilizing the unique physicochemical properties of rice waste to derive valuable carbon nanomaterials such as carbon nanotubes, graphene, and carbon dots. Particularly, we will focus on the role of these materials to discuss the environmental solutions such as water purification, pesticides and microbial removals, VOCs sensing, and gas storage (Fig. 7.1).

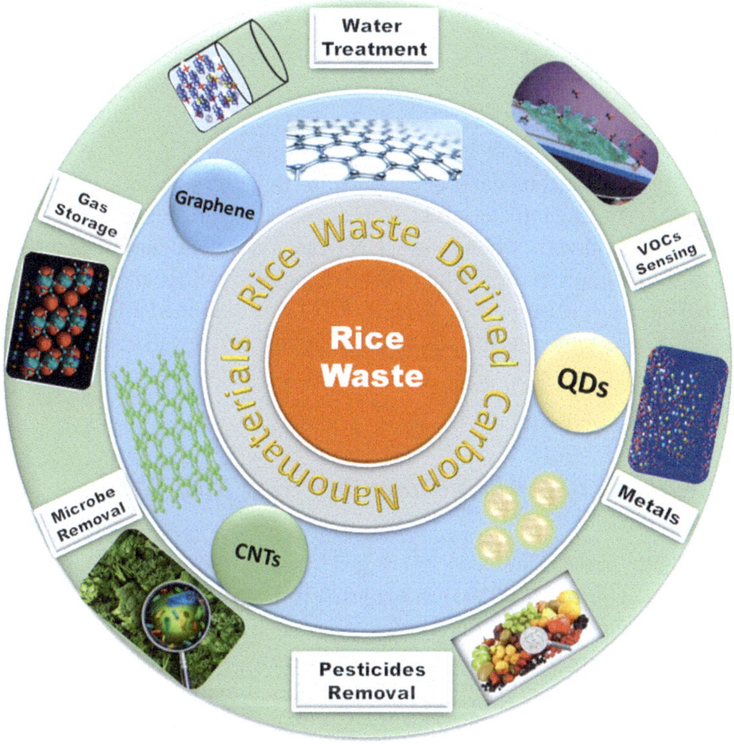

Fig. 7.1 An overview of carbon nanomaterials derived from rice waste and their environmental applications

7.2 Carbon Nanomaterials

Carbon nanomaterials are widely applied in several applications based on their excellent physiochemical properties [10, 11]. Based on their morphological structures, these materials are often categorized into zero-dimensional (0-D), one-dimensional (1-D), and two-dimensional (2-D) materials. The examples of 0-D, 1-D, and 2-D carbon nanomaterials include carbon dots and fullerenes, carbon nanotubes, and graphene, respectively [12, 13]. All of these materials possess differential characteristics that make the carbon nanomaterials a high-valued material to be applied in a variety of applications. For instance, CNTs are excellent materials with remarkable conductivity and superior chemical and mechanical stability. On the other hand, graphene possesses excellent feasibility of doping and superior thermal conductivity. These multifunctional nanomaterials have been extensively applied across different applications in electronics, composite materials, energy storage, biomedical research, and environmental sensing [14, 15].

7.2.1 Carbon Nanotubes

After the discovery of carbon nanotubes by Lijima in 1991, it became one of the most renowned 1-D carbon nanomaterials [16]. This nanostructure consists of a hexagonal network of sp^2 hybridized carbon atoms. CNTs are cylindrical nanostructures with a high length-to-diameter ratio, holding promising electrical, mechanical, electronic, thermal, and mechanical properties. Broadly, CNTs are classified into single-wall and multiple-wall CNTs with tuneable diameters ranging from few to several nanometres in diameter.

7.2.2 Synthesis of Carbon Nanotubes from Rice Waste

Several synthetic approaches have been developed to synthesize CNTs; however, chemical vapor deposition (CVD) is extensively used to synthesize CNTs. CVD involves the formation of a thin film as a product over the exposed surface of the substrate by the interaction of volatile precursors. The CVD method depends on the choice of precursor, catalyst, temperature, and required energy. Fathy et al. reported the synthesis of CNT bundles by combining two processes. The foremost step involves the hydrothermal treatment of Fe-impregnated rice straw at 250 °C for 2 h to produce hydro-char which acts as a substrate. Next, camphor is decomposed in a heating tube under nitrogen to decompose into gases. The CVD involves joining two separate chambers as a single work unit [17]. Another approach widely used in recent years was to synthesize CNTs by using pyrolysis. Pyrolysis involves heating the precursors in a cylindrical tube in the presence of H_2, N_2, and Ar for a few hours

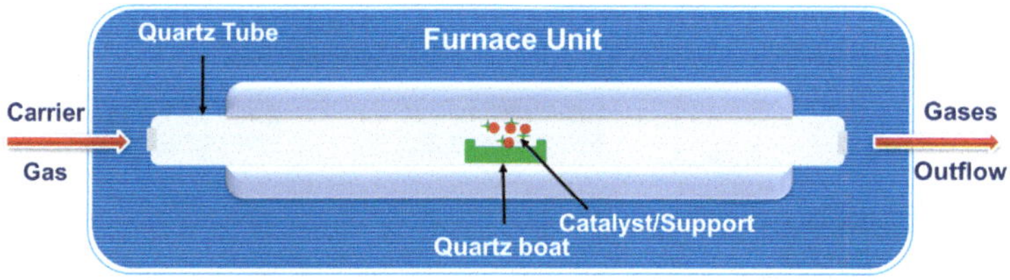

Fig. 7.2 A schematic illustration of the setup used for the catalytic degradation of rice waste

(Fig. 7.2) [18]. However, the attention of researchers has been in recent years to adopt green technology for mass production of CNTs. Several studies have reported the synthesis of CNTs by exploiting agri-waste as a source of carbon nanomaterials. A suitable precursor should have a high content of carbon (40–50%) to be exploited to constitute the carbon backbone of CNTs. A very simple approach is to utilize microwaves as an energy source to convert rice husk into CNTs. Another approach is to utilize microwave plasma irradiation in the presence of air and argon to convert rice husk into graphene–CNTs composite material [19].

7.2.3 Environmental Applications of CNTs

Adsorption techniques are widely applied to deal with both organic and inorganic environmental contaminants from water bodies. Typically, activated carbons are ideal candidates for removing environmental contaminants based on their chemical inertness, thermal stability, and superior surface area [20]. However, a few limitations still exist that linger on their effectiveness at a mass scale. The foremost issue is the challenges in regeneration and recycling of these materials. Another issue that hampers the effectiveness is the slow sorption kinetics of activated carbon that leads to time consumption and ultimately the cost of the process. Therefore, alternate sources are highly desirable, particularly renewable sources. Nowadays, CNTs derived from rice waste are promisingly applied to deal with water purification. CNTs have the potential to extract organic, inorganic, and biological contaminants from water bodies. Figure 7.3 illustrates the range of water contaminants removed by applying CNTs.

Before industrialization, the contaminants in drinking water were mostly organic pollutants and microbes that were treated by coagulation and disinfectant procedures. In recent times, rapid industrialization and urbanization have led to a rapid rise in inorganic pollutants, dyes, and pesticides in water bodies [21]. These contaminants are hazardous to human health and therefore must be removed from water entities. CNTs-based contaminants removal by using adsorption techniques offers a valid solution to the aforesaid issues. CNTs hold several promising advantages like high surface area, porosity, low density, entrapment ability, and tailoring functionalities that make them an ideal candidate for water cleaning. Another important factor to consider is the interstitial channels, grooves, internal surface, and outer surface [22].

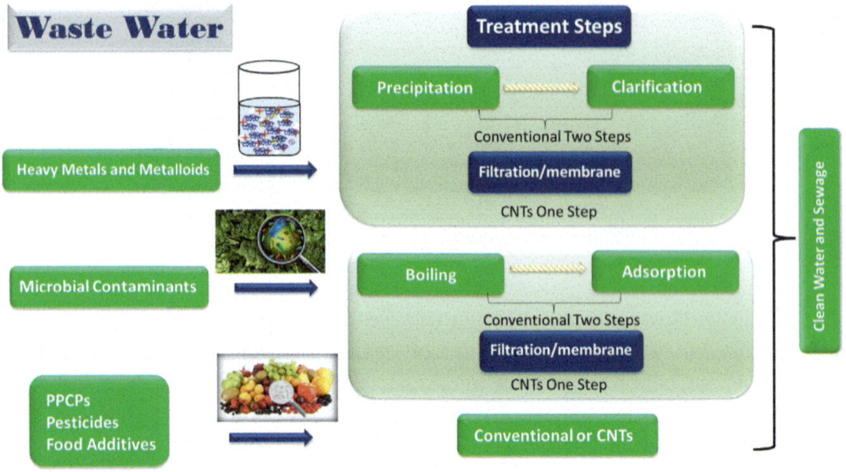

Fig. 7.3 Schematic illustration of different strategies used for wastewater treatment using CNTs

A recent study has reported that interstitial channels are foremost important in the removal of impurities. The notable inorganic impurities in water bodies include Zn, Pb, Cu, Hg, Fe, and Cd ions [23]. Many studies have recently been presented to report the removal of these ions from water bodies by using CNTs and compared with other methods. Overall, it has been noted that the capability of CNTs to remove heavy metals from water bodies is far better than other traditional adsorbents under similar conditions [24].

Recently, various methods have been established to enhance the contaminant removal efficiency for CNTs. One strategy is to utilize surface-functionalized CNTs (f-CNTs) for the efficient removal of heavy metals (Fig. 7.4). Another advancement is to utilize f-CNTs with magnetic metal oxides for the removal of heavy elements. However, once the metals are removed by f-CNTs, the separation of these magnetic nanoparticles from the surface of CNTs is never an easy task. The workable solution proposed by the researchers was to utilize CNTs/metal oxide nanocomposites, because a magnetic field may help to easily separate the magnetic nanoparticles from the composite. Gupta et al. developed CNTs/Fe_2O_3 nanocomposite to anchor the Fe ions from the polluted water [25]. In another report, CNTs/Al_2O_3 were prepared to excavate the Pb^{2+} ions from the contaminated water. Overall, several other factors like the chemistry of contaminated water (pH, temperature, hardness, salt concentration, and nature of contaminants) are always considered when choosing proper CNTs as adsorbents for water cleaning [26].

Microbial contaminants (bacteria, viruses, and protozoa) are the major disease-causing pathogens in fresh and waste water effluents. The traditional contaminant removal materials, namely polymers and activated carbon, were not productive enough to get pure water. On the contrary, CNTs have shown high potential to remove pathogenic microbes from water bodies. For instance, CNTs' ability to

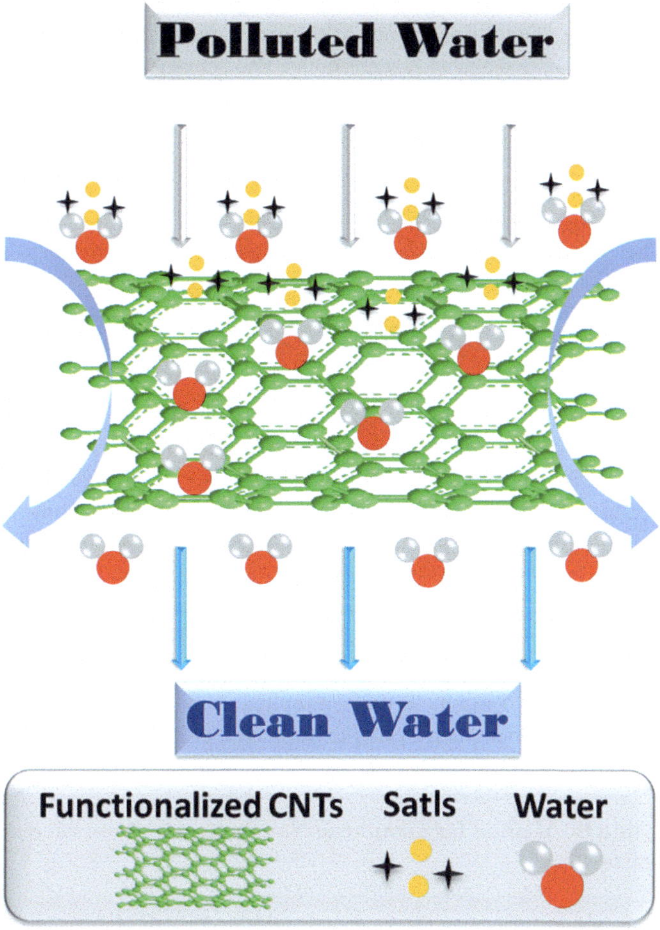

Fig. 7.4 Schematic demonstration of f-CNTs membranes used for wastewater treatment

remove *Bacillus subtilis* spores is 36 times higher as compared to activated carbon [27]. CNTs also possess a remarkable ability to deactivate a range of water pollutants including bacteria (*Escherichia coli*, *Streptococcus mutans*, *Micrococcus lysodeikticus*, and *Salmonella*), viruses (MS2 bacteriophage), and protozoa (*Tetrahymena pyriformis*) [28, 29]. SWCNTs are also advantageous in terms of preferential selectivity for certain pathogens over others. For instance, the removal of *Staphylococcus aureus* was almost 100 times that of *E. coli* [30]. In addition, the removal efficiency was successfully enhanced by functionalizing the CNTs with metal oxides and polymeric substances.

Pesticides are increasingly applied to combat the harmful microorganisms and control the plant growth. It is estimated that approximately 5 million tons of pesticides are sprayed annually which causes about 3 million cases of pesticide poisoning.

These pesticides are difficult to restrict to the target and enter the ecosystem through agricultural waste and landfill leaching. Additionally, pesticides enter the ecosystem from industrial waste, leakage of storage tanks, and sewage leaks. It is necessary to control the pesticides entering the water bodies and afterward remove these species from the water sources. To this end, both SWCNTs and semiconductor SWCNTs are applied for pesticide removal. Rocha et al. reported that semiconductor SWCNTs possess a remarkably higher ability (70%) to remove 2,4-dichlorophenoxy acetic acid, diquat dibromide, and 1-pyrine butyric acid [31].

7.2.4 *Graphene Nanomaterials*

Graphene is a monolayer of sp^2 hybridized carbon atoms packed in a two-dimensional honeycomb network with unique physical and electronic properties. This material acts as the basis of several carbon nanomaterials. For example, it can be stacked to produce three-dimensional graphite, rolled to one-dimensional nanotubes, and wrapped to zero-dimensional fullerenes. After its separation in 2004 by the micromechanical method, it has been extensively investigated across different disciplines, such as physics, chemistry, materials science, and engineering. Up till now, researchers have been widely exploiting the chemistry of graphene and exploring potential applications in different fields. Generally, graphene can be categorized into two types: functionalized graphene (e.g., graphene oxide, reduced graphene) and non-functionalized graphene (e.g., pristine graphene, graphene sheets).

7.2.4.1 Synthetic Method for Graphene Nanomaterials from Rice Waste

In this section, we will discuss the rice waste-derived methods used to synthesize graphene nanomaterials. As we know huge amount of rice waste is usually dumped or burned which adversely affects the environment. So, utilizing the rice waste for valuable carbon nanomaterials could be the best solution to deal with rice waste. Some important methods utilized to produce graphene and graphene oxide are chemical activation and microwave-assisted methods.

Chemical activation involves treating the rice waste with activating agents, such as KOH, H_2SO_4, $ZnCl_2$, etc. After chemical activation, the resulting material is carbonized at high temperatures to produce activated carbon with high surface area and porosity. Rice waste-derived graphene can be produced by using different chemical activating agents. However, the most effective activating agent is KOH because it not only provides a high surface of the continuous 3D network but also produces pure graphene by etching the impurities [32]. Figure 7.5 illustrates the steps involved in producing graphene from rice waste. It involves the following three steps: pre-treatment, chemical treatment, and post-treatment. Initially, rice husk feedstock is pre-treated to get the carbonaceous matter. The second step involves the chemical

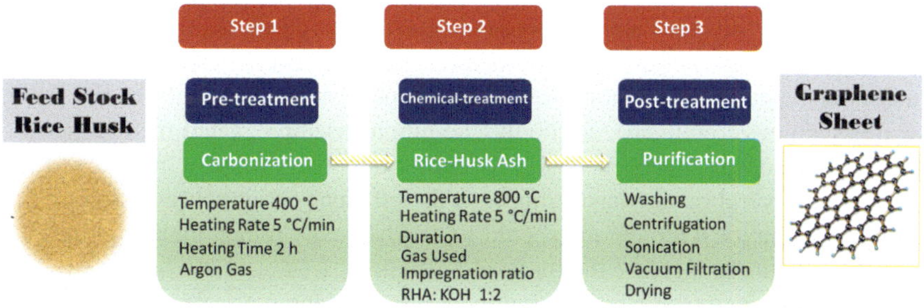

Fig. 7.5 Flow sheet chart for the rice waste-derived graphene sheet synthesis

treatment with an activating agent in a furnace at high temperatures. This step produces high surface area graphene which is further purified in a post-treatment step. This step involved washing with different solvents and filtration to remove the impurities. In a modified synthetic method, Singh et al. omitted the carbonization step and carried out the graphene synthesis in two steps. He directly annealed the rice waste at a high temperature after KOH activation and obtained a few layers of thick graphene [33].

Another facile method used to synthesize graphene nanomaterials is microwave-assisted heating. This method can be used to directly convert rice waste into graphene under microwave-assisted radiation. Recently, Kumar et al. prepared graphene from rice husks by using the following sequential steps [34]. The rice husk was washed with distilled water by ultrasonication and dried the sample. Next, the samples were grounded with different concentrations of ferrocene (catalyst) and ethanol as a solvent. Finally, microwave-assisted heating was provided at three different temperatures of 650, 750, and 800 °C. Finally, it was concluded that the prime factor for obtaining the minimal size of the sample was microwave irradiation temperature.

7.2.4.2 Environmental Applications of Graphene Nanomaterials

Rice waste-derived graphene has been widely used across different environmental applications like desalination of water, gas storage, and bactericidal action. According to World Health Organization (WHO) reports, about 1/5th of the world population is facing potable water shortage, these numbers are expected to increase in coming years because of climate changes and population growth. Shortage of potable water is posing the risk of fatal diseases like diarrhea and acute respiratory infections to a huge number of masses. Therefore, a big challenge is to manage clean drinking water at less cost and investigate new methods to clean the drinking water. Desalination of water by using membrane technology could be a promising strategy to fulfill the shortage of fresh water [35]. Several types of graphene such as pristine graphene, reduced graphene, and graphene oxide are often employed in membrane technology based on their efficiency and low cost. Graphene being a one-atom thick material having high surface area and nanopores is capable of desalinating water contaminants. These graphene-impregnated membranes have several advantages over the traditional methods: the foremost being the inexpensive raw material like rice waste used to synthesize graphene. Secondly, the graphene-impregnated membranes are easy to charge due to functionalized graphene oxide [36]. Nowadays, researchers are working on several parameters to improve the efficiency of graphene-impregnated membranes. An ideal membrane should be mechanically tough, highly permeable, and fine structure to increase the removal efficiency.

Another important application of graphene-based nanomaterials is in the storage of gases. Among the hydrocarbons, natural gas is one the most widely used energy sources across the globe. In comparison with other fossil fuels, natural gas is not only environmentally friendly but also provides a high value of energy per unit mass. The spatial distribution of reservoirs requires an efficient transportation of this energy source across the world. The conventional methods used to transport natural gas were to compress the gas in the form of liquefied natural gas (LNG) and transport it in containers. However, the storage of compressed natural gas (CNG) requires a high pressure of ~ 250 bar and more space. To overcome the issue, researchers are looking to develop alternate technologies. One possible solution is to use adsorbed natural gases (ANG). The ANG requires a highly porous material to store the gas at less pressure of ~ 3–4 MPa only [37]. Carbon nanomaterials are considered suitable for ANG because of their adsorption capacity, low cost, regenerative ability, and microporosity. Several researches have been reported to utilize graphene for CH_4 adsorption. Surface functionalization is an important advancement to improve the adsorption capacity of graphene. De la Hoz et al. reported the synthesis of graphene-impregnated activated carbon nanofibers for the adsorption of CH_4 gas [38]. This material exhibited a high surface area of 384 m^2/g with a micropore volume of 0.158 cm^3/g which was 88% of the total pores volume.

7.2.5 Carbon Dots

7.2.5.1 Synthetic Methods for Carbon Dots from Rice Waste

In recent years, various methods have been devised to synthesize rice waste-derived carbon dots and explore their potential applications in environmental sensing and energy direction. CDs are quasi-spherical nanomaterials with a size of less than 10 nm. The commonly applied methods to prepare waste rice-derived CDs include thermal calcination, microwave hydrothermal, and hydrothermal carbonization. Herein, we will take a brief overview of commonly applied methods to synthesize carbon dots.

Thermal calcination is a green, economical, and simple route to synthesize CDs. In a report, Chaudhry et al. synthesized a series of fluorescent CDs by exploiting seeds of edible crops of sorghum, wheat, rice, and millet. He carried out synthesis at 400 °C in a muffle furnace for 2 h in a nitrogen atmosphere and employed CDs for

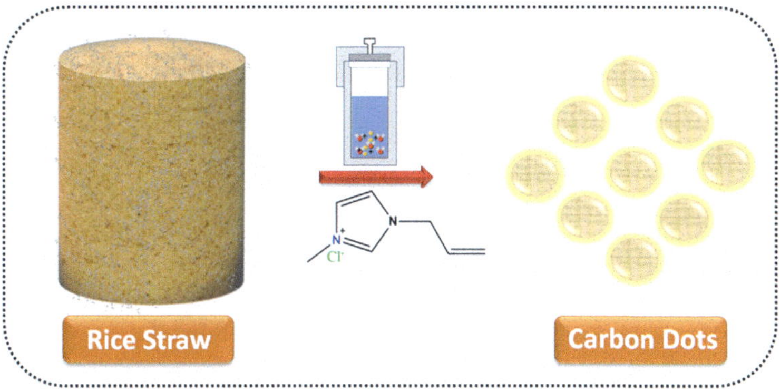

Fig. 7.6 Scheme illustrating the ion-liquid-based hydrothermal synthesis of carbon dots from rice straw as starting precursor

the identification of Cr^{3+} ions in water [39]. In another report, Zhu et al. reported the rice husk-derived Si-CQDs with blue emission with a quantum yield of 8.1%. The hydrothermal approach has also been widely applied to synthesize CDs and functionalized CDs (f-CDs) [40]. Wongso et al. reported a facile approach to synthesizing hydrophilic f-CDs with avg. size of 4–5 nm and quantum yield of 3%. These f-CDs were effective as electronic nose systems for sensing ethanol, methanol, and other volatile organic compounds [41]. The microwave-assisted hydrothermal method is another effective approach to synthesizing rice waste-derived CDs. The biochemistry of rice waste reveals that it has major components of cellulose and hemicellulose which have strong H-bonding that inhibits its dissolution in conventional solvents like water, ethanol, acetone, and tetrahydrofuran; therefore, ionic liquids were supposed to an alternate green solvent with high stability, better recyclability, and excellent dissolution for raw components in rice waste. Recently, Liu et al. reported the synthesis of heteroatom-doped CDs by using rice waste in the microwave-assisted hydrothermal approach for the detection of Fe^{3+} ions (Fig. 7.6). Briefly, he employed a mixture of water and an ionic liquid 1-allyl-3-methylimidazolium chloride, AMIM-Cl. The AMIM-Cl not only helps to dissolve the cellulose components but also imparts nitrogen to CDs [42].

7.2.5.2 Environmental Applications of Carbon Dots

An important application of CDs is to detect volatile organic compounds (VOCs). VOCs are a big threat to human health in laboratories and industries. Therefore, an efficient method is required to estimate VOCs in labs and other places. CDs have great potential to be applied as sensors for VOCs because of their binding capacities, great surface area, high polarity, and photostability. An electronic nose is the artificial replica of a human nose that can sense smell. Recently, various types of VOCs such as diethyl ether, hexane, ketones, toluene, and alcohols have been detected by using this

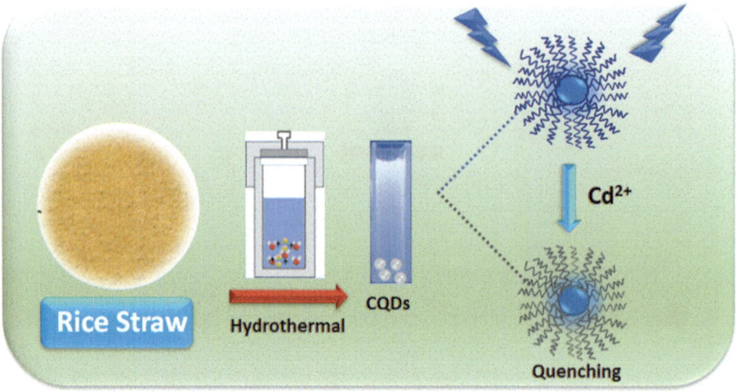

Fig. 7.7 An illustration of the preparation of fluorescent carbon dots derived from rice waste and application in sensing of Cd^{2+} ions

type of system in laboratories [43, 44]. The diagrammatic presentation of the sensing system for VOCs by using CDs film and an electronic nose is shown in Fig. 7.7. In 2019, Thongsai et al. developed a method to detect alcohol vapors and VOCs by using the carbon dot-impregnated electronic nose. The difference in sensor response by a change in transmitted light intensity through carbon dots film was estimated. This sensor developed better stability and more sensitivity for alcohol. This type of sensor can differentiate VOCs based on their polar modifications [45].

Metal ions are an important constituent of biological systems. However, the higher uptake of these metals affects the biochemistry of the human body and causes adverse effects on it. One heavy metal that pollutes the environment and has an adverse effect on human health is cadmium (Cd^{2+}). Fluorescent carbon quantum dots (CQDs) wide range of absorption spectra and adjustable emission wavelengths making them useful for both metal ion detection and removal. Abidin et al. (2020) described the use of CQDs made from rice husk to detect and remove cadmium ions (Fig. 7.7) [46]. The appropriate concentration of ascorbic acid and EDA as functionalizing agents are utilized in the adsorption and extraction of cadmium ions from water. To test the functionalized CQDs ability to detect metal ions, cadmium nitrate solutions of different concentrations were made. The removal of the CQDs was evaluated by calculating the difference in concentration between the pre-and post-addition times of the functionalized CQDs. The amount of cadmium solution added to carbon quantum dots decreases the photoluminescent intensity (PL) of functionalized CQDs. The expansion of heavy metal ions caused the CQDs structure to be quenched. Electron transfer from the carbon dots to heavy metal ions is triggered by the interaction between Cd^{2+} ions and CQDs. When evaluating PL quenching, the electrostatic interaction between negatively charged CQDs and positively charged metal ions plays a major role. Because the functionalization agent EDA for amino-functionalized CQDs is greater than ascorbic acid for carboxyl-functionalized CQDs during the production of CQDs, amino-functionalized CQDs may remove a higher proportion of cadmium than carboxyl-functionalized CQDs. Positively charged metal ions and negatively charged carbon quantum dots can work very well at the adsorption sites. As a result, metal–CQD binding energy increases. Finally, metal clusters will eventually disperse across CQDs' surface, facilitating the easy removal of the metal particles [47].

7.3 Future Outlook

The increasing energy crisis and related environmental issues have drawn the attention of researchers to adopt innovative technologies to produce high-performance carbon nanomaterials. Although various technologies are employed to produce carbon nanomaterials, however, green technology to produce valuable carbon nanomaterials from low-value bio-waste is particularly attractive. Among the various bio-waste, rice waste-derived nanomaterials possess unique physicochemical properties like high surface area, porosity, and unique functionalities which make them a source of graphene, CDs, and CNTs. Although these materials are applied across several fields, there are still several significant obstacles that must be addressed before rice waste-based nanomaterials can be used for practical applications. Following are some of the suggestions to further boost up rice waste-based nanomaterials:

- In many cases, rice waste-derived carbon nanomaterials essentially require a pre-treatment step which requires additional reaction conditions that ultimately undermine its potential applications. Therefore, new research methods are needed to design that can help to omit the pre-treatment step.
- Applications for carbon-based nanomaterials formed from rice husks are not restricted to the treatment of water pollution; many protocols may be developed in this way to broaden the usage of carbon-based nanomaterials obtained from biomass for the treatment of soil contamination and air pollution.
- There exists a potential to research other carbon nanomaterials like fullerenes to address the environmental challenges.

7.4 Conclusion

To conclude, different techniques have been investigated to produce CNMs, but more environment-friendly approaches have attracted wide attention. Therefore, researchers are focusing on developing new methods for producing CNMs using easy, inexpensive, sustainable, and environmentally friendly processes that make use of plentiful and naturally renewable resources, especially agricultural waste. The main topic of this chapter is rice waste, a major agricultural waste that may be utilized as a raw material to produce carbon nanomaterials (CNMs) such as graphene, CNTs, and CDs. Since carbon nanostructures made from rice waste have unique physicochemical properties like high surface area, porosity, and unique functionalities that have excellent potential to be applied across a range of disciplines from material science to life science, however, we have focused on the prevailing environmental issues. CNMs plentiful use has been seen in water purification technology like desalination, pesticides and microbial removals, VOCs sensing, and gas storage. It is expected that environmental researchers will focus more on novel agri-waste-derived CNMTs to deal with rising environmental issues in the coming years and benefit society.

References

1. Ouyang J, Zhou L, Liu Z et al (2020) Biomass-derived activated carbons for the removal of pharmaceutical mircopollutants from wastewater: a review. Sep Purif Technol 117536
2. Wang Y, Sun J, He B et al (2020) Synthesis and modification of biomass derived carbon dots in ionic liquids and their application. A mini review. Green Chem Eng 1:94–108
3. Boruah A, Saikia M, Das T et al (2020) Blue-emitting fluorescent carbon quantum dots from waste biomass sources and their application in fluoride ion detection in water. J Photochem Photobiol B Biol 209:111940
4. Wang Z, Shen D, Wu C et al (2018) State-of-the-art on the production and application of carbon nanomaterials from biomass. Green Chem 20:5031–5057
5. Gu S, Zhou J, Yu C et al (2015) A novel two-staged thermal synthesis method of generating nanosilica from rice husk via pre-pyrolysis combined with calcination. Ind Crop Prod 65:1–6
6. Wanninayake P, Rathnayake M, Thushara D et al (2021) Conversion of rice straw into 5-hydroxymethylfurfural: review and comparative process evaluation. Biomass Convers Bior 12:1013–1047
7. Pode R (2016) Potential applications of rice husk ash waste from rice husk biomass power plant. Renew Sustain Energy Rev 53:1468–1485
8. Masulili A, Utomo W, Syechfani M (2010) Rice husk biochar for rice based cropping system in acid soil 1. The characteristics of rice husk biochar and its influence on the properties of acid sulfate soils and rice growth in West Kalimantan, Indonesia. J Agric Sci 2:39–47
9. Umeda J, Kondoh K (2010) High-purification of amorphous silica originated from rice husks by combination of polysaccharide hydrolysis and metallic impurities removal. Ind Crop Prod 32:539–544
10. Nath N, Chakroborty S, Pal K et al (2023) Recent Advances in Plasmonic Enhanced Nanocatalyst for Oxidation of Alcohol. Top Catal 1:1–11
11. Nath N, Chakroborty S, Vishwakarma DP et al (2023) Recent advances in sustainable nature-based functional materials for biomedical sensor technologies. Environ Sci Pollut Res 1:1–25
12. Villarreal CC, Pham T, Ramnani P et al (2017) Carbon allotropes as sensors for environmental monitoring. Curr Opin Electrochem 3:106–113
13. Jiang J, Leng J, Li J et al (2017) Twin graphene: a novel two-dimensional semiconducting carbon allotrope. Carbon 118:370–375
14. Feng Y, Jiang J, Xu Y et al (2023) Biomass derived diverse carbon nanostructure for electrocatalysis, energy conversion and storage. Carbon 211:118105
15. Tiwari S, Bystrzejewski M, Adhikari A et al (2022) Methods for the conversion of biomass waste into value-added carbon nanomaterials: Recent progress and applications. Prog Energy Combust Sci 92:101023
16. Iijima S (1991) Helical microtubules of graphitic carbon. Nature 354:56–58
17. Fathy N (2017) Carbon nanotubes synthesis using carbonization of pretreated rice straw through chemical vapor deposition of camphor. RSC Adv 7:28535–28541
18. Lotfy V, Fathy N, Basta A (2018) Novel approach for synthesizing different shapes of carbon nanotubes from rice straw residue. J Environ Chem Eng 6:6263–6274

19. Alslaibi T, Abustan I, Ahmad M (2013) A review: Production of activated carbon from agricultural byproducts via conventional and microwave heating. J Chem Technol Biotechnol 88:1183–1190
20. Khan F, Mubarak N, Khalid M et al (2021) Comprehensive review on carbon nanotubes embedded in different metal and polymer matrix: fabrications and applications. Crit Rev Solid State Mater Sci 47:837–864
21. Betancourt W, Rose J (2004) Drinking water treatment processes for removal of cryptosporidium and giardia. Vet Parasitol 126:219–234
22. Yang J, Hou B, Wang J et al (2019) Nanomaterials for the removal of heavy metals from wastewater. Nanomaterials 9:424–463
23. Lu C, Su F (2007) Adsorption of natural organic matter by carbon nanotubes. Sep Purif Technol 58:113–121
24. Ren X, Chen C, Nagatsu M et al (2011) Carbon nanotubes as adsorbents in environmental pollution management. A review. Chem Eng J 170:395–410
25. Gupta V, Agarwal S, Saleh T (2011) Chromium removal by combining the magnetic properties of iron oxide with adsorption properties of carbon nanotubes. Water Res 45:2207–2212
26. Smith S, Rodrigues D (2015) Carbon-based nanomaterials for removal of chemical and biological contaminants from water: a review of mechanisms and applications. Carbon 91:122–143
27. Upadhyayula V, Deng S, Mitchell M et al (2009) Application of carbon nanotube technology for removal of contaminants in drinking water. A review. Sci Total Environ 408:1–13
28. Sarkar B, Mandal S, Tsang Y et al (2019) Sustainable sludge management by removing emerging contaminants from urban wastewater using carbon nanotubes. In: Industrial and municipal sludge. Elsevier, Amstredam, pp 553–571
29. Ali S, Rehman S, Luan H et al (2019) Challenges and opportunities in functional carbon nanotubes for membrane-based water treatment and desalination. Sci Total Environ 646:1126–1139
30. Deng S, Upadhyayula V, Smith G (2008) Adsorption equilibrium and kinetics of microorganisms on single-wall carbon nanotubes. IEEE Sens J 8:954–962
31. Rocha JDR, Rogers R, Dichiara A et al (2017) Emerging investigators series: highly effective adsorption of organic aromatic molecules from aqueous environments by electronically sorted single-walled carbon nanotubes. Environ Sci: Water Res Technol 3:203–212
32. Arifin N, Yusof N, Ismail A et al (2020) Graphene from waste and bioprecursors synthesis method and its application: a review: Malays. J Fundam Appl Sci 16:342–350
33. Singh P, Bahadur J, Pal K (2017) One-step one chemical synthesis process of graphene from rice husk for energy storage applications. Graphene 6:61–71
34. Kumar M, Sachdeva A, Garg R et al (2020) Synthesis and characterization of graphene prepared from rice husk by a simple microwave process. Nano Hybrids Compos 29:74–83
35. Water U (2014) The United nations world water development report 2014: water and energy. United Nations, Paris, France
36. Han Y, Xu Z, Gao C (2013) Ultrathin graphene nanofiltration membrane for water purification. Adv Funct Mater 23:3693–3700
37. Policicchio A, Filosa R, Abate S et al (2017) Activated carbon and metal organic framework as adsorbent for low-pressure methane storage applications: an overview. J Porous Mater 24:905–922
38. De la Hoz A, Diaz-Ortiz A, Moreno A (2005) Microwaves in organic synthesis. Thermal and non-thermal microwave effects. Chem Soc Rev 34:164–178
39. Chaudhary S, Kumar S, Kaur B et al (2016) Potential prospects for carbon dots as a fluorescence sensing probe for metal ions. RSC Adv 6:90526–90536
40. Zhu L, Shen D, Wu C et al (2020) State-of-the-art on the preparation, modification, and application of biomass-derived carbon quantum dots. Ind Eng Chem Res 59:22017–22039
41. Wongso V, Sambudi N, Sufian S (2019) The effect of pH in the synthesis of carbon quantum dots from rice husk on their photoluminescence properties. IOP Conf Ser Earth Environ Sci 268:012087

42. Liu R, Gao M, Zhang J et al (2015) An ionic liquid promoted microwave-hydrothermal route towards highly photoluminescent carbon dots for sensitive and selective detection of iron (III). RSC Adv 5:24205–24209
43. Canhoto O, Pinzari F, Fanelli C et al (2004) Application of electronic nose technology for the detection of fungal contamination in library paper. Int Biodeterior Biodegrad 54:303–309
44. Thaler E, Kennedy D, Hanson C (2001) Medical applications of electronic nose technology: review of current status. Am J Rhinol 15:291–295
45. Thongsai N, Tanawannapong N, Praneerad J et al (2019) Real-time detection of alcohol vapors and volatile organic compounds via optical electronic nose using carbon dots prepared from rice husk and density functional theory calculation. Colloids Surf A Physicochem Eng Asp 560:278–287
46. Abidin N, Wongso V, Hui K et al (2020) The effect of functionalization on rice-husks derived carbon quantum dots properties and cadmium removal. J Water Process Eng 38:101634
47. Shtepliuk I, Caffrey N, Iakimov T et al (2017) On the interaction of toxic Heavy Metals (Cd, Hg, Pb) with graphene quantum dots and infinite graphene. Sci Rep 7:1–17

Chapter 8
Nutshell-Derived Efficient Carbon Nanomaterials as a Potential Smart Electrode Material for Electrocatalytic Hydrogen Production

Monika Sindhu, Arshjeet Singh, Yarramsetti Praveen Kumar, Neetu Vishwakarma, Afshan Hassan Wani, and Ajit Sharma

Abstract As part of ongoing exertions by scientists to develop alternative solutions to curtail the issue of continuous emissions of greenhouse gases by the use of fossil fuels, attention has been directed toward biomass (organic matter that can be turned into useful energy). Green nanotechnology is advancing to address society's global sustainability disquiets by recycling a variety of industrial and bio-wastes to generate functional carbonaceous nanomaterials that include biochar and activated carbon. Biomass can be the primary source of bio-based carbon nanomaterials for electrochemical energy production and storage technologies. Carbon nanomaterials derived from biomass possess diverse chemical, structural, and mechanical characteristics that have been investigated as a carbon source due to their high surface area, configurable structure, abundance, cost-effectiveness, and cyclic and environmentally friendly nature. In this chapter, various nutshells are used as a green carbon source for the electrode to produce the electrocatalytic hydrogen. This chapter aims to provide a thorough understanding and perspective on all of the potential uses of nutshell biomass for energy generation. We explore the latest advancements and current obstacles and possibilities for biomass materials in energy generation.

M. Sindhu
Department of Chemistry, University Centre for Research and Development, Chandigarh University, Mohali, Punjab, India

A. Singh · Y. P. Kumar · A. Sharma (✉)
Department of Chemistry, School of Chemical Engineering and Physical Science, Lovely Professional University, Phagwara, Punjab, India
e-mail: ajitsharma2003@gmail.com

N. Vishwakarma
Department of Physics, School of Chemical Engineering and Physical Science, Lovely Professional University, Phagwara, Punjab, India

A. H. Wani
Department of Computer Science and Engineering, Lovely Professional University, Jalandhar, Punjab, India

N. Talreja et al. (eds.), *Waste-Derived Carbon Nanostructures*, Nanostructure Science and Technology, https://doi.org/10.1007/978-3-031-75247-6_8

Keywords Biomass · Nanotechnology · Carbon nanomaterials · Hydrogen evolution

8.1 Introduction

Nowadays, energy is the most essential need for us to continue living our lives, and as the population expands, rises the energy demand. Hydrocarbon-based fossil fuels such as crude oil, natural gas, and coal provide the vast majority of the energy required in the world [1]. As a result of increased public awareness about the depletion of fossil fuels and their environmental effects, researchers and industry have invented extremely effective and environmentally friendly alternative energy solutions. Furthermore, excessive fossil fuel consumption results in global warming, abrupt climate change, and significant ecological contamination [2]. Energy sources should have characteristics like high reserve, renewable, high-energy content, and eco-friendly to work as an alternative to fossil fuels. Alternative sources of energy, like solar, wind, biomass, hydro, and geothermal, are becoming more and more significant as environmentally friendly alternatives as shown in Fig. 8.1 [3]. Notably, variations in the seasons, the period of the day, and geographic location may all affect how much energy is generated from these renewable sources [4]. In spite of these circumstances, clean-burning hydrogen gas (H_2) has become an efficient and clean energy carrier with a high density of energy. H_2, having these characteristics, has become the fuel of the next generation and is predicted to be the primary alternative energy source that substitutes fossil fuel systems by the end of this century. H_2 can be produced by mainly two sources, renewable and conventional non-renewable as shown in Fig. 8.2. The electro- or photocatalytic splitting of water into H_2 and O_2 is a promising way to reduce our current reliance on fossil fuels, and it is inspired by nature. Two half-reactions are involved in the splitting of water: the reduction of protons to H_2 and an oxidation of the water to O_2, protons, and electrons [5–7].

The development of novel catalysts for solar H_2 production hinges on a thorough comprehension of the two reactions. Because it is simple, inexpensive, and safe for the environment, electrochemical (EC) water splitting is a reliable and effective process [8]. Scientists have been working extensively to develop superior electrode materials to improve the electrochemical water-splitting process for hydrogen evolution reaction (HER) [9]. An increasing number of electrocatalysts are constructed from noble metals and their alloys, including platinum, palladium, iridium, and ruthenium. As an example, platinum (Pt) is the ideal electrocatalyst for hydrogen evolution reactions (HER), but large-scale uses of noble metal catalysts are not feasible due to their high cost and rarity [10]. Therefore, it is crucial to investigate and develop electrocatalysts that are highly efficient for HER, using non-noble metals. Therefore, it is essential to explore electrocatalysts for the hydrogen evolution reaction (HER) that are free of Pt. Developing cost-effective and sustainable nonprecious metal catalysts is a major area of interest.

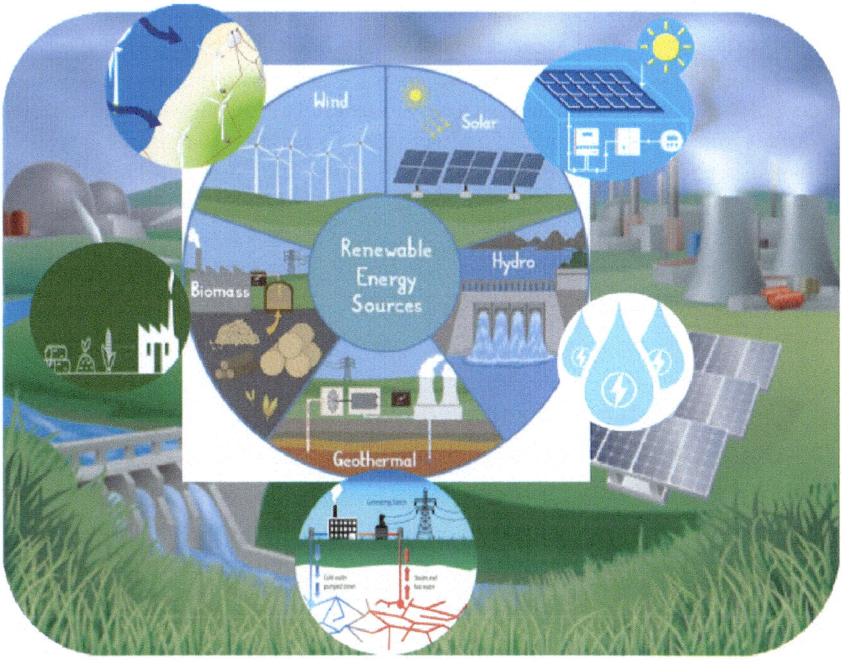

Fig. 8.1 Renewable energy sources. Reproduced with permission from Ref. [3], 2021, renewable and sustainable energy reviews

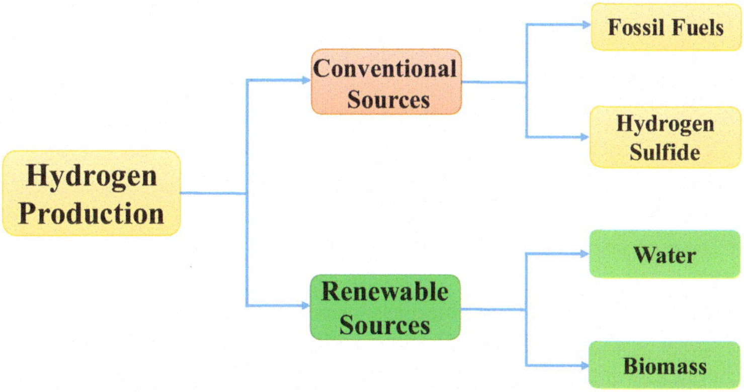

Fig. 8.2 The production of hydrogen using conventional and renewable energy sources

According to these research endeavors, significant attention has been directed toward the development of biomaterials that are abundant in nature and environmentally friendly, while also possessing sturdy catalytic activity. These materials are chosen for their inherent characteristics, including superior electrical conductivity, an extensive surface area, and ease of synthesis. Over the years, there has been a rising attraction to the practical uses of renewable biomaterials, mainly because of their cost-effectiveness. These biomaterials have shown great potential for a wide range of applications [11]. Meanwhile, by taking advantage of the evenly distributed heteroatom sources observed in biomass cells, it is feasible to directly obtain heteroatom self-doped biomass carbon via a straightforward biomass pyrolysis process [12]. By transforming biomass waste into efficient bio-carbon materials, the cost of electrode materials can be significantly reduced, leading to significant improvements in sustainable energy technologies.

8.2 Carbon Materials Produced from Biomass for Energy Storage and Conversion Systems

Carbon-based materials are essential in various clean energy conversion and storage technologies. It is commonly acknowledged that the choice of electrode materials significantly impacts the electrochemical performance. It has been recognized that the key to developing such new energy technologies is the fabrication of innovative electrode materials to power these devices. In recent studies, various carbon-based hybrids such as carbon nanotubes, graphene, and other carbon-based composite materials have shown significant potential [13]. However, conventional precursors and procedures have drawbacks such as high cost, inadequate production, elaborate production procedures, and lack of sustainability, which greatly hinder their widespread use [14]. Furthermore, the starting materials are even hazardous, which pose a risk to public health and the environment. Thus, efforts to synthesize a substitute carbon material for biomass through ecologic synthesis methods have the potential to revolutionize the field of renewable energy. As depicted in Fig. 8.3, the number of biomass sources is steadily increasing each year, causing up more than a third of all renewable energy sources. Several sources of biomass include agricultural waste (which includes vegetables, fruit, wheat, and rice), industrial waste, resulting from the utilization of microorganisms, and domestic food waste like animal waste and coffee [15].

Carbon materials derived from biomass have gained significant attention as electrocatalysts for several applications such as fuel cells, supercapacitors, batteries, and electrochemical water splitting as shown in Fig. 8.4.

These materials are renewable, environmentally friendly, and cost-effective, facilitating accelerated ion transportation from the electrolyte to the electrode surface while minimizing the distance over which ions diffuse [16]. It is well known that electrode materials are essential for analyzing the electrochemical performance. Pine needle-derived microporous nanoparticles were produced by Liu et al. and demonstrated outstanding electrocatalytic activity [17].

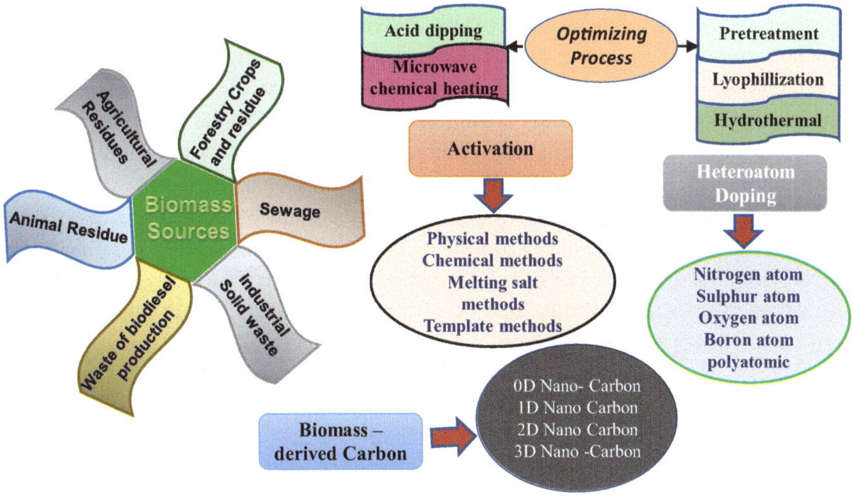

Fig. 8.3 Schematic diagram of the sources of biomass on Earth and biomass-derived carbon

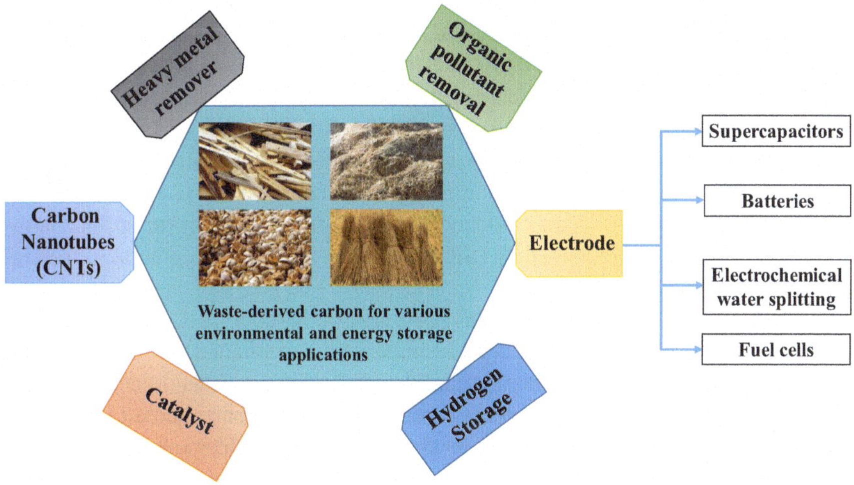

Fig. 8.4 Carbon derived from biomass for various environmental and energy conversion/storage applications

According to study findings by Feng et al., biomass resources can be utilized to synthesize hierarchically porous carbons, which have a large surface area and have functional groups consisting of nitrogen and oxygen [18]. According to Moreno et al., the unique porous properties of biomass materials can be utilized to develop electrode materials for supercapacitors [19]. To promote ion transport and electrolyte penetration, Lu et al. found that structures that are porous and linked can be produced from biomass carbons [20]. Research on catalysts derived from biomass has been ongoing ever since the development of silk-derived activated carbon as a highly effective ORR electrocatalyst in 2010 [21]. Meanwhile, it presents exciting possibilities for the advancement of carbon obtained from biomass in alternative energy technologies, thanks to its numerous advantages such as its abundance, environmental friendliness, and ease of functioning. Therefore, it is evident that the choice of electrode materials greatly impacts the efficiency of the electrochemical process. Efficient electrode materials with enhanced surface area are crucial for optimizing the performance of electrochemical water-splitting devices. Efficient carbon-based electrodes derived from nutshells are not only low cost, but also have a naturally porous structure, excellent biocompatibility, and eco-friendly properties. These qualities make them excellent for various applications, including electrochemical energy production systems. Manoharan et al. examined the hydrogen uptake capacity of imogolite and activated carbon nanocomposites made from tangerine peel. To synthesize the imogolite (IMO), solutions of sodium orthosilicate and aluminum chloride were mixed, and then NaOH was added gradually until the pH level reached 6. Tangerine peel biomass was ground, combined with KOH, and then pyrolyzed at 800 °C to produce Tangerine peel biomass-derived activated carbon (TPAC).

According to Fig. 8.5 [22], the nanocomposite (IMO/TPAC) was synthesized by ultrasonication, subsequently washed, and dried. There are numerous examples of these materials being effectively utilized for particular applications. In microbial fuel cells, Chen et al. found that using chestnut shells instead of carbon fiber anodes resulted in a significant 2.3-fold increase in power output [23].

Huggins et al. used wood-derived carbon as the cathode and anode materials in an MFC to study the effects of treating wastewater and recovering nutrients. The results indicated a 95% removal of chemical oxygen demand (COD) [24]. Abbas et al. fabricated porous structures and induced structural flaws in the form of SiC nanospecies by using the abundant silica (SiO_2) found in bamboo as a natural sacrificial template and dopant (Fig. 8.6 [25]).

Additionally, it was found that the pyrolysis of bamboo N-functional groups produced pyrrole-N, which enhanced the material's capacitance. By incorporating SiC and pyrrole-N, the composite material demonstrated improved rate ability even at higher current densities. The porous carbon material obtained had an impressive specific surface area of 1789 m^2g^{-1} and showed a remarkable specific capacitance of 369 F g^{-1} at 0.5 A g^{-1}, indicating excellent electrochemical stability. These findings strongly indicate that carbonizing biomass is an excellent technique to generate conductive materials suitable for inexpensive electrodes. Investigating other naturally abundant biomass sources is necessary, and there is still significant room for improvement in the preparation method. It is still a difficult task to understand the causes and mechanisms of activity. Therefore, further endeavors are required to create simple techniques utilizing cost-effective biomass precursors in order to manufacture long-lasting electrode materials for applications in energy.

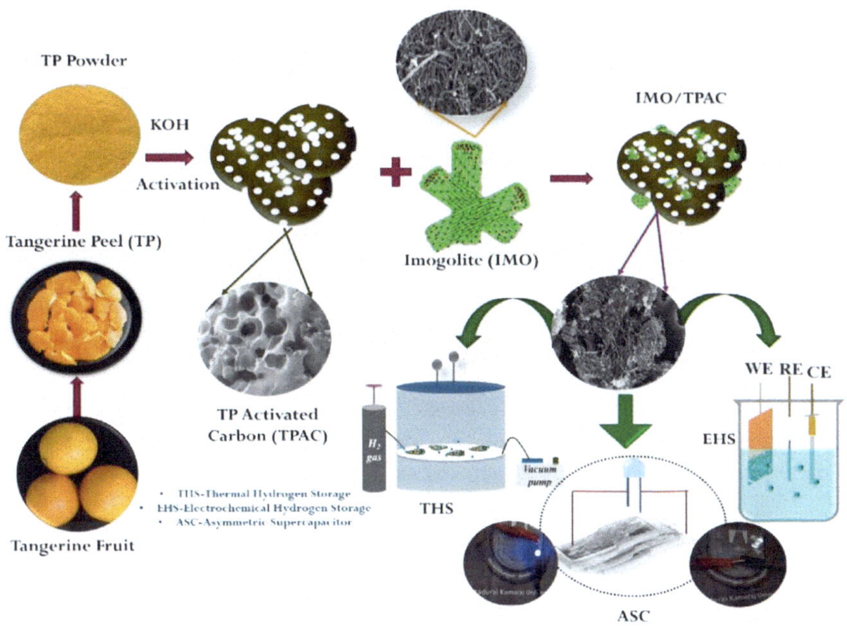

Fig. 8.5 The synthesis process of IMO/TPAC nanocomposite, reproduced with permission [22]

8.2.1 Activated Biochar Derived from Peanut Shell

Several materials based on carbon, like carbon nanotubes, activated carbon, and graphene, are mainly derived from non-renewable resources such as coal, petrochemicals, or natural gas [26]. These resources are becoming limited as technology and society advancement. To produce carbon-based materials in such circumstances, it is essential to look for abundant and sustainable carbon resources. Peanut is a highly significant leguminous plant and a major oil-bearing crop that has been extensively cultivated worldwide. PS, commonly used crop residue from peanut harvesting, is a clean, abundant, cost-effective, and renewable biomass resource with a high global annual yield of 7.44 million tons. This makes it a promising feedstock for the production of carbon-based materials, ensuring a practically infinite supply. PS has the potential to be employed as raw materials for extremely efficient electrode production because of its relatively high carbon content, which makes it a good substrate for activated carbon (AC) [27, 28]. In recent years, peanut shells have been utilized in various applications, such as developing adsorbents that can eliminate heavy metal ions and organic dyes from wastewater. It has also been utilized as electrode materials in various energy conversion and storage devices such as supercapacitors, electrocatalytic water splitting, lithium-ion batteries, as well as metal-air batteries [29]. However, the catalysts made up of carbon fabricated through the process of pyrolyzing peanut shells as precursors at temperatures ranging from 500

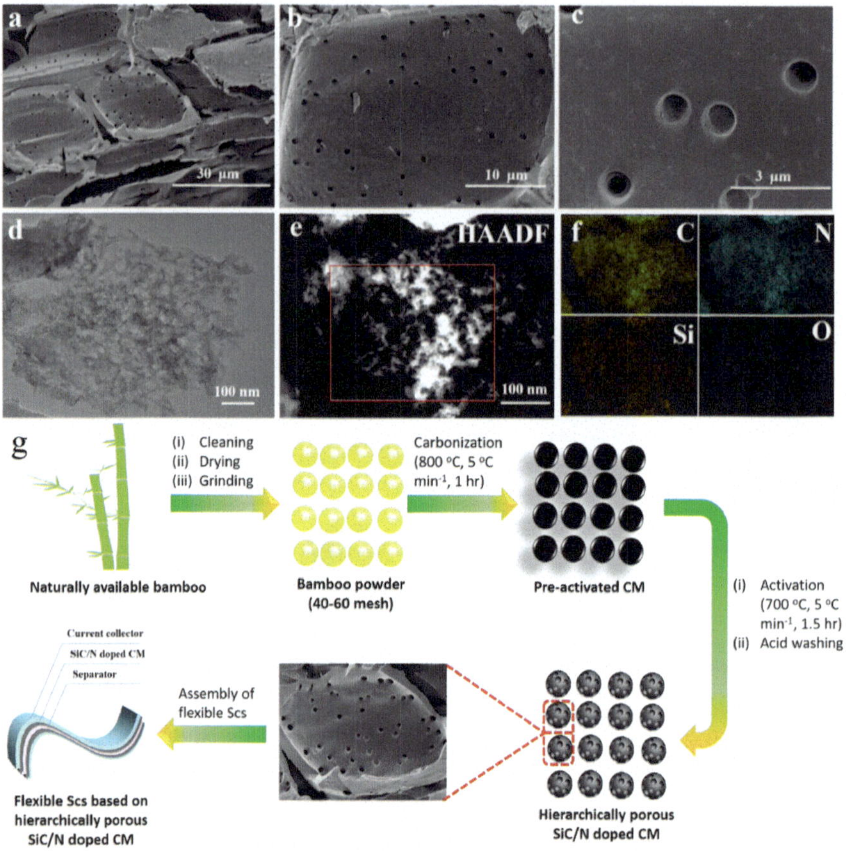

Fig. 8.6 **a–c** FESEM images of SNAC-1, **d** TEM image of SNAC-1, and **e**, **f** HAADF; HAADF-EDS mapping of SNAC-1 for C, N, Si, and O, **g** Scheme for the synthesis of SNAC-1 and SNAC-2-based electrodes via natural self-activating sacrificial template methods for flexible solid-state supercapacitors. Adopted with permission ref [25]

to 800 °C in an N_2 or CO_2 environment exhibited a half-wave capability of the oxygen reduction reaction (ORR) that was 50–123 mV lower compared to the one observed on commercial Pt/C [30]. The NC-900 catalyst, synthesized by pyrolyzing a mixture of melamine and peanut shells, exhibited a half-wave potential of 0.81 V for ORR, slightly lower than that of Pt/C [26]. Furthermore, materials derived from peanut shells that are carbon-based regularly exhibit specific capacitances that are less than 340 F g^{-1} [31]. Thus, it is crucial to develop an innovative carbon-based catalyst that exhibits exceptional performance in oxygen reduction reactions (ORR) and remarkable capacitive properties. Lotus root starch was used by Zhong et al. as the biomass material, and as illustrated in Fig. 8.7a [32], they mixed it with urea, magnesium chloride ($MgCl_2$), and lotus root starch carbon material in a 30 mL water solution by using a pot stew method. After heating the mixture to 90 °C for one hour, a homogeneous slurry was formed. This was followed by a 24-h drying process at 120 °C to make a mixed powder. Subsequently, porous carbon materials were prepared by activating them at 800 °C, resulting in a specific surface area of 844.82 m^2 g^{-1}. These materials exhibited a specific capacitance of 331.80 F g^{-1}. In Fig. 8.7b, Gang et al. used Ulva lactuca as a biomass material and nickel sulfate (Ni_2SO_4) as an activator to submerge 0.8 g of Ulva lactuca powder in 50 mL solutions of nickel sulfate ($NiSO_4$ $6H_2O$) with different concentrations (0.2, 0.6, 0.8, and 1.0 M) for 48 h [33]. After collection, the.

powder was freeze-dried for 12 h. It was then carbonized and activated in a tube furnace at 850 °C. The resultant porous carbon material has a specific capacitance of 167 F g^{-1} and a significant degree of graphitization. In Fig. 8.7c, Guo et al. used $FeCl_3$, $ZnCl_2$, and $MgCl_2$ as activators, combining the three salt activators via impregnation [34]. This catalyst can be derived from peanut shells, serving a dual purpose of waste utilization and cost-effectiveness. Such a development would contribute to ecological environmental protection and the sustainable advancement of society.

8.2.2 *Activated Biochar Derived from Almond Shell*

The use of biomass to produce bio-carbon expands the potential uses of nutshells as a natural filler. Additionally, advancements in technologies like activated bio-carbon enable its application in supercapacitors, absorbents, hydrogen production, and storage. Among the different agricultural residues, nutshells are highly valued not only for their large production volume but also for their chemical composition. Tree nut fruits are processed to extract the edible nut kernel by removing the shell and/or hull. In the 2020/21 season, the global production of tree nuts attained a staggering 5.3 million metric tons [35]. It is preferable to valorize nut biomass since it is cheap and simple to obtain, and it is abundant in a nutshell and hull material. The almond shell is a plentiful byproduct from the almond cultivation process and is a valuable renewable resource in almond-producing nations like Spain and the USA. The global yearly harvest of almonds is estimated to be 1.97 million tons (MT), with Spain contributing around 196,700 tons [36]. Annually, approximately 0.68–1.47 million metric tons of almond shells are generated. This material is a widespread and inexpensive biomass that can be converted into a conductive material through the activation procedure. Researchers are exploring ways to make use of agricultural waste products to support the concept of a circular economy. Recent studies have been exploring the potential of nut biomass for value-added use as biofuel and as natural filler substitutes for bio-composites. Through the process of torrefaction and pyrolysis, these bio-constituents are transformed into carbon-rich materials. Torrefaction is typically carried out at temperatures ranging from 200 to 300 °C, whereas pyrolysis occurs at temperatures above 300 °C [37].

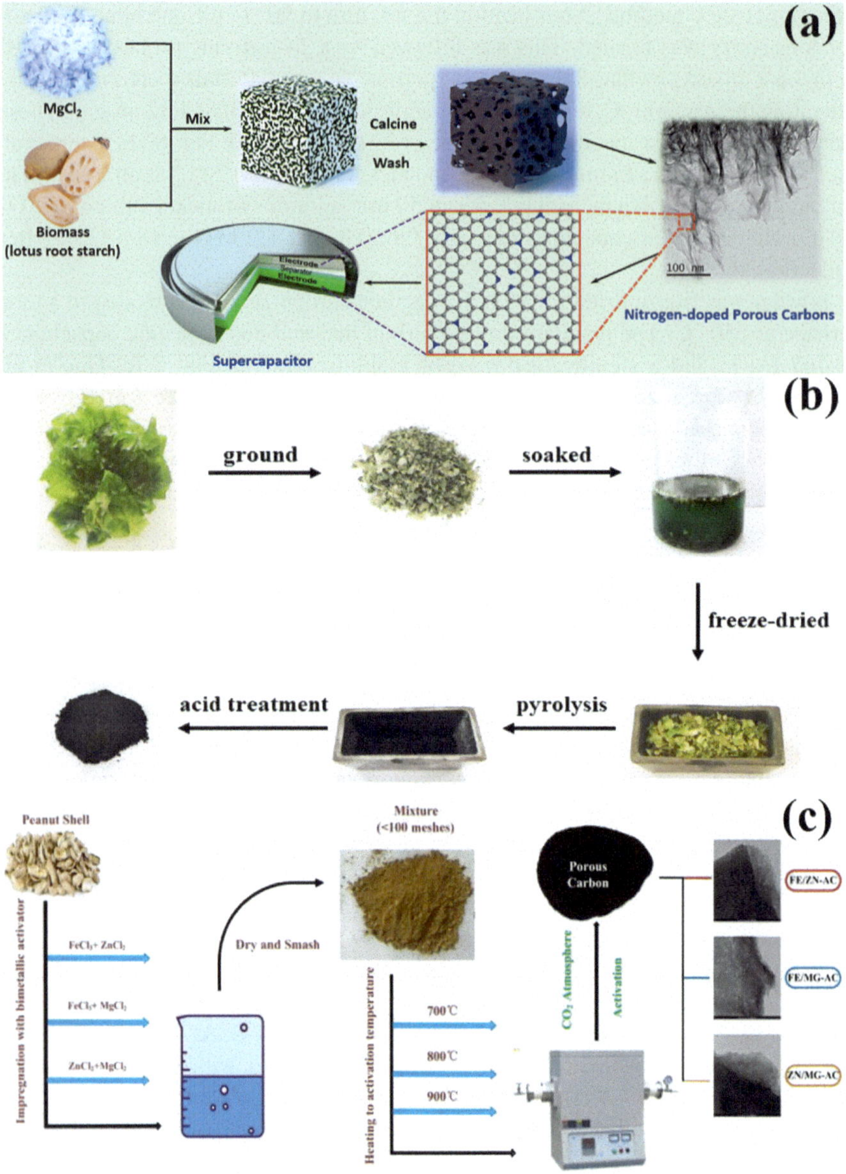

Fig. 8.7 Scheme for the synthesis of biomass-activated carbons **a** lotus root starch [32], **b** Ulva lactuca [33], **c** peanut shell [34]. Adopted with permission ref [28]

Understanding the electrical conductivity of bio-carbon is crucial for its various applications, such as storage and conversion of energy, electrochemical sensing, and electronic systems. The electrically conductive property of carbon derived from biomass is largely dependent on the pyrolysis temperature. At higher temperatures, functional groups are lost through dehydration, leading to an increase in carbon content and, subsequently, electrical conductivity [38]. Research has been done on almond nutshell bio-carbon for different applications, but there is still a need for more investigation into the impact of pyrolysis temperature on the properties and electrical conductivity to expand its possibilities of application. The porous carbon derived from almonds exhibits excellent properties, such as high conductivity, expanding surface area, an interconnected stacked porous structure, and strong biocompatibility. These characteristics render it an ideal choice for anode materials in electrocatalytic hydrogen production. Almond shells are a readily available agricultural by-product that can be used to produce microporous activated carbons. This is achieved by first carbonizing the shells and then activating them [39]. Activated carbon production enables the utilization of agricultural wastes and their integration into the economy. The researchers successfully synthesized porous carbon materials by utilizing almond shells as a precursor. These materials have a remarkable Brunauer–Emmett–Teller (BET) surface area of 616.04 $m^2\ g^{-1}$ and possess a complex porous structure that includes both macropores and mesopores [40]. The MFCs developed with the PC-800 to be the anode exhibit exceptional performance, reaching an ideal power density of 4346 ± 12 mW m^{-2}. Considering the almond shell's affordability and eco-friendliness as a resource for carbon materials, along with the straightforward and effective fabrication process and impressive electrochemical performance of MFCs, the carbon derived from almond shells shows great potential as an anode for MFC applications.

Since pyrolyzed almond shells function similarly to carbon-felt electrodes, it is possible to produce sustainable and affordable electrodes for electrochemical systems using them. The microbiological study showed that the Pyrolyzed Almond Shells (PAS) electrode and the CF electrode had a similar structure and taxonomic composition. This confirms the strong electrical connection between bacteria and the electrode, leading to improved electron transfer efficiency with the low-cost material [41]. These results indicate that PAS is a highly suitable choice for anode material in MFCs. There has been a growing interest in using biomass-derived materials as cost-effective and high-performance options to develop electrode materials in fuel cells. Because of the growing need for sustainable energy solutions, the utilization of carbonaceous materials as electrodes has gained significant attention. Thus, because of their affordable and reusable properties, carbon derivatives obtained from bio-waste are seen as a highly promising substitute for conventional electrodes in electrochemical applications. These derivatives should exhibit excellent conductivity, biocompatibility, chemical stability, specific surface area, and a well-distributed pore structure, while also being cost-effective [42].

8.2.3 Activated Biochar Derived from Walnut Shell

The walnut tree, scientifically known as Juglans regia, is a popular plant in the Juglandaceae family. It is highly sought after for its delicious dry fruits known as walnuts. In 2019, the global production of walnuts reached an incredible 3.7 million tons. Unfortunately, this massive amount of production also resulted in the generation of millions of tons of waste due to the discarding of shells. In addition, walnuts have a high protein content of about 15%. This protein content is the main source of nitrogen (N) element. As a result, we can conclude that there is also a small amount of nitrogen element present in the shell structure of walnuts. This finding is significant because it helps improve the electrochemical activity of the corresponding carbon materials [43]. The edible part of a walnut is a highly nutritious food, while the shell is often discarded as waste. However, by adopting a waste-to-wealth approach, researchers have found a valuable use for walnut shells. They have discovered that walnut shells can be used as a precursor material to produce activated carbon, which can then be used in the fabrication of electrodes for energy storage. This innovative approach has led to the investigation of the energy storage capabilities of these walnut shell-derived electrodes. The carbon produced from walnut shells has shown remarkable results, boasting a significant micropore area and volume, and could be used in electrocatalysts and supercapacitors. The activation temperature plays a crucial role in determining the surface structure, including surface area, pore texture, and morphology of the resulting walnut shell-derived carbon series. In addition, a thorough analysis is conducted on the correlation between surface structure and electrochemical performance for SCs and HER. This analysis highlights the potential of WS-derived carbon materials with exceptional surface structure as promising biomass electrode materials [44]. In a recent study, Huafang Yang and colleagues discovered that AC derived from walnut shells can be used as an electrode material for supercapacitors. They found that the specific capacitance reached 255 and 168.68 F/g with a current density of 0.5 A/g, demonstrating the potential of this material for energy storage applications [45]. Xiya Li et al. used the AC produced from walnut shells for applications such as Zn (II) removal and the adsorption of methylene blue dye [46].

8.3 Synthesis of Biomass-Derived Carbon and Fabrication of Electrode

Over time, the synthesis and use of carbon produced from biomass for energy applications have garnered significant attention in studies. This might be a result of its affordability, sustainability, and simplicity of manufacture from different bio-precursors. Carbon derived from biomass is generated through various thermal processes, including gasification, carbonization, liquefaction, and pyrolysis [47].

Typically, carbon derived from bio-waste can be activated through various activation methods, including both physical and chemical approaches.

8.3.1 Thermal Carbonization and Chemical Activation

Thermal carbonization is a commonly employed method for developing biochar from biomass. It involves heating the biomass at high temperatures in a muffle furnace, which eliminates lighter compounds. By optimising the specific conditions such as temperature and time the resulting biochar has the desired characteristics. For example a slower heating rate promotes the production of greater quantities of biochar [48].

8.3.2 Fabrication of Electrode

To improve the electrochemical performance, appropriate electrodes must be fabricated. In the past decade, there has been significant progress in the advancement of electrode materials, resulting in highly efficient choices. When developing the electrode, the necessary quantity of carbon derived from biomass was mixed into the solvent. The mixture was sonicated until it formed a slurry, which was then applied to both sides of a substrate such as carbon cloth or stainless-steel mesh sheet with an optimal loading amount. The porous carbon electrodes were air-dried at room temperature [49].

8.4 Application of Biomass-Derived Carbon Materials

8.4.1 Electrocatalysis (Carbon-Based Electrode for Electrocatalytic Hydrogen Production)

The technologies associated with electrochemistry have emerged as a possible alternative to address the environmental problems and the rapid global warming caused by the increasing usage of fossil fuels. These reactions, such as HER, ORR, OER, and others, have gained global recognition as crucial processes in the conversion and storage of clean energy technologies. They play an essential role in various applications such as fuel cell technology, MFCs, splitting of water, and supercapacitors [50]. Carbon materials derived from nutshells have shown promising potential in meeting the requirements for efficient electrocatalytic performance. To produce hydrogen and oxygen by water splitting, one such substance, biochar, has been used as an electrocatalyst and photocatalyst as shown in Fig. 8.8. The introduction of heteroatoms into

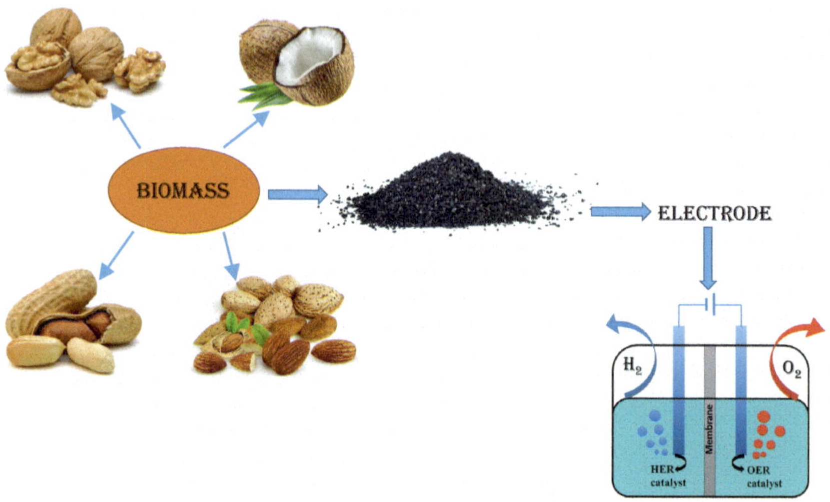

Fig. 8.8 Schematic of biomass-derived carbon electrode for electrochemical water splitting

biochar creates active sites that enhance the HER. For instance, biochar doped with sulfur and nitrogen, derivatives from peanut root nodules, has been found to be highly effective electrocatalysts for HER [51]. These biochars have been doped to enhance their properties, resulting in a high porosity and an extensive electrochemical area of 27.4 mF cm^{-2}. As a result, they exhibit an impressive onset potential of 27 mV compared to the reversible hydrogen electrode (RHE) for HER. This performance is similar to the performance of a commercial Pt/C catalyst which has a 20 wt% loading [52].

A novel carbon-based catalyst generated from porphyra was synthesized by Zhang and colleagues. Significant ORR performance and an impact on OER activity are demonstrated by using Fe, N co-doping bio-carbon [53]. S, N-co-doped Structure present in porphyra is capable to improve the efficiency of OER and ORR activities. Recently, Yuan et al. achieved a successful synthesis of a porous tar-based carbon material that was co-doped with Fe–N. It demonstrated exceptional ORR electrocatalytic efficiency [54]. The results of these investigations can be used as a blueprint for the production of high-tech electrocatalyst materials from biomass. Liu et al. synthesized nitrogen, cobalt-co-doping bio-carbon ORR catalysts using corn stover for Al-air batteries, and these catalysts perform outstandingly in ORR and metal-air battery experiments [55]. Direct pyrolysis of bio-carbon fiber cloth electrodes, as proposed by Xiao et al., is a straightforward method for synthesizing such electrodes. In terms of electrocatalytic performance toward HER, the BCF/Mo_2C electrode is ideal [56].

8.4.2 Application in MFCs

Microbial fuel cells (MFCs) are a new kind of bioenergy generator and wastewater treatment system that can biodegrade both organic and inorganic materials at the same time. In these bio-electrochemical reactors, microorganisms catalyze chemical processes that transfer chemical energy from organic substrates to electrical energy [57]. The electrochemical reactions that take place on the electrode surfaces determine the overall performance of the MFCs, developing the electrodes a crucial component of the device. A large surface area, good conductivity, low cost, stability, and biocompatibility are important considerations for choosing electrode materials. For a sustainable, carbon–neutral society, nutshell leftovers from nuts including peanuts, almonds, coconuts, and walnuts could be used as a carbon electrode [58]. Due to its low cost and low environmental impact, biomass-produced carbon is being used for fabricating electrodes and electrocatalysts for MFCs [59]. As a result, microbial fuel cells can make use of carbon from a variety of nutshell biomass as an electrocatalyst, anode, or cathode material [60]. There are three main reasons why MFCs utilize electrodes fabricated from biochar: Compared to traditional electrodes made of carbon or platinum, those made of biochar are more cost-effective. Progress toward a sustainable future is safe for the environment and can even be fabricated from waste. Power generation produces a power density similar to that of traditional electrodes [61].

8.4.3 Application in Supercapacitor

Supercapacitors (SCs), with their outstanding power density, quick charge/discharge rate, extended cycle life, reversibility, and high coulombic efficiency, have garnered a lot of interest these days to store clean and renewable energy to meet the world's energy demands [62]. Carbon-based porous materials derived from biomass show great potential as electrodes for supercapacitors. These materials possess remarkable physicochemical stability, a large surface area, good electrical conductivity, and are abundant, cost-effective, renewable, environmentally friendly, and non-toxic [63]. Furthermore, when compared to non-renewable activated carbon and graphene, biochar is known for its sustainable cycle, as it adsorbs CO_2 onto its carbon matrix, resulting in less carbon being released back into the atmosphere.

8.5 Advantages of Using Biomass-Derived Carbon Electrodes in Electrocatalytic Hydrogen Production

Nutshell-derived carbon materials offer numerous advantages compared to regular biochar when utilized in electrocatalytic hydrogen energy production. Here are some of the advantages:

- Increased surface area: It increases the surface area, which enhances the efficacy of transporting electrons and improves overall electrochemical performance.
- Enhanced electrical conductivity: It boosts electrical conductivity, leading to improved electron transmission in electrocatalytic water splitting. This results in improved electrochemical efficiency and a higher yield of hydrogen.
- Cost-effective: Utilising nutshell biomass as a material for electrochemical applications offers an economical and sustainable solution due to its low cost.
- Chemical stability: The efficient carbon-based electrode exhibits remarkable chemical stability, ensuring its durability and suitability for prolonged use in electrocatalytic water splitting.
- Environmental benefits: The utilization of biomass-derived carbon materials in electrocatalytic processes not only transforms waste into energy but also helps reduce greenhouse gas emissions. This has the potential to minimize the environmental consequences of waste management and energy generation.
- Flexibility: Biomass-derived carbon electrodes can be developed from different sources such as agricultural residue, forestry, and municipal waste. This makes it a versatile material for electrocatalytic green energy generation. It offers several advantages in electrochemical applications, including a higher surface area, better electrical conductivity, chemical stability, cost-effectiveness, and environmental benefits. Due to these advantages, carbon derived from nutshells is a promising material for electrocatalytic hydrogen generation. This technology can make a substantial impact on the production of renewable energy and waste management [64–66].

8.6 Conclusion and Future Perspectives

In summary, recent studies have shown that carbon electrodes derived from nutshells are shown to be highly effective in electrochemical energy storage and conversion devices. This is attributed to their unique porous structure and the ability to adjust their surface chemistry, resulting in excellent performance. This chapter discusses the various electrochemical characteristics and the several application of biomass derived carbon such as electrocatalytic hydrogen production, supercapacitors, and fuel cells. Extensive research has revealed that carbon derived from biomass has the potential to replace traditional raw materials for electrochemical energy storage and conversion. Nutshell-derived carbon has interesting properties that could prevent biomass from being burned or buried while also opening the door to the possibility of using it to make inexpensive electrode catalysts for real-world uses. This chapter provides a thorough overview of the most recent advancements in the study of carbon electrode materials derived from biomass. The main objective is to develop a method of production that is both energy-efficient and environmentally friendly, while also being easy to operate, in order to produce high-performance carbon electrodes for potential electrochemical energy storage and conversion technologies. The utilization of carbon electrodes derived from nutshells in electrocatalytic water splitting shows promising prospects for improving performance and energy conversion efficiency. Carbon derived from nutshells can improve electrode performance, advance electron transmission, and enhance the production of hydrogen. Scientists are currently conducting ongoing research in this field, tackling obstacles like production costs and structural stability. Despite the obstacles, the future of hydrogen production using biomass-derived carbon electrodes seems promising due to its environmentally friendly sources and noteworthy efficiency. Nevertheless, before these advantages may be fully experienced, a few obstacles must be solved. Efforts should be focused on enhancing fabrication techniques to ensure the flexibility and affordability of biomass carbon-based electrodes for electrocatalytic energy applications on a large scale. In addition, it is essential to optimize the interface between the carbon-based electrode derived from biomass and the electrolyte to enhance the transportation of electrons and enhance the overall efficacy of the device. By meticulous examination of production processes, flexibility, cost-effectiveness, stability, and interface optimization, biomass-derived carbon holds immense potential in enhancing sustainable energy generation and establishing electrochemical devices as an exciting technology for the future.

References

1. Kapsalyamova Z, Paltsev S (2020) Use of natural gas and oil as a source of feedstocks. Energy Economics 92:104984
2. Nema P, Nema S, Roy P (2012) An overview of global climate changing in current scenario and mitigation action. Renew Sustain Energy Rev 16(4):2329–2336
3. Siwal SS, Zhang Q, Devi N, Saini AK, Saini V, Pareek B, Gaidukovs S, Thakur VK (2021) Recovery processes of sustainable energy using different biomass and wastes. Renew Sustain Energy Rev 150:111483
4. Widén J, Carpman N, Castellucci V, Lingfors D, Olauson J, Remouit F, Bergkvist M, Grabbe M, Waters R (2015) Variability assessment and forecasting of renewables: a review for solar, wind, wave and tidal resources. Renew Sustain Energy Rev 44:356–375
5. Veziroğlu TN, Şahi S (2008) 21st Century's energy: hydrogen energy system. Energy Convers Manag 49(7):1820–1831
6. Puga A (2023) Photocatalytic hydrogen production in the context of sustainable energy. In: Photocatalytic hydrogen production for sustainable energy, pp 1–18
7. Qi J, Zhang W, Cao R (2018) Solar-to-hydrogen energy conversion based on water splitting. Adv Energy Mater 8(5):1701620
8. Zhai Y, Ren X, Yan J, Liu S (2021) High density and unit activity integrated in amorphous catalysts for electrochemical water splitting. Small Structures 2(4):2000096
9. Verma J, Goel S (2022) Cost-effective electrocatalysts for hydrogen evolution reactions (HER): challenges and prospects. Int J Hydrogen Energy 47(92):38964–38982

10. Li C, Baek J-B (2019) Recent advances in noble metal (Pt, Ru, and Ir)-based electrocatalysts for efficient hydrogen evolution reaction. ACS Omega 5(1):31–40
11. Zhang M, Zhang J, Ran S, Sun W, Zhu Z (2022) Biomass-derived sustainable carbon materials in energy conversion and storage applications: status and opportunities. A mini review. Electrochem Commun 138:107283
12. Feng Y, Jiang J, Xu Y, Wang S, W An, Chai Q, Prova UH, Wang C, Huang G (2023) Biomass derived diverse carbon nanostructure for electrocatalysis, energy conversion and storage. Carbon, p 118105
13. Zhang Z, Yang S, Li H, Zan Y, Li X, Zhu Y, Dou M, Wang F (2019) Sustainable carbonaceous materials derived from biomass as metal-free electrocatalysts. Adv Mater 31(13):1805718
14. Han J, Huang G, Wang Z, Lu Z, Du J, Kashani H, Chen M (2018) Low-temperature carbide-mediated growth of bicontinuous nitrogen-doped mesoporous graphene as an efficient oxygen reduction electrocatalyst. Adv Mater 30(38):1803588
15. Bilandzija N, Voca N, Jelcic B, Jurisic V, Matin A, Grubor M, Kricka T (2018) Evaluation of Croatian agricultural solid biomass energy potential. Renew Sustain Energy Rev 93:225–230
16. Wang Q, Guo R, Wang Z, Shen D, Yu R, Luo K, Wu C, Gu S (2021) Progress in carbon-based electrocatalyst derived from biomass for the hydrogen evolution reaction. Fuel 293:120440
17. Liu D, Li X, Ma J, Li M, Ren F, Zhou L (2021) Metal-organic framework modified pine needle-derived N, O-doped magnetic porous carbon embedded with Au nanoparticles for adsorption and catalytic degradation of tetracycline. J Clean Prod 278:123575
18. Feng D, Guo D, Zhang Y, Sun S, Zhao Y, Shang Q, Sun H, Wu J, Tan H (2021) Functionalized construction of biochar with hierarchical pore structures and surface O-/N-containing groups for phenol adsorption. Chem Eng J 410:127707
19. Moreno-Castilla C, García-Rosero H, Carrasco-Marín F (2017) Symmetric supercapacitor electrodes from KOH activation of pristine, carbonized, and hydrothermally treated Melia azedarach stones. Materials 10(7):747
20. Lu B, Hu L, Yin H, Mao X, Xiao W, Wang D (2016) Preparation and application of capacitive carbon from bamboo shells by one step molten carbonates carbonization. Int J Hydrogen Energy 41(41):18713–18720
21. Iwazaki T, Yang H, Obinata R, Sugimoto W, Takasu Y (2010) Oxygen-reduction activity of silk-derived carbons. J Power Sources 195(18):5840–5847
22. Manoharan K, Sundaram R, Raman K (2023) Preparation and characterization of hydrogen storage medium (IMO/TPAC) and asymmetric supercapacitor (IMO/TPAC|| TPAC) using imogolite (IMO) and biomass derived activated carbon from tangerine peel (TPAC) for renewable energy storage applications. Int J Hydrogen Energy 48(74):28694–28711
23. Chen Q, Pu W, Hou H, Hu J, Liu B, Li J, Cheng K, Huang L, Yuan X, Yang C, Yang J (2018) Activated microporous-mesoporous carbon derived from chestnut shell as a sustainable anode material for high performance microbial fuel cells. Bioresour Technol 249:567–573
24. Huggins TM, Latorre A, Biffinger JC, Ren ZJ (2016) Biochar based microbial fuel cell for enhanced wastewater treatment and nutrient recovery. Sustainability 8(2):169
25. Abbas SC, Lin C, Hua Z, Deng Q, Huang H, Ni Y, Cao S, Ma X (2022) Bamboo-derived carbon material inherently doped with SiC and nitrogen for flexible supercapacitors. Chem Eng J 433:133738
26. Qiao Y, Zhang C, Kong F, Zhao Q, Kong A, Shan Y (2021) Activated biochar derived from peanut shells as the electrode materials with excellent performance in Zinc-air battery and supercapacitance. Waste Manage 125:257–267
27. Saini S, Chand P, Joshi A (2021) Biomass derived carbon for supercapacitor applications. J Energy Storage 39:102646
28. Guo F, Jiang X, Li X, Peng K, Guo C, Rao Z (2019) Carbon electrode material from peanut shell by one-step synthesis for high performance supercapacitor. J Mater Sci Mater Electron 30:914–925
29. Ramos R, Abdelkader-Fernández VK, Matos R, Peixoto AF, Fernandes DM (2022) Metal-supported biochar catalysts for sustainable biorefinery, electrocatalysis, and energy storage applications: a review. Catalysts 12(2):207

30. Liu J, Wang Z, Li H, Hu C, Raymer P, Huang Q (2018) Effect of solid state fermentation of peanut shell on its dye adsorption performance. Biores Technol 249:307–314
31. Ke L, Chen Y, Wang D, Wang W, Jia S, Mitsuzakic N, Chen Z (2023) Post-modified biomass derived carbon materials for energy storage supercapacitors: a review. Sustain Energy Fuels 7(15):3541–3559
32. Zhong G, Xu S, Chao J, Fu X, Liao W, Xu Y, Liu Z, Cao Y (2020) Biomass-derived nitrogen-doped porous carbons activated by magnesium chloride as ultrahigh-performance supercapacitors. Ind Eng Chem Res 59(50):21756–21767
33. Gang B, Zhang F, Li X, Zhai B, Wang X, Song Y (2021) A ulva lactuca-derived porous carbon for high-performance electrode materials in supercapacitor: synergistic effect of porous structure and graphitization degree. J Energy Storage 33:102132
34. Guo F, Jiang X, Jia X, Liang S, Qian L, Rao Z (2019) Synthesis of biomass carbon electrode materials by bimetallic activation for the application in supercapacitors. J Electroanal Chem 844:105–115
35. Debevc S, Weldekidan H, Snowdon MR, Vivekanandhan S, Wood DF, Misra M, Mohanty AK (2022) Valorization of almond shell biomass to biocarbon materials: Influence of pyrolysis temperature on their physicochemical properties and electrical conductivity. Carbon Trends 9:100214
36. Arenas C, Sotres A, Alonso RM, González-Arias J, Morán A, Gómez X (2022) Pyrolysed almond shells used as electrodes in microbial electrolysis cell. Biomass Convers Biorefinery 12(2):313–321
37. Basu P (2018) Biomass gasification, pyrolysis and torrefaction: practical design and theory. Academic press
38. Tiwari SK, Bystrzejewski M, De Adhikari A, Huczko A, Wang N (2022) Methods for the conversion of biomass waste into value-added carbon nanomaterials: recent progress and applications. Prog Energy Combust Sci 92:101023
39. Plaza MG, Pevida C, Martín CF, Fermoso J, Pis JJ, Rubiera F (2010) Developing almond shell-derived activated carbons as CO2 adsorbents. Sep Purif Technol 71(1):102–106
40. Li M, Ci S, Ding Y, Wen Z (2019) Almond shell derived porous carbon for a high-performance anode of microbial fuel cells. Sustain Energy Fuels 3(12):3415–3421
41. Molahalli V, Sharma A, Bijapur K, Soman G, Chattham N, Hegde G (2024) Low-cost bio-waste carbon nanocomposites for sustainable electrochemical devices: a systematic review. Mater Today Commun, p 108034
42. Chudhary Z, Khera RA, Hanif MA, Ayub MA, Hamrouni L (2020) Walnut. In: Medicinal plants of South Asia. Elsevier, pp 671–684
43. Wang Y, Jiang H, Ye S, Zhou J, Chen J, Zeng Q, Yang H, Liang T (2019) N-doped porous carbon derived from walnut shells with enhanced electrochemical performance for supercapacitor. Funct Mater Lett 12(03):1950042
44. Fu H, Chen L, Gao H, Yu X, Hou J, Wang G, Yu F, Li H, Fan C, Shi Y, Guo X (2020) Walnut shell-derived hierarchical porous carbon with high performances for electrocatalytic hydrogen evolution and symmetry supercapacitors. Int J Hydrog Energy 45(1):443–451
45. Yang H, Tang Y, Huang X, Wang L, Zhang Q (2017) Activated porous carbon derived from walnut shells with promising material properties for supercapacitors. J Mater Sci Mater Electron 28:18637–18645
46. Li X, Qiu J, Hu Y, Ren X, He L, Zhao N, Ye T, Zhao X (2020) Characterization and comparison of walnut shells-based activated carbons and their adsorptive properties. Adsorpt Sci Technol 38(9–10):450–463
47. Wang Y, Zhang M, Shen X, Wang H, Wang H, Xia K, Yin Z, Zhang Y (2021) Biomass-derived carbon materials: controllable preparation and versatile applications. Small 17(40):2008079
48. Yang X, Wang H, Strong PJ, Xu S, Liu S, Lu K, Sheng K, Guo J, Che L, He L, Ok H, Yuan G, Shen Y, Chen X (2017) Thermal properties of biochars derived from waste biomass generated by agricultural and forestry sectors. Energies 10(4):469
49. Yang Z, Ren J, Zhang Z, Chen X, Guan G, Qiu L, Ye Z, Peng H (2015) Recent advancement of nanostructured carbon for energy applications. Chem Rev 115(11):5159–5223

50. Hu C, Qu J, Xiao Y, Zhao S, Chen H, Dai L (2019) Carbon nanomaterials for energy and biorelated catalysis: recent advances and looking forward. ACS Cent Sci 5(3):389–408
51. Rahman MZ, Edvinsson T, Kwong P (2020) Biochar for electrochemical applications. Curr Opin Green Sustain Chem 23:25–30
52. Zhou Y, Leng Y, Zhou W, Huang J, Zhao M, Zhan J, Feng C, Tang Z, Chen S, Liu H (2015) Sulfur and nitrogen self-doped carbon nanosheets derived from peanut root nodules as high-efficiency non-metal electrocatalyst for hydrogen evolution reaction. Nano Energy 16:357–366
53. Zhang J, Zhang M, Zeng Y, Chen J, Qiu L, Zhou H, Sun C, Yu Y, Zhu C, Zhu Z (2019) Single Fe atom on hierarchically porous S, N-codoped nanocarbon derived from porphyra enable boosted oxygen catalysis for rechargeable Zn-air batteries. Small 15(24):1900307
54. Yuan H, Chen H, Li D, Deng L, Chen J, Fan Y, He M, Sun F (2019) Catalytic synthesis and simultaneous co-doping of hierarchically porous carbon with in-situ coated graphene from biomass tar as efficient catalyst for ORR. Electrochem Commun 100:52–59
55. Liu Z, Li Z, Ma J, Dong X, Ku W, Wang M, Sun H, Liang S, Lu G (2018) Nitrogen and cobalt-doped porous biocarbon materials derived from corn stover as efficient electrocatalysts for aluminum-air batteries. Energy 162:453–459
56. Xiao J, Zhang Y, Zhang Z, Lv Q, Jing F, Chi K, Wang S (2017) Self-supported biocarbon-fiber electrode decorated with molybdenum carbide nanoparticles for highly active hydrogen-evolution reaction. ACS Appl Mater Interfaces 9(27):22604–22611
57. Sevda S, Singh S, Garlapati VK, Sharma S, Pandey L, Sreekrishnan TR, Singh A (2019) Sustainability assessment of microbial fuel cells. In: Waste to sustainable energy. CRC Press, pp 313–330
58. Pandey G (2019) Biomass based bio-electro fuel cells based on carbon electrodes: an alternative source of renewable energy. SN Appl Sci 1:1–10
59. Cao TND, Mukhtar H, Yu C-P, Bui X-T, Pan S-Y (2022) Agricultural waste-derived biochar in microbial fuel cells towards a carbon-negative circular economy. Renew Sustain Energy Rev 170:112965
60. Kalinke C, de Oliveira PR, Bonacin JA, Janegitz BC, Mangrich AS, Marcolino-Junior LH, Bergamini MF (2021) State-of-the-art and perspectives in the use of biochar for electrochemical and electroanalytical applications. Green Chem 23(15):5272–5301
61. Sangrulkar P, Gupta S, Kandasubramanian B (2023) Advancements in biochar-based electrodes for improved performance of microbial fuel cells. Bioresour Technol Rep, p 101684
62. Olabi AG, Abbas Q, Al Makky A, Abdelkareem MA (2022) Supercapacitors as next generation energy storage devices: properties and applications. Energy 248:123617
63. Li Z, Guo D, Liu Y, Wang H, Wang L (2020) Recent advances and challenges in biomass-derived porous carbon nanomaterials for supercapacitors. Chem Eng J 397:125418
64. Orooji Y, Han N, Nezafat Z, Shafiei N, Shen Z, Nasrollahzadeh M, Karimi-Maleh H, Luque R, Bokhari A, Klemeš JJ (2022) Valorisation of nuts biowaste: prospects in sustainable bio (nano) catalysts and environmental applications. J Clean Prod 347:131220
65. Haghighi Poudeh L, Berktas I, Ali HQ, Okan BS, Yıldız M (2020) Toward next-generation carbon-based materials derived from waste and biomass for high-performance energy applications. Energy Technol 8(12):2000714
66. Mojapelo NA, Seroka NS, Khotseng L (2023) Macadamia nut bio-waste: an agricultural waste with potential to be used as carbon support material in fuel cell applications. Coatings 13(9):1545

Chapter 9
Agricultural Waste-Derived Carbon Nanomaterials for Biomedical Applications

Jaison Jeevanandam and Michael K. Danquah

Abstract Agricultural waste has emerged as a pressing concern for numerous countries worldwide. While these wastes may be less toxic compared to other forms of waste, their volume poses a significant challenge. They can encroach upon valuable agricultural lands, encumber forests, and even infiltrate residential areas due to their extensive production. Moreover, efforts to recycle these agricultural wastes have, in some cases, resulted in the creation of additional hazardous byproducts. Thus, there arises a compelling need for innovative approaches to harness these agricultural residues for potential material fabrication, particularly with a focus on their applicability in biomedical contexts. In recent times, agricultural waste materials have gained considerable attention among researchers as a valuable resource for the synthesis of carbon-based nanomaterials. Their richness in carbon content, coupled with the need for recycling, presents distinct advantages over conventional methods of carbon-based nanomaterial synthesis, which can have pronounced environmental impacts. This chapter offers a comprehensive overview of various agricultural waste sources and their potential utility in the synthesis of carbon nanomaterials. The discussion extends to exploring the diverse applications of these biosynthesized carbon nanomaterials within the field of biomedicine. Repurposing agricultural waste into advanced nanomaterials with biomedical significance can potentially address both environmental concerns and the growing demand for innovative materials in the healthcare sector. This integration of sustainability and cutting-edge science underscores the transformative potential of agricultural waste in shaping a more sustainable and technology-driven future for various applications, particularly in the biomedicine space.

Keywords Agriculture wastes · Carbon nanomaterials · Carbon quantum dots · Graphene · Biomedical applications

J. Jeevanandam
CQM—Centro de Química da Madeira, Universidade da Madeira, Campus da Penteada, 9020-105 Funchal, Portugal

M. K. Danquah (✉)
Chemical Engineering Department, University of Tennessee, Knoxville, TN 37403, USA
e-mail: mdanquah@utk.edu

N. Talreja et al. (eds.), *Waste-Derived Carbon Nanostructures*, Nanostructure Science and Technology, https://doi.org/10.1007/978-3-031-75247-6_9

9.1 Introduction

Carbon nanomaterials are gaining significance among researchers to be utilized in several applications [1]. The high chemical resistance and excellent mechanical properties with light weight make carbon nanomaterials a potential candidate for biomedical applications [2], such as implants [3], drug delivery [4], and biosensors [5]. These carbon nanoparticles are classified into zero (0D), one (1D), two (2D), and three-dimensional (3D), based on their morphology and dimensions [6]. Fullerene, carbon nanotubes (CNTs), graphene, and carbon nanostructures are examples of 0, 1, 2, and 3D carbon nanoparticles, respectively, and are considered allotropes of carbon [7–10]. Furthermore, fullerenes (hollow 0D structures with hexagonal carbon arrangement) are sub-classified into C60 (a spherical shape with 60 carbon atoms in sp^2 orbitals) [11], C70 (cage shaped with 70 carbon atoms in sp^2 hybridization) [12], and fullerenols (water-soluble fullerenes) [13]. Furthermore, CNTs are classified into single and multi-walled CNTs [14], whereas there are several types of graphene, such as ribbons, sheets, platelets, and flakes, depending on their morphology [15]. It is noteworthy that microwave-assisted exfoliation [16], chemical vapor deposition to unzip CNT [17], Hummer [18], ball milling [19], and green approaches [20] were used to fabricate distinct morphologies of graphene nanostructures. In general, fullerenes are fabricated using arc discharge [21] and laser ablation method [22], whereas CNTs are synthesized via chemical vapor deposition, electric arc discharge, laser ablation, pulsed laser deposition, combined with purification steps, such as gas-phase, liquid-phase, and intercalation procedures [23]. Moreover, common physical and chemical approaches, such as sol–gel [24], co-precipitation [25], hydrothermal [26] and sonoelectrodeposition [27], were also used for the formation of carbon nanoparticles. Even though these synthesis approaches can lead to stable carbon nanostructures, these nanoparticles cannot be utilized in biomedical applications due to the existence of toxic byproducts or functional groups [28]. Hence, there is a high demand for an alternative approach for the synthesis of non-toxic carbon nanostructures to be utilized in biomedical applications.

Green synthesis is a common term for the formation of nanostructures using natural sources of heat/light [29], reduced energy [30], or biological sources [31]. The natural sources are not reliable for a long time, dependent on several external environmental factors, and the resultant nanoparticles are sparsely stable. Therefore, green synthesis of nanostructures utilizing biological organisms, either whole organism or their extract, is termed a biosynthesis approach and is introduced as a potential green reducing/stabilizing source [32]. Bacteria, fungi, algae, or plants are widely used biosynthesis sources [33], where it is difficult to purify the resultant nanoparticle while utilizing whole live microbe or plants for biosynthesis [34]. Biomolecular extracts of microbes and plants are the common biosynthesis agents in recent times, due to their easiness in extraction procedure, and can be stored for a long time to be utilized in future [35]. However, the longer reaction time and specific growth conditions for microbes are the limitations of microbe-mediated synthesis of nanoparticles [36]. Hence, phytochemical extracts from plants have been introduced as a significant biosynthesis source for nanoparticle fabrication [37]. However, the utilization of native or medicinal plants for large-scale nanoparticle synthesis will lead to bioprospecting issues or lead to several ecological imbalances [38]. It can be noted that agricultural waste is a menace to the environment and about one billion tons of agricultural waste are produced every year, globally [39]. It has been established in various reports that careful extraction of potential phytochemicals from these agricultural wastes can be utilized as an alternative to fresh/native plant extracts for nanoparticle synthesis [40]. Thus, this chapter aims to provide an overview of agricultural wastes that are used for the synthesis of specific carbon nanostructures. In addition, the mechanism of nanostructure formation and the future of agriculture waste in nanoparticle synthesis were also discussed.

9.2 Agricultural Waste for Carbon Nanostructure Synthesis

All the dimensional classifications of carbon nanostructures have been reported to be synthesized via extracts of agricultural wastes.

9.2.1 Carbon Quantum Dots

Tan et al. [41] synthesized carbon dots via thermal pyrolysis of sago industrial waste. Generally, starch food industries produce sago wastes, which consist of wastewater, woody bark, and fibrous sago pith, that are usually dumped in nearby water sources and cause an environmental population. Hence, these wastes are used for the fabrication of carbon dots to reduce their environmental impact and also to produce less toxic carbon nanostructures. In this study, the sago wastes were pre-dried and grounded, and heated for 1 h at a temperature range of 250–450 °C without gas flow. Later, 10 g/L of dried sago wastes were dispersed in 10 ml of ultrapure water and were homogenized via sonication process for 2h followed by centrifugation for 20 min at 13,400 rotation per min (rpm), where the carbon dots were present in the supernatant. The resultant carbon dots were 6–17 nm in size and in uniform spherical shape. Similarly, Al-Qahtani et al. [42] utilized rice straw waste for the development of carbon dot-based ink. In this study, the nitrogen-doped carbon dots were prepared using a hydrothermal approach by initially drying the rice straw waste at 80 °C for 1 h. Later, ammonium hydroxide and deionized water were mixed with dried rice straw waste, which was heated at 240 °C for 15 h and centrifuged for 10 min at 10,000 rpm to form carbon dots. The resultant carbon dots were ~ 45 nm in size with spherical morphology and amorphous carbon phase. Likewise, Alfi et al. [43] prepared nitrogen-doped carbon dots using the agricultural waste of sugarcane bagasse. In this study, ammonium hydroxide and cellulose diacetate were mixed with the waste and heated at 240 °C for 14 h, which was later centrifuged for 10 min and dialyzed for 22 h for the formation of carbon dots. The resultant nanodots were about 5–8 nm in size and in spherical shape with an amorphous carbon phase. Further, Omran et al. [44] listed several agricultural peel wastes for the synthesis of fluorescent carbon quantum dots as shown in Fig. 9.1. It has been reported that watermelon, orange, pomelo, grapefruit, lemon, pineapple, onion, mango, and sweet lemon peels can be used for the synthesis of spherical, quasi-spherical, or irregular carbon dots with sizes of 2, 2–7, 2–4, 4.2 ± 0.11, 4.5, 2–3, 15, 3, and 1.5–6 nm, respectively. Furthermore, litchi pulp and banana pseudo-stem wastes were also utilized for the formation of 2–3 and 4.1 nm sized spherical/irregular carbon dots, respectively. It is noteworthy that the carbon dots were synthesized widely using a hydrothermal approach, followed by combination of carbonization-oxygenolysis, heating, and simple filtration process. Moreover, Bandi et al. [45] utilized onion waste, which was rich in dietary fiber, glucose, sucrose, fructose, alk(en)yl cysteine sulphoxide, and fructose-oligosaccharides, for the fabrication of novel fluorescent carbon dots. In this study, onion wastes were treated with ethylene diamine at 120 °C and 15 lbs of pressure in an autoclave for the carbonization and passivation of non-structural carbohydrates to form carbon dots. The results revealed that the carbon dots were ~ 15 nm in size with spherical morphology and a maximum quantum yield of 28%.

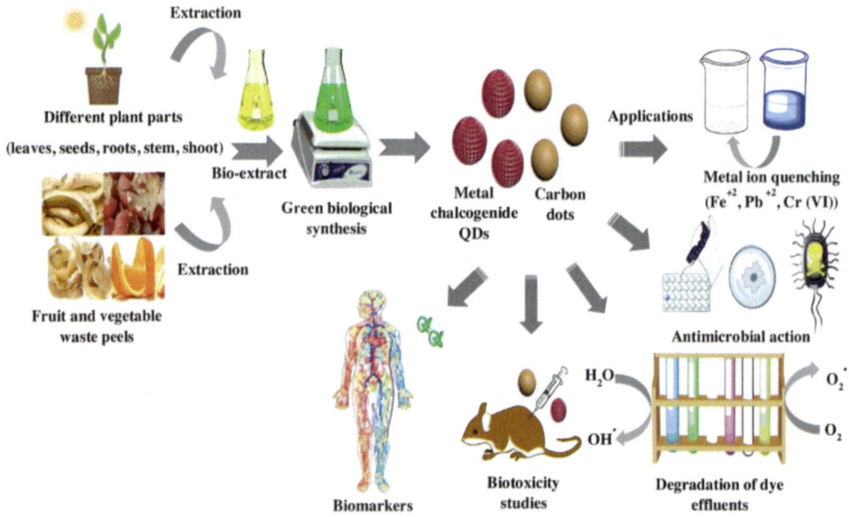

Fig. 9.1 Scheme of carbon quantum dot synthesis using fruit and vegetable waste peels. Reproduced with permission from [44], ©Elsevier, 2021

9.2.2 Carbon Nanotubes

Similar to carbon dots, carbon nanotubes (CNTs) were also widely synthesized using agricultural wastes. Recently, Singh et al. [46] fabricated fluorescent multi-walled CNTs using chickpea peel waste via a pyrolysis approach. In this study, 5 g of grounded chickpea peel was carbonized at 400 °C for 4 h at a heating rate of 3 °C per minute via a pyrolysis process to decompose oxygen, carbon, and hydrogen species in the waste. The precipitation of CNTs was initiated by the oxygen in the nucleation stage, which eventually led to the subsequent growth of multi-walled CNTs. Later, 1 g of pyrolyzed carbon from chickpea peel waste was dispersed in water, sonicated in a probe sonicator for 10 min, centrifuged for 15 min at 11,000 rpm, and dialyzed using a dialysis bag for 24 h. The results revealed that the multi-walled CNTs were 7 nm of diameter and 114 nm of average length with 0.28 nm of lattice spacing, which was identified to be formed with ~ 20 layers of stacked concentric graphene layers. Further, Osman et al. [47] produced multi-walled CNTs and activated carbon using potato peel wastes of lignocellulosic biomass as displayed in Fig. 9.2. In this study, potassium hydroxide was mixed with iron oxalate precursor and melamine via stirring in methanol solution for 4 h and dried for 24 h at 120 °C. Later, the dried samples were subjected to pyrolysis at 600 °C with lateral subsequent pyrolysis at 900 °C at a heating of 2 °C per minute and maintained at a static temperature for 1 h. Finally, the samples were washed several times at room temperature for the removal of impurities and were dried at 80 °C for 24 h to form CNTs. The results emphasized the formation of hydrophilic multi-walled CNTs with a surface area of 52 m^2/g and an average of 40 walls, which adds to their enhanced surface area. Furthermore, Roquia et al. [48] demonstrated the fabrication of carbon nanotubes using walnut shell waste of *Juglans regia.* Initially, activated carbon was prepared by burning walnut shell waste in an electric Bunsen burner, which has led to the formation of charcoal, which is grinned in an electric grinder. The activated charcoal was mixed with zinc chloride under continuous stirring for 24 g, which was later heated for 30 min at 300 °C and then incubated at 600 °C for 60 min. The resultant activated carbon was washed in hydrochloric acid, transferred onto filter paper, and dried in an oven for 5 h at 120 °C. The activated carbon was mixed with ferrocene and benzene, which was later pyrolyzed for 4 h at 700 °C to form multi-walled CNTs. The resultant CNTs were identified to be multi-walled with tubes of closed-end and dome-like structures. It can be noted that single-walled CNTs were difficult to synthesize via agriculture waste, whereas aggregation of CNTs due to biomolecules has led to the easier formation of multi-walled CNTs.

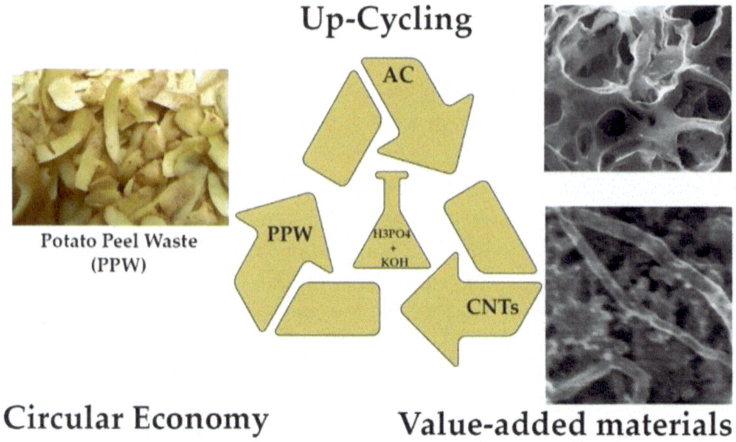

Fig. 9.2 Scheme representing synthesis of multi-walled CNTs with potato peel waste. Reproduced with permission from [47], ©Springer, 2019 with Creative Common Attributions 4.0 (Open access)

9.2.3 Graphene Nanostructures

Graphene is not only a 2D thin film, but it is also fabricated as 3D nanostructures, using agricultural wastes. Hashmi et al. [49] prepared graphene oxide nanoparticles using a mixture of orange peel, rice bran, and sugarcane bagasse agriculture waste. In this study, ferrocene was mixed with carbonaceous powder from standalone wastes and heated for 15 min at 300 °C, whereas the tri-composite agro-wastes of equal weight were mixed with ferrocene and were heated for 15 min at 400 °C for graphene nanoparticle synthesis. The results emphasized that the orange peel, sugarcane bagasse, rice bran wastes, and tri-composite agro-wastes were utilized for the fabrication of 23.99, 15.5, 2.235, and 1.77 nm sized graphene oxide, where the graphite is identified to be 17.17 nm in size. Similarly, Tewari et al. [50] utilized oak (*Quercus ilex*) fruit waste for the fabrication of photoluminescent potassium-doped 2D graphene oxide nanomaterials. In this study, a solvothermal process was utilized for the formation of potassium-doped graphene oxide, where graphitic sheets were grown by mixing oak fruit waste, ethanol, and distilled water, and heated for 4 h at 120 °C. The resultant brown solid residue was crushed, dispersed in water for 30 min via magnetic stirring, and centrifuged for 15 min at 7000 rpm at room temperature for the elimination of impurities and larger particles. The final step is to filter the brown solution using 0.2 μm of nylon paper and dried at 100 °C for graphene oxide nanoparticle fabrication. The results showed the formation of graphene nanosheets which contain 6.8 wt% of uniform potassium arrangement. Likewise, Goswami et al. [51] prepared nanoplatelets of graphene oxide using biomass of rice straw waste as a potential carbon source as shown in Fig. 9.3. In this study, the seven-day sundried straw was washed with water to remove impurities and moisture was eliminated by heating at 80 °C for 48 h in a hot air oven, which was further annealed for 45 min

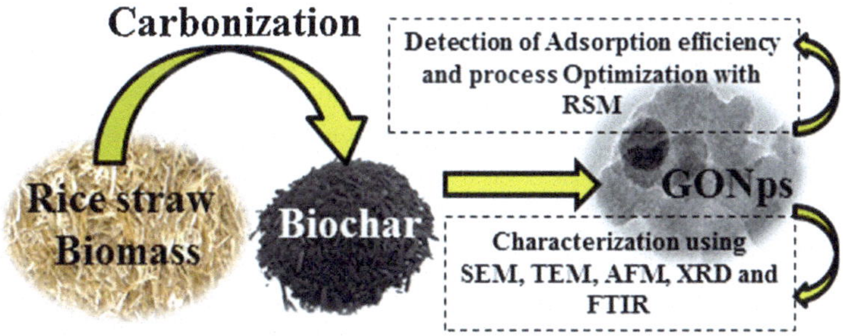

Fig. 9.3 Scheme showing rice straw biomass waste for the formation of graphene oxide nanoplatelets. Reproduced with permission from [51], ©Elsevier, 2017

at 450 °C in nitrogen atmosphere to form biochar via carbonization. The resultant biochar was sieved and crushed to obtain homogenized powder, which was mixed with concentrated sulfuric acid under continuous stirring in an ice tray along with potassium permanganate at room temperature for 30 min. Later, the resultant mixture was incubated in a shaker for 3 h at 45 °C with distilled water and 30% of hydrogen peroxide for the formation of graphene oxide nanoplatelets. The results revealed that the nanoplatelets possess smoothening of crumpled edges in the graphene sheets of 60–100 nm (diameter).

9.2.4 Three-Dimensional Carbon Nanostructures

Apart from 0, 1, and 2D carbon nanostructures, 3D carbon nanoparticles were also fabricated using agriculture waste. Khalafallah et al. [52] listed several biomass wastes, such as orange peel, rice waste, and several other biomass for the formation of porous carbon nanostructures doped with heteroatoms. Furthermore, Boyrazli and Guler [53] prepared carbon nanostructures containing fibers, nanotubes, nano-onion, and amorphous carbon using corn stalk waste. Initially, corn stalk was dried for 24 h at 80 °C and was carbonized for 1–8 h at 673–1073 K, after which it was subjected to heating at 750 ± 10 °C for 6 h. Further, the corn stalk was heated to 950 °C at 50 °C per min of ramp heating rate for 7 min and cooled to room temperature. The carbonized corn stalk was milled using a planetary ball mill for 20 h at 450 rpm in a normal air atmosphere with a ball-to-powder ratio of 40:1, where 5% of nickel was included to increase graphitization via the milling process. Later, the milled powders were annealed for 2 h at 1400 °C with an argon gas atmosphere at 1×10^{-1} l/min to form mixed carbon nanostructures. The results showed that the resultant nanostructures were in the size range of 50–500 nm with agglomeration and distinct morphologies. Furthermore, Zhang et al. [54] fabricated hierarchical silicon/nitrogen-doped carbon/carbon nanotube spheres using rice husk wastes as shown in Fig. 9.4. Initially, the rice husks were refluxed for 2 h with 10% hydrochloric acid for the elimination of metal ion impurities, which were washed in water and dried at 700 °C for 2 h in air with 2 °C per min of ramp heating rate for organic component removal process. Later, pretreated multi-walled CNTs with nitric acid, sodium citrate, and rice husk-obtained silicon were dispersed in N, N-dimethylformamide via ultrasonic treatment for an hour to form CNTs. Also, polyacrylonitrile (PAN) was mixed with the pretreated CNT mixture for 12 h and the mixture was electrosprayed with 0.8 mL/h under 20 kV of working voltage. The resultant electrospray product was heated for 3 h in argon flow at 230 °C with 1 °C per min of ramp heating rate, followed by heating the sample for 2 h at 800 °C for the formation of carbon/CNT spheres. The resultant spheres were identified to be homogeneous as microspheres with 3.2 ± 0.8 μm of size, where silicon nanoparticles were encapsulated on the surface of nitrogen-doped carbon matrix (~ 4 nm) with CNT insertion to form a hierarchical sphere structure. Moreover, Magagula [55] utilized corncob waste biomass for the synthesis of carbon quantum dots along with cellulose nanocrystals and activated carbon. In this study, dried cellulose nanocrystals extracted from corncob were mixed with urea and water and heated at 180 °C for 10 min, for the preparation of nitrogen-doped carbon quantum dots. The results revealed the formation of 1–5 nm sized spherical nitrogen-doped carbon quantum dots. Likewise, Karaman [56] prepared carbon dots, which are co-doped with nitrogen and sulfur, using orange peel waste. In this study, acidic oxidation pretreatment and hydrothermal carbonization processes were used for carbon dot fabrication. Initially, waste orange peels were cleaned with water, ball milled, and filtered, which were dried at 70 °C in a vacuum oven for 24 h. The dried peels were mixed with sulfuric acid, ultra-sonicated for 2 h, and maintained for 24 h at 25 °C for acidic oxidation pretreatment, followed by centrifugation for 10 min at 5000 rpm, neutralized with sodium hydroxide and dried in vacuum oven at 70 °C for overnight. Later, the resultant peels were ultra-sonicated for an hour, mixed with thiourea and water, heated at 200 °C for 12 h in an autoclave reactor to form yellowish brown solution in the supernatant after 15 min of centrifugation at 15,000 rpm. Finally, the solution was filtered, dialyzed, and freeze-dried for two days to form carbon dots. The resultant carbon dots were identified to be quasi-spherical in shape with an average diameter of 4 ± 0.2 nm in size.

9.2.5 Other Novel Carbon Nanostructures

Recently, Chugh and Kaur [56] fabricated nanocomposites of carbon dots and silver using orange peel wastes. In this study, the orange peel waste was collected, cleaned, sun-dried for 48 h, sliced, and ground into fine powder. The peel powder was dissolved in the mixture of ethylene glycol and water to be heated at 180–200 °C in a microwave reactor for a few minutes to form carbon dots. The resultant carbon dot solution was centrifugated at 10,000 rpm, precipitated using acetone, and freeze-dried to yield hydrophilic fluorescent carbon dots. Later, the carbon dot powders were mixed with silver nitrate via constant stirring for 35–40 min at 50–60 °C, which was further centrifuged for 10–15 min at 12,000 rpm for silver-carbon dot nanocomposite

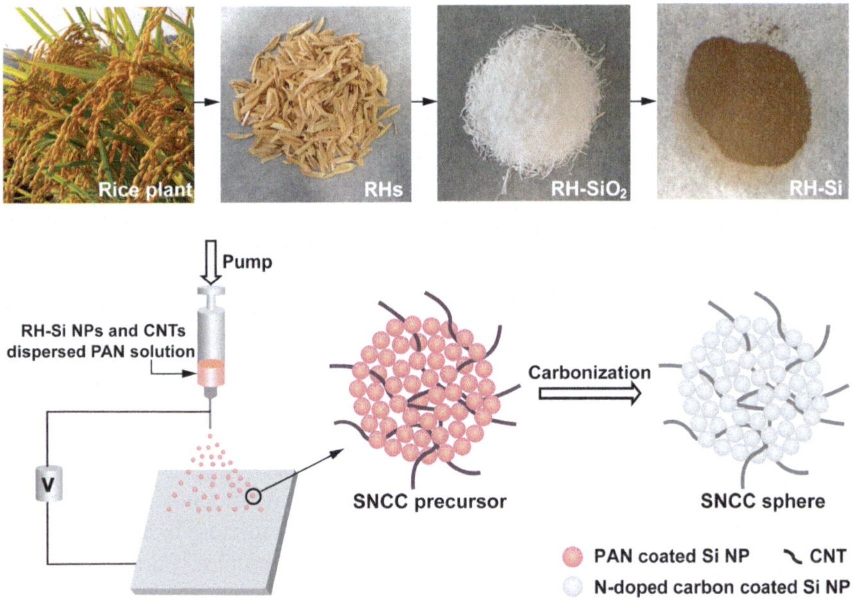

Fig. 9.4 Scheme of silicon/nitrogen-doped carbon/carbon nanotube spheres synthesis using rice husk. Reproduced with permission from [54], ©Elsevier, 2016

synthesis. The results showed that the carbon dots were 5–8 nm in size, whereas the nanocomposites were ~ 15–20 nm in size, which are well dispersed and spheroidal in shape. Similarly, Naeimi et al. [57] synthesized bio-nanocomposite using multi-walled CNTs and zinc oxide nanoparticles using cellulose fibers extracted from *Quinoa* waste. In this study, wastes of *Quinoa* were mixed with sodium chlorite, heated for 2.5 h at 80 °C at pH 4, and filtered to obtain residue, which was washed in water–ethanol mixture, dried at room temperature, mixed with potassium hydroxide for 8 h at 25 °C, washed to make final pH as 7, and freeze-dried to extract cellulose. Later, the cellulose was functionalized with a carboxyl group and immobilized manganese complex on its surface. On the other hand, *Quinoa* waste was boiled for 10 min by mixing them in water, where multi-walled CNTs were mixed for 30 min by adding zinc nitrate at pH 7 and room temperature for 24 h to form nanocomposites. Finally, the manganese complex with oxidized cellulose was mixed with nanocomposites by heating them for 8 h at 80 °C, which was dried for 7 h at 70 °C. The results emphasized that the multi-walled CNTs were 10 nm in size and their size increased to 90 nm after combining with zinc oxide and cellulose. Likewise, Rai et al. [58] demonstrated the preparation of tin-iron oxide-activated carbon magnetic nanocomposite using castor seed pod husk. In this study, activated carbon was synthesized using castor seed pod husk, which was initially washed to remove impurities, dried at 80 °C, crushed into powder, and sieved. The husk powder was mixed with 1:5 w/v of concentrated sulfuric acid overnight, where the activated carbon was obtained by decanting the waste liquid, and dried for 24 h at 120 °C. Later, this activated carbon was used to form nanocomposite via co-precipitation approach, where the tin chloride was dissolved in water and mixed with iron chloride at 80 °C for 4 h along with sodium hydroxide to maintain pH at 8. This mixture was collected as a thick suspension, which was collected, washed with ethanol, and heated at 120 °C for 24 h to form tin-iron oxide particles. Finally, the tin-iron oxide and activated carbon were dissolved in water along with a few nitric acid drops, which were heated for 1 h at 80 °C and dried to form a magnetic nanocomposite. The results showed that the activated carbon possesses several pores, tin-iron oxide with rough surfaces, and nanocomposite with high corrugation and agglomeration. Moreover, Hossein et al. [59] produced a novel iron oxide-silica-activated carbon nanocomposite using *Nigella sativa* oil waste. In this study, oil waste was washed in water, dried for 4 h at 110 °C, grounded, and sieved, which was heated for 3 h at 500 °C, immersed in nitric acid for 4 h at 40 °C, and dried for 2 h at 110 °C to obtain activated carbon. Later, the stopper method was used for the iron oxide nanoparticle, where the resultant nanoparticles were mixed with ethanol, water, and concentrated ammonia for 3 h in a three-necked flask under a nitrogen atmosphere at room temperature. Finally, tetraethoxysilane was added dropwise into the flask with mixture in nitrogen atmosphere via vigorous stirring for an hour to form nanoparticles. Furthermore, the nanoparticles were combined with activated carbon by adding ethanol under ultrasonication for 5 min at temperatures ranging from 30 to 50 °C for 8–24 h, for the formation of nanocomposite. The results showed that the nanoparticles were 16–27 nm in range, whereas the activated carbons were in the size range of 2–50 nm and the nanocomposites were in the shape of spheres with mesoporous cavities. Additionally, Nthwane et al. [60] fabricated carbon hollow nanosphere using orange peels via reflux approach, where gamma-aluminum oxide nanoparticles were coated on its surface using hydrothermal approach as shown in Fig. 9.5. The peel waste was washed, blended as powder, mixed with nitric acid for overnight, and refluxed for 1 h at 105 °C, which was washed in water and ethanol mixture, dried for 12 h at 60 °C for carbon hollow sphere preparation. Later, alumina nanoparticles were mixed with green synthesized carbon hollow nanosphere and water via stirring at room temperature for 24 h, which was washed in the ethanol and water mixture, and dried for 24 h at 70 °C to form a nanocomposite. The resultant nanocomposites were identified to be of porous structures, irregular shapes with almost spherical nanostructures, and 1–2 nm in size.

9.3 Biological Applications of Agriculture Waste-Synthesized Carbon Nanostructures

Recently, various carbon nanostructures were synthesized via agriculture wastes to be utilized for specific biological applications as summarized in Table 9.1 [61]. Roy et al. [62] prepared mesoporous carbon nanospheres using garlic peel wastes via a novel pyrolysis approach. In this study, 31 ± 6.3 nm sized nanospheres were

Fig. 9.5 Scheme of carbon hollow nanosphere synthesis using orange peel waste. Reproduced with permission from [60], ©Springer, 2023 (CC BY 4.0; Open access)

synthesized with a surface charge of -25.2 ± 1.75 mV, indicating their high stability, 0.032 cm^3/g of total pore volume, isotherm for type IV, and 15.2 nm of mean pore diameter. The resultant nanoparticles were identified to possess enhanced gram-selective (gram-negative) antibacterial activity against *Escherichia coli* (*E. coli*) with 480 ± 0.5 μg/ml of minimum inhibitory concentration and 495 ± 0.5 μg/ml of minimum bacterial concentration. Similarly, Selvaraju et al. [63] prepared carbon dots using fruit juice without pulp from *Citrus medica* via hydrothermal process by heating in a reactor for 5 h at 240 °C as shown in Fig. 9.6. In this study, 4.5 nm sized carbon quantum dots with nitrogen as dopants, fluorescent property at 340 nm, maximum intensity at 412 nm of wavelength, and 16.2% of relative quantum yield were fabricated. The resultant carbon dots were hydrophilic with high stability and excellent dispersibility in water along with the ability to inhibit 1.25% (v/v) *Pseudomonas aeruginosa* bacterial growth of lowest concentration level and 0.07% (v/v) of biofilm formation concentration level.

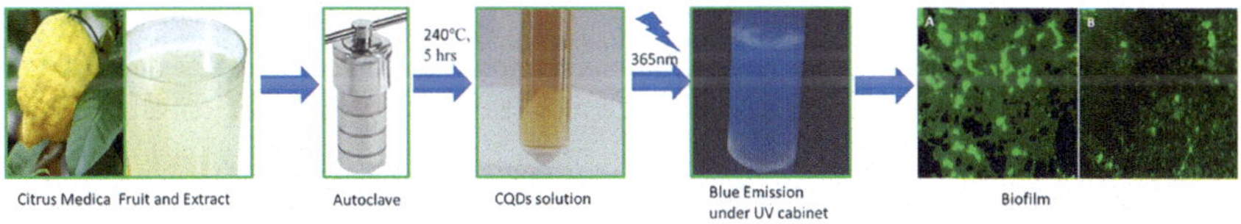

Fig. 9.6 Scheme representation formation of carbon dots using *Citrus medica* fruit juice via hydrothermal approach for blue emission to inhibit bacterial biofilm growth. Reproduced with permission from [63], ©American Chemical Society (ACS), 2022

Table 9.1 List of carbon nanostructures synthesized using agricultural waste for biological applications

Carbon nanostructures	Agriculture wastes	Properties	Biological applications	References
Mesoporous carbon nanospheres	Garlic peel	Size: 31 ± 6.3 nm	Antibacterial activity against *Escherichia coli*	[62]
Carbon dots	*Citrus medica* fruit juice	Size: 4.5 nm	Antibacterial activity against *Pseudomonas aeruginosa*	[63]
Multi-walled CNTs	Chickpea peel	Size: 7 nm	Fluorescence to detect 22RV1 human prostate carcinoma	[46]
Carbon dots	*Actinidia deliciosa* peel	Size: 5 nm	Fluorescence to detect breast cancer, thyroid cancer cells, and mesenchymal stem cells	[64]
Carbon dots	*Punica granatum* fruit waste	Size: 3.5 nm	*Pseudomonas aeruginosa* bacteria and *Fusarium avenaceum* fungi detection	[65]
Carbon dots	*Manilkara zapota* fruit wastes	Size: 1.9 ± 0.3, 4.5 ± 1.25, and 2.9 ± 0.7 nm	Bioimaging of *E. coli*, *Fomitopsis* species, and *Aspergillus aculeatus*	[66]
Carbon dots	*Carica papaya* leaf waste	Size: 0–20 nm	Antioxidant and anti-inflammatory activity	[67]
Carbon dots	*Actinidia deliciosa* fruit peels	Size: 3–9 nm	Fluorescence detection of ferric ions	[68]
Mesoporous carbon nanospheres	Onion peel waste	Size: 2–50 nm	Oral feeding with less toxicity in *D. melanogaster* for 25 days	[69]
Carbon dot-silver nanohybrids	Orange, mango, and *Colocasia* peel	Size: 4–10 nm	Bio-larvicidal activity against mosquito vectors such as *Anopheles stephensi*, *Aedes albopictus*, and *Culex quinquefasciatus*	[70]

Singh et al. [46] recently utilized chickpea peel waste for the fabrication of fluorescent multi-walled CNTs using pyrolysis in an aqueous medium at 400 °C. The study showed the formation of multi-walled CNTs with a diameter size of 7 nm and graphitic spacing of 0.28 nm with functional carboxylic acid, carbonyl, and hydroxyl groups. The resultant nanotubes were identified to possess a 10% blue fluorescence quantum yield, especially in an aqueous medium with a triple exponential decay average fluorescence lifetime of 4.7 ns. These nanotubes demonstrated consistency in exhibiting fluorescence in 22RV1 human prostate carcinoma cell cytoplasm without any cytotoxicity, that are cyto-compatible in immortalized human prostate RWPE1 epithelial cells. Likewise, Atchudan et al. [64] prepared fluorescent carbon dots from kiwi fruit (*Actinidia deliciosa*) peel wastes using a hydrothermal-carbonization approach. The carbon dots were identified to be spherical with an average size of 5 nm, a quantum yield of 18%, and monodispersed homogeneously in an aqueous medium. The resultant nanodots demonstrated low cytotoxicity and high biocompatibility with excellent cell labeling ability in breast cancer, thyroid cancer cells, and mesenchymal stem cells via an in vitro imaging approach. Furthermore, Kasibabu et al. [65] synthesized fluorescent carbon dots using pomegranate (*Punica granatum*) fruit waste via a hydrothermal approach. In this study, the carbon dots were identified to be 3.5 nm in diameter, spherical in shape, well dispersed with amorphous carbon and turbostratic graphitic nanostructures with densely packed alkyl chains. The quantum yield of the resultant carbon dots was 7.6% with the ability to detect *Pseudomonas aeruginosa* bacteria and *Fusarium avenaceum* fungi at laser emission of 405 and 458 nm to emit green and red fluorescence. Furthermore, Bhamore et al. [66] fabricated multi-color emissive carbon dots using *Manilkara zapota* fruit wastes with the help of oxidizing agents, such as sulfuric acid and dihydrogen phosphate, which fine-tune their emission. The resultant nanodots were identified to emit blue, yellow, and green color at 443, 563, and 515 nm with excitation of 350, 440, and 420 nm, and quantum yield of 5.7, 5.2, and 7.9% with average sizes of 1.9 ± 0.3, 4.5 ± 1.25, and 2.9 ± 0.7 nm, respectively. These nanodots possess high biocompatibility to act as a potential bioimaging agent to identify bacterial and fungal cells, such as *E. coli*, *Fomitopsis* species, and *Aspergillus aculeatus*.

Recently, Gudimella et al. [67] utilized leaf waste of *Carica papaya* for the effective synthesis of fluorescent carbon dots with the help of a sand bath approach. In this study, the formation of carbon dots was identified to be due to the presence of cyanogenic glucosides, ascorbic acid, flavonoids, papain, glucosinolates, and tocopherol as functional groups. Additionally, the synthesized carbon dots were monodispersed, uniformly spherical with an average particle size distribution of 0–20 nm along with maximum emission at 530 nm at 440 nm of excitation. The resultant carbon dots were demonstrated to possess free radical scavenging activity with maximal effective concentration (EC50) at 27.6 μg/ml, antioxidant activity of 23 μg/ml, and in vitro anti-inflammatory activity of 15.52 μg/ml, which can be used as an effective therapeutic agent. Moreover, Atchudan et al. [68] established the synthesis of carbon dots with and without ammonium hydroxide using leftover *Actinidia deliciosa* (kiwi) fruit peels using the hydrothermal-carbonization route. In this study, the carbon dots were identified to possess smooth surfaces with monodispersity and particle size of 3–9 nm with average diameters of 5.1 and 5.6 nm and 0.21 nm of d-spacing values, lattice fringes, and graphitic nature. The study also showed that the presence of amine, hydroxyl, and carboxyl functional groups from kiwi fruit peel helps in the formation of carbon dots. The results revealed that the carbon dots possess a quantum yield of 14 and 19% with increased emission intensity from 300 to 360 nm of excitation wavelength, while the emission intensity decreased from 370 to 450 nm. The selective fluorescence of carbon dots helped in the quenching of ferric ions for the detection of iron at a concentration range of 5–25 μM with 0.95 and 0.85 μM detection limits, which will be beneficial to detect iron deficiency and improve human health.

Recently, Bhat et al. [69] prepared mesoporous carbon nanospheres using onion peel waste via pyrolysis at 500–1000 °C and evaluated their toxicological profile using in vivo *Drosophila melanogaster* as a genetic animal model. The study showed that the samples pyrolyzed at higher temperatures have led to the formation of porous spherical-shaped carbon nanomaterials with pore diameters of 2–50 nm and trace quantities of sodium, potassium, magnesium, and calcium. Furthermore, the survival assay revealed that the oral feeding of mesoporous carbon nanoparticles does not exhibit significant toxicity in *D. melanogaster* for 25 days without any behavioral and developmental defects. In addition, Raul et al. [70] fabricated a nanohybrid of carbon dot and silver using peels of orange (*Citrus limetta*), mango (*Mangifera indica*) and *Colocasia* (*Colocasia esculenta* var. *antiquorum Schott.*). The results showed that smaller diameter of carbon dots with 4–10 nm of size, 0.25 nm of layer spacing, and 8–20 nm diameter of silver nanoparticles along with high stability. The resultant nanohybrids were identified to possess bio-larvicide activity with minimum concentrations of 0.40–0.70 mg/L against *Anopheles stephensi, Aedes albopictus,* and *Culex quinquefasciatus* mosquito vectors.

9.4 Future Perspective

It has been reported in various literature that agricultural wastes have been gradually gaining attention among researchers for the formation of carbon nanostructures [71]. In particular, these agricultural wastes have been identified to be a potential reducing and stabilizing agent, which can be a significant alternative for conventional green or plant-mediated synthesis of nanoparticles [72]. The major limitation of agriculture waste is the degradation of biomolecules, due to decomposition over a period of time [73]. These decomposed biomolecules, that contribute to the formation of nanostructures, may affect the stability and morphology of the resultant nanostructures [74]. Until now, there has been no standard deterioration period for the usage of agricultural wastes for the preparation of nanostructures [75]. Hence, there must be a standard protocol in the future for the preparation of nanostructures using agricultural waste [76]. Further, crude extracts of plants or agricultural waste are the main components of the biosynthesis approach to form any nanoparticles [77]. Until now, the specific phytochemicals responsible for nanoparticle formation have not been identified by researchers [78]. The identification of specific phytochemicals required for nanoparticle systems will reduce the usage of large crude phytochemical volumes and extraction of such phytocompounds will reduce the time and effort for nanoparticle production in the future [79]. Furthermore, waste management is the key to the success of agriculture waste-mediated nanoparticle fabrication [72]. Large-scale collection of specifically segregated agriculture waste based on their carbon content must be the priority of waste management in the future to make these alternative biosynthesis a primary approach for nanostructure synthesis in the future [80].

In recent times, computational approaches have been widely employed in the synthesis of nanoparticles [81]. Computational modeling of nanoparticle complex synthesis systems using software such as MATLAB® will help in the characterization and optimization of numerous variables at the same time [82]. It is noteworthy that MATLAB® can be beneficial for the conversion of grayscale to binary images, which helps in the better visualization of nanoparticles [83–85]. Moreover, machine learning is currently gaining significant attention among researchers to be utilized for the formation of nanostructures [86]. Recently, Krasnikov et al. [87] utilized a machine learning approach via an artificial neural network in a semi-industrial aerosol reactor for the synthesis of CNTs. Also, Li et al. [88] utilized machine learning models for the automation of CNT synthesis. Numerous machine learning approaches have been introduced in recent times for the effective synthesis of carbon nanostructures [89–91]. In addition, computation optimization of parameters using design expert® was also under consideration by various researchers for the fabrication of smaller-sized carbon nanostructures with distinct morphologies [92–94]. However, these computational approaches are currently in their infancy stage and are only utilized for lab-scale applications. Thus, these computational methods can be beneficial for the agriculture waste-mediated commercial production of carbon nanostructures in the future.

9.5 Conclusion

In this chapter, we have summarized various agricultural waste sources and their potential utility in the synthesis of carbon nanomaterials. Further, we have extended the discussion on the diverse applications of biosynthesized carbon nanomaterials within the field of biomedicine. Further, repurposing agricultural waste into advanced nanomaterials with biomedical significance can potentially address both environmental concerns and the growing demand for innovative materials in the healthcare sector. However, these agriculture waste-mediated carbon nanostructure formations have been utilized in lab-scale synthesis. Thus, the emergence of agriculture waste management and computational approaches will help to achieve smaller carbon nanostructures and make this approach a potential alternative to the conventional biosynthesis method.

References

1. Speranza G (2021) Carbon nanomaterials: synthesis, functionalization and sensing applications. Nanomaterials 11:967

2. Simon J, Flahaut E, Golzio M (2019) Overview of carbon nanotubes for biomedical applications. Materials 12:624
3. Holmannova D, Borsky P, Svadlakova T, Borska L, Fiala Z (2022) Carbon nanoparticles and their biomedical applications. Appl Sci 12:7865
4. Misra SK, Ohoka A, Kolmodin NJ, Pan D (2015) Next generation carbon nanoparticles for efficient gene therapy. Mol Pharm 12:375–385
5. Asadian E, Ghalkhani M, Shahrokhian S (2019) Electrochemical sensing based on carbon nanoparticles: a review. Sens Actuators, B Chem 293:183–209
6. Roy D, Srivastava AK, Mukhopadhyay K, Namburi EP (2024) 0D, 1D, 2D & 3D Nano materials: synthesis and applications. In: Novel defence functional and engineering materials (NDFEM) Volume 1: functional materials for defence applications. Springer, pp 73–91
7. Fiorito S, Serafino A, Andreola F, Togna A, Togna G (2006) Toxicity and biocompatibility of carbon nanoparticles. J Nanosci Nanotechnol 6:591–599
8. Negri V, Pacheco-Torres J, Calle D, López-Larrubia P (2020) Carbon nanotubes in biomedicine Surface-modified. In: Nanobiomaterials for electrochemical and biomedicine applications, pp 177–217
9. Wang G, Yang J, Park J, Gou X, Wang B, Liu H, Yao J (2008) Facile synthesis and characterization of graphene nanosheets. J Phys Chem C 112:8192–8195
10. Liu H, Zhang L, Yan M, Yu J (2017) Carbon nanostructures in biology and medicine. J Mater Chem B 5:6437–6450
11. Panda J, Patra T, Panda PK, Sahu R, Tripathy BC, Biswal A (2024) Fullerenes: synthesis and applications. In: New forms of carbon. Apple Academic Press, pp 93–122
12. Xie F-F, Chen Z-C, Zhang M, Xie X-M, Chen L-F, Tian H-R, Deng S-L, Xie S-Y, Zheng L-S (2022) Capturing nonclassical C 70 with double heptagons in low-pressure combustion. Chem Commun 58:9814–9817
13. Sharoyko VV, Ageev SV, Podolsky NE, Petrov AV, Litasova EV, Vlasov TD, Vasina LV, Murin IV, Piotrovskiy LB, Semenov KN (2021) Biologically active water-soluble fullerene adducts: Das Glasperlenspiel (by H. Hesse)? J Molec Liquids 323:114990
14. Jeevanandam J, Sundaramurthy A, Sharma V, Murugan C, Pal K, Kodous MHA, Danquah MK (2020) Sustainability of one-dimensional nanostructures: fabrication and industrial applications. In: Sustainable nanoscale engineering. Elsevier, pp 83–113
15. Sattar T (2019) Current review on synthesis, composites and multifunctional properties of graphene. Top Curr Chem 377:1–45
16. Skorupska M, Ilnicka A, Lukaszewicz JP (2021) N-doped graphene foam obtained by microwave-assisted exfoliation of graphite. Sci Rep 11:2044
17. Kang Y-R, Li Y-L, Deng M-Y (2012) Precise unzipping of flattened carbon nanotubes to regular graphene nanoribbons by acid cutting along the folded edges. J Mater Chem 22:16283–16287
18. Zaaba N, Foo K, Hashim U, Tan S, Liu W-W, Voon C (2017) Synthesis of graphene oxide using modified hummers method: solvent influence. Procedia Eng 184:469–477
19. Mondal O, Mitra S, Pal M, Datta A, Dhara S, Chakravorty D (2015) Reduced graphene oxide synthesis by high energy ball milling. Mater Chem Phys 161:123–129
20. Rai S, Bhujel R, Biswas J, Swain BP (2020) A green approach for synthesis of graphene. In: AIP conference proceedings. AIP Publishing, vol 2273
21. Kyesmen PI, Onoja A, Amah AN (2016) Fullerenes synthesis by combined resistive heating and arc discharge techniques 5:1–7
22. Itagaki H, Fujiwara Y, Minowa Y, Ikehara Y, Kaneko T, Okazaki T, Iizumi Y, Kim J, Sakakita H (2019) Synthesis of endohedral-fullerenes using laser ablation plasma from solid material and vaporized fullerenes. In: AIP advances, vol 9
23. Jagadeesan AK, Thangavelu K, Dhananjeyan V (2020) Carbon nanotubes: synthesis, properties and applications. IntechOpen
24. Verma S, Padha B, Singh A, Khajuria S, Sharma A, Mahajan P, Singh B, Arya S (2021) Sol-gel synthesized carbon nanoparticles as supercapacitor electrodes with ultralong cycling stability. Fullerenes, Nanot Carb Nanostruct 29:1045–1052

25. Salarizadeh N, Sadri M (2017) Preparation and characterization of a carbon-based magnetic nanostructure via co-precipitation method: peroxidase-like activity assay with 3, 3′, 5, 5′-tetramethylbenzidine. Nanomed Res J 2:150–157
26. Sedira S, Mendaci B (2020) Hydrothermal synthesis of spherical carbon nanoparticles (CNPs) for supercapacitor electrodes uses. Mater Renew Sustain Energy 9:1
27. Costa JM, de Almeida Neto AF (2020) Ultrasound-assisted electrodeposition and synthesis of alloys and composite materials: a review. Ultrason Sonochem 68:105193
28. Patil SS, Lekhak UM (2020) Toxic effects of engineered carbon nanoparticles on environment. In: Carbon nanomaterials for agri-food and environmental applications. Elsevier, pp 237–260
29. Srikar SK, Giri DD, Pal DB, Mishra PK, Upadhyay SN (2016) Light induced green synthesis of silver nanoparticles using aqueous extract of Prunus amygdalus. Green Sustain Chem 6:26–33
30. Madani M, Hosny S, Alshangiti DM, Nady N, Alkhursani SA, Alkhaldi H, Al-Gahtany SA, Ghobashy MM, Gaber GA (2022) Green synthesis of nanoparticles for varied applications: green renewable resources and energy-efficient synthetic routes. Nanotechnol Rev 11:731–759
31. Shanker U, Jassal V, Rani M, Kaith BS (2016) Towards green synthesis of nanoparticles: from bio-assisted sources to benign solvents: a review. Int J Environ Anal Chem 96:801–835
32. Jeevanandam J, Kiew SF, Ansah SB, Lau SY, Barhoum A, Danquah MK, Rodrigues J (2022) Green approaches for the synthesis of metal and metal oxide nanoparticles using microbial and plant extracts. Nanoscale 14:2534–2571
33. de Jesus RA, de Assis GC, de Oliveira RJ, Costa JAS, da Silva CMP, Iqbal HM, Ferreira LFR (2024) Metal/metal oxide nanoparticles: a revolution in the biosynthesis and medical applications. Nano-Struct Nano-Obj 37:101071
34. Sivakumar A, Suresh V, Sethuraman S, Sivaperumal P (2024) Biosynthesis of zinc nanoparticles from actinobacterium streptomyces species and their biological potential. Cureus 16
35. Qamar SUR, Ahmad JN (2021) Nanoparticles: mechanism of biosynthesis using plant extracts, bacteria, fungi, and their applications. J Molecul Liquids 334:116040
36. Koul B, Poonia AK, Yadav D, Jin J-O (2021) Microbe-mediated biosynthesis of nanoparticles: applications and future prospects. Biomolecules 11:886
37. Ovais M, Khalil AT, Islam NU, Ahmad I, Ayaz M, Saravanan M, Shinwari ZK, Mukherjee S (2018) Role of plant phytochemicals and microbial enzymes in biosynthesis of metallic nanoparticles. Appl Microbiol Biotechnol 102:6799–6814
38. Chaudhary P, Sharma R, Rawat S, Janmeda P (2023) Antipyretic medicinal plants, phytocompounds, and green nanoparticles: an updated review. Curr Pharmaceut Biotechnol 24:23–49
39. Peng X, Jiang Y, Chen Z, Osman AI, Farghali M, Rooney DW, Yap P-S (2023) Recycling municipal, agricultural and industrial waste into energy, fertilizers, food and construction materials, and economic feasibility: a review. Environ Chem Lett 21:765–801
40. Goswami P, Mathur J (2022) Application of agro-waste-mediated silica nanoparticles to sustainable agriculture. Bioresour Bioproces 9:9
41. Tan XW, Romainor ANB, Chin SF, Ng SM (2014) Carbon dots production via pyrolysis of sago waste as potential probe for metal ions sensing. J Anal Appl Pyrolys 105:157–165
42. Al-Qahtani SD, Hameed A, Snari RM, Shah R, Alfi AA, El-Metwaly NM (2022) Development of fluorescent carbon dots ink from rice straw waste toward security authentication. J Mol Liq 354:118927
43. Alfi AA, Alamrani NA, Azher OA, Snari RM, Abumelha HM, Al-Ahmed ZA, El-Metwaly NM (2022) Development of carbon dots sensor dipstick from sugarcane bagasse agricultural waste toward all-cellulose-derived tetracycline sensor. J Market Res 19:4697–4707
44. Omran BA, Whitehead KA, Baek K-H (2021) One-pot bioinspired synthesis of fluorescent metal chalcogenide and carbon quantum dots: applications and potential biotoxicity. Colloids Surf, B 200:111578
45. Bandi R, Gangapuram BR, Dadigala R, Eslavath R, Singh SS, Guttena V (2016) Facile and green synthesis of fluorescent carbon dots from onion waste and their potential applications as sensor and multicolour imaging agents. RSC Adv 6:28633–28639

46. Singh V, Chatterjee S, Palecha M, Sen P, Ateeq B, Verma V (2021) Chickpea peel waste as sustainable precursor for synthesis of fluorescent carbon nanotubes for bioimaging application. Carbon Lett 31:117–123
47. Osman AI, Blewitt J, Abu-Dahrieh JK, Farrell C, Al-Muhtaseb AaH, Harrison J, Rooney DW (2019) Production and characterisation of activated carbon and carbon nanotubes from potato peel waste and their application in heavy metal removal. Environ Sci Pollut Res 26: 37228–37241
48. Roquia A, Khalfan Hamed Alhashmi A, Hamed Abdullah Alhasmi B (2021) Synthesis and characterisation of carbon nanotubes from waste of *Juglans regia* (walnut) shells. Fullerenes Nanotubes Carb Nanostruct 29:860–867
49. Hashmi A, Singh AK, Jain B, Singh A (2020) Muffle atmosphere promoted fabrication of graphene oxide nanoparticle by agricultural waste. Fullerenes Nanotubes Carb Nanostruct 28:627–636
50. Tewari C, Tatrari G, Karakoti M, Pandey S, Pal M, Rana S, SanthiBhushan B, Melkani AB, Srivastava A, Sahoo NG (2019) A simple, eco-friendly and green approach to synthesis of blue photoluminescent potassium-doped graphene oxide from agriculture waste for bio-imaging applications. Mater Sci Eng 104:109970
51. Goswami S, Banerjee P, Datta S, Mukhopadhayay A, Das P (2017) Graphene oxide nanoplatelets synthesized with carbonized agro-waste biomass as green precursor and its application for the treatment of dye rich wastewater. Process Saf Environ Prot 106:163–172
52. Khalafallah D, Zhi M, Hong Z (2021) Heteroatoms doped porous carbon nanostructures recovered from agriculture waste for energy conversion and storage. In: Waste recycling technologies for nanomaterials manufacturing, pp 465–512
53. Boyrazlı M, Güler SH (2020) Synthesis of carbon nanostructures from corn stalk using mechano-thermal method. J Mol Struct 1199:126976
54. Zhang Y-C, You Y, Xin S, Yin Y-X, Zhang J, Wang P, Zheng X-S, Cao F-F, Guo Y-G (2016) Rice husk-derived hierarchical silicon/nitrogen-doped carbon/carbon nanotube spheres as low-cost and high-capacity anodes for lithium-ion batteries. Nano Energy 25:120–127
55. Magagula LP (2022) Synthesis and characterization of agricultural waste carbon-based structures for application in sensing. University of the Witwatersrand, Johannesburg
56. Karaman C (2021) Orange peel derived-nitrogen and sulfur Co-doped carbon dots: a nano-booster for enhancing ORR electrocatalytic performance of 3D graphene networks. Electroanalysis 33:1356–1369
57. Naeimi A, Khoshkam S, Eslaminejad T (2022) Natural cellulose fibers from Quinoa wastes reinforced carbon nanotube/ZnO bio-nanocomposite as a novel recyclable catalyst for oxidation reaction. Polym Bull 79:7795–7808
58. Rai P, Gautam RK, Banerjee S, Rawat V, Chattopadhyaya M (2015) Synthesis and characterization of a novel $SnFe_2O_4$@ activated carbon magnetic nanocomposite and its effectiveness in the removal of crystal violet from aqueous solution. J Environ Chem Eng 3:2281–2291
59. Fekri MH, Tousi F, Heydari R, Razavi Mehr M, Rashidipour M (2022) Synthesis of magnetic novel hybrid nanocomposite (Fe_3O_4@ SiO_2/activated carbon (by a green method and evaluation of its antibacterial potential. Iranian J Chem Chem Eng 41:767–776
60. Nthwane Y, Fouda-Mbanga B, Thwala M, Pillay K (2024) Novel synthesis of γ-Al_2O_3/carbon hollow nanosphere nanocomposite for the optimal removal of Ni^{2+} ions from wastewater and reuse in blood fingerprint enhancement. J Iran Chem Soc 21:101–115
61. Thippeswamy BH, Maligi AS, Hegde G (2021) Roadmap of effects of biowaste-synthesized carbon nanomaterials on carbon nano-reinforced composites. Catalysts 11:1485
62. Roy P, Bhat VS, Saha S, Sengupta D, Das S, Datta S, Hegde G (2021) Mesoporous carbon nanospheres derived from agro-waste as novel antimicrobial agents against gram-negative bacteria. Environ Sci Pollut Res 28:13552–13561
63. Selvaraju N, Ganesh PS, Palrasu V, Venugopal G, Mariappan V (2022) Evaluation of antimicrobial and antibiofilm activity of *Citrus medica* fruit juice based carbon dots against Pseudomonas aeruginosa. ACS Omega 7:36227–36234

64. Atchudan R, Kishore SC, Gangadaran P, Edison TNJI, Perumal S, Rajendran RL, Alagan M, Al-Rashed S, Ahn B-C, Lee YR (2022) Tunable fluorescent carbon dots from biowaste as fluorescence ink and imaging human normal and cancer cells. Environ Res 204:112365
65. Kasibabu BSB, D'souza SL, Jha S, Singhal RK, Basu H, Kailasa SK (2015) One-step synthesis of fluorescent carbon dots for imaging bacterial and fungal cells. Anal Methods 7:2373–2378
66. Bhamore JR, Jha S, Park TJ, Kailasa SK (2019) Green synthesis of multi-color emissive carbon dots from *Manilkara zapota* fruits for bioimaging of bacterial and fungal cells. J Photochem Photobiol, B 191:150–155
67. Kanthi Gudimella K, Gedda G, Kumar PS, Babu B, Yamajala B, Rao BV, Singh PP, Kumar D, Sharma A (2022) Novel synthesis of fluorescent carbon dots from bio-based Carica Papaya leaves: optical and structural properties with antioxidant and anti-inflammatory activities. Environ Res 204:111854
68. Atchudan R, Edison TNJI, Perumal S, Vinodh R, Sundramoorthy AK, Babu RS, Lee YR (2021) Leftover kiwi fruit peel-derived carbon dots as a highly selective fluorescent sensor for detection of ferric ion. Chemosensors 9:166
69. Bhat VS, Kudva AK, Naik HV, Raghu SV, De Padova P, Hegde G (2022) Toxicological profiling of onion-peel-derived mesoporous carbon nanospheres using in vivo drosophila melanogaster model. Appl Sci 12:1528
70. Raul PK, Santra P, Goswami D, Tyagi V, Yellappa C, Mauka V, Devi RR, Chattopadhyay P, Jayaram RV, Dwivedi SK (2021) Green synthesis of carbon dot silver nanohybrids from fruits and vegetable's peel waste: applications as potent mosquito larvicide. Curr Res Green Sustain Chem 4:100158
71. Abioye AM, Ani FN (2015) Recent development in the production of activated carbon electrodes from agricultural waste biomass for supercapacitors: a review. Renew Sustain Energy Rev 52:1282–1293
72. Saratale RG, Saratale GD, Shin HS, Jacob JM, Pugazhendhi A, Bhaisare M, Kumar G (2018) New insights on the green synthesis of metallic nanoparticles using plant and waste biomaterials: current knowledge, their agricultural and environmental applications. Environ Sci Pollut Res 25:10164–10183
73. Capanoglu E, Nemli E, Tomas-Barberan F (2022) Novel approaches in the valorization of agricultural wastes and their applications. J Agric Food Chem 70:6787–6804
74. Aslam M, Abdullah AZ, Rafatullah M (2021) Recent development in the green synthesis of titanium dioxide nanoparticles using plant-based biomolecules for environmental and antimicrobial applications. J Ind Eng Chem 98:1–16
75. Zamani A, Marjani AP, Mousavi Z (2019) Agricultural waste biomass-assisted nanostructures: synthesis and application. Green Proces Synth 8:421–429
76. Rao P, Rathod V (2019) Valorization of food and agricultural waste: a step towards greener future. Chem Rec 19:1858–1871
77. Aswathi V, Meera S, Maria CA, Nidhin M (2023) Green synthesis of nanoparticles from biodegradable waste extracts and their applications: a critical review Nanotechnology for. Environ Eng 8:377–397
78. Shaikh WA, Chakraborty S, Owens G, Islam RU (2021) A review of the phytochemical mediated synthesis of AgNP (silver nanoparticle): the wonder particle of the past decade Appl Nanosci 11:2625–2660
79. Yousaf Z, Saleh N (2018) Advanced concept of green synthesis of metallic nanoparticles by reducing phytochemicals. Nanobotany 17–36
80. Debroy A, Joshi S, Yadav M, George N (2023) Green synthesis of nanoparticles from bio-waste for potential applications: current trends, challenges, and prospects. In: Bio-based materials and waste for energy generation and resource management, pp 431–466
81. Nathanael K, Pico P, Kovalchuk NM, Lavino AD, Simmons MJ, Matar OK (2022) Computational modelling and microfluidics as emerging approaches to synthesis of silver nanoparticles—a review. Chem Eng J 436:135178
82. Shabanzadeh P, Yusof R, Shameli K, Hajalilou A, Goudarzi S (2019) Computational modeling of biosynthesized gold nanoparticles in black *Camellia sinensis* leaf extract. J Nanomater

83. Hunter CN, Check MH,Bultman JE, Voevodin AA (2008) Development of matrix-assisted pulsed laser evaporation (MAPLE) for deposition of disperse films of carbon nanoparticles and gold/nanoparticle composite films. Surf Coatings Technol 203:300–306
84. Cheng X, Li P, Zhu X, Luo C, Zheng L, Hou C, Wu D, Liang H (2021) Role of different dimensional carbon nanoparticles in catalytic oxidation of organic pollutants and alleviating membrane fouling during ultrafiltration of surface water. Sep Purif Technol 270:118804
85. Nasiri N, Mosayebi A (2018) Prediction of resulting properties from adding nanoparticles to polymers by the aid of MATLAB software specialty. J Chem 3:1–9
86. Bamidele EA, Ijaola AO, Bodunrin M, Ajiteru O, Oyibo AM, Makhatha E, Asmatulu E (2022) Discovery and prediction capabilities in metal-based nanomaterials: an overview of the application of machine learning techniques and some recent advances. Adv Eng Inform 52:101593
87. Krasnikov DV, Khabushev EM, Gaev A, Bogdanova AR, Iakovlev VY, Lantsberg A, Kallio T, Nasibulin AG (2023) Machine learning methods for aerosol synthesis of single-walled carbon nanotubes. Carbon 202:76–82
88. Li Y, Wang S, Lv Z, Wang Z, Zhao Y, Xie Y, Xu Y, Qian L, Yang Y, Zhao Z (2024) Transforming the synthesis of carbon nanotubes with machine learning models and automation. arXiv preprint arXiv:2404.01006
89. Han Y, Tang B, Wang L, Bao H, Lu Y, Guan C, Zhang L, Le M, Liu Z, Wu M (2020) Machine-learning-driven synthesis of carbon dots with enhanced quantum yields. ACS Nano 14:14761–14768
90. Ji Z-H, Zhang L, Tang D-M, Chen C-M, Nordling TE, Zhang Z-D, Ren C-L, Da B, Li X, Guo S-Y (2021) High-throughput screening and machine learning for the efficient growth of high-quality single-wall carbon nanotubes. Nano Res 1–6
91. Vivanco-Benavides LE, Martínez-González CL, Mercado-Zúñiga C, Torres-Torres C (2022) Machine learning and materials informatics approaches in the analysis of physical properties of carbon nanotubes: a review. Comput Mater Sci 201:110939
92. Mehta P, Shende P (2024) Collation of fullerenes and carbon nanotubes with genistein for synergistic anti-Alzheimer's activity by amyloid-β deaggregation. J Drug Deliv Sci Technol 91:105205
93. Mohammadian N, Ghoreishi SM, Hafeziyeh S, Saeidi S, Dionysiou DD (2018) Optimization of synthesis conditions of carbon nanotubes via ultrasonic-assisted floating catalyst deposition using response surface methodology. Nanomaterials 8:316
94. Yousefi M, Gholami M, Oskoei V, Mohammadi AA, Baziar M, Esrafili A (2021) Comparison of LSSVM and RSM in simulating the removal of ciprofloxacin from aqueous solutions using magnetization of functionalized multi-walled carbon nanotubes: process optimization using GA and RSM techniques. J Environ Chem Eng 9:105677

Chapter 10
Synthesis and Characterization of Agricultural Waste-Derived Carbon Nanomaterials (AWD-CNMs) for Tissue Engineering Application

Mohamed Abdelmonem, Rathi Devi Nair, Mohamed Ahmed Ibrahim, Romesa Soomro, Abubakar Dantani Meli, Ariyati Retno Pratiwi, Wessam I. Elsisi, and Che Azurahanim Che Abdullah

Abstract The increasing surge in agricultural waste materials (AWMs) presents both environmental challenges and opportunities for sustainable resource utilization. Recent trends have seen a massive transformation of discarded resources into value-added products that could be catered for various applications. One such pioneering avenue looks into the exploration of carbon nanomaterials (CNMs) using agricultural waste residues. This chapter delves into the synthesis and characterization aspects of agricultural waste-derived CNMs (AWD-CNMs) with remarkable biomedical potential. The recent developments in the synthesis methodologies provide insight into the efficient conversion of these waste feedstocks into high-quality CNMs. The significance of customized synthesis procedures in regulating the dimensions, structure, and

---

M. Abdelmonem · R. Soomro · A. D. Meli · C. A. Che Abdullah (✉)
Department of Physics, Faculty of Science, Universiti Putra Malaysia, UPM Serdang, 43400 Selangor, Malaysia
e-mail: azurahanim@upm.edu.my

R. D. Nair
Nanomaterial Synthesis and Characterization Laboratory, Institute of Nanoscience and Nanotechnology, Universiti Putra Malaysia, 43400 Serdang, Malaysia

Nutrition, Metabolism and Cardiovascular Research Centre, Institute for Medical Research, National Institute of Health, Ministry of Health Malaysia, Seksyen U13 Setia Alam, No 1, Jalan Setia Murni U13/52, 40170 Shah Alam, Malaysia

M. A. Ibrahim
Natural Medicines and Products Research Laboratory, Institute of Bioscience, Universiti Putra Malaysia, 43400 Serdang, Selangor Darul Ehsan, Malaysia

W. I. Elsisi
Department of Therapeutic Chemistry, Pharmaceutical and Drug Industries Research Institute, National Research Centre, El-Bohouth Street, P. O. Box 12622, Dokki, Cairo, Egypt

A. R. Pratiwi
Department of Oral Biology, Faculty of Dentistry, Universitas Brawijaya, Malang, Indonesia

N. Talreja et al. (eds.), *Waste-Derived Carbon Nanostructures*, Nanostructure Science and Technology, https://doi.org/10.1007/978-3-031-75247-6_10

surface characteristics of these nanomaterials (NMs) is also emphasized. The characterization techniques are also detailed, emphasizing the importance of the structural and functional attributes that dictate their biomedical applicability. In biomedical research, CNMs hold significant potential for tissue engineering. They support tissue growth by forming structures that encourage cell development and regeneration. In conclusion, this chapter showcases the promising biomedical prospects of agricultural waste-derived CNMs (AWD-CNMs) as a sustainable and environmentally conscious solution.

Keywords Agricultural waste · Carbon nanomaterials · Green synthesis · Characterization · Drug delivery · Tissue engineering

10.1 Introduction

In 1991, S. Iijima conducted a breakthrough discovery of carbon nanoparticles in the form of nanofibers. The nanofibers can be found in a diameter of at least 2.2 nm, with the majority having diameters in the range of 4–30 nm [1]. Ebbesen and Ajayan proposed a method for a large-scale synthesis of carbon nanotubes using the arc-discharge process. Interestingly, the proposal directed all researcher's attention to designing the methods for the mass production of carbon nanotubes (CNTs) and etched a way for various production methods to be developed [2]. The milestone discovery within the domain of carbon nanoparticles was graphene, which was ascribed to the innovative efforts conducted by Geim and Novoselov. The discovery heralded new frontiers and implications within the domain, emphasizing the scientific community's understanding of two-dimensional materials [3]. Graphene is defined as a single-atom-thick layer formed by the hexagonal configuration of the sp2 carbon atoms. The distinctive characteristics of the carbon lattice and sub-lattice make the material unique. Indeed, the effective mass of graphene amounts to 0.77 mg/m^2, demonstrating nearly no mass, and outstanding physical properties, holding the potential to be used in nanotechnology and material science [4]. The latter material has demonstrated the capacity to sustain current densities sixfold greater than copper, marking a significant advancement in material science. The past decade enabled the rapid development of carbon-based nanoparticles, largely thanks to their unique properties and implications [5].

As carbon is the biogeosphere's most abundant element, it appears to offer great potential for manufacturing materials with far-reaching applications [6, 7]; graphene (a two-dimensional layer of hexagonal carbon atoms) is certainly one compared to other materials. Depending on the application, this material can take very different forms from zero-dimensional fullerenes and one-dimensional CNTs to three-dimensional graphite stacks. Figure 10.1 exemplifies the structural diversity of CNMs concerning their dimensionality configurations. Moreover, the modification of graphene with various functional groups improves its properties which

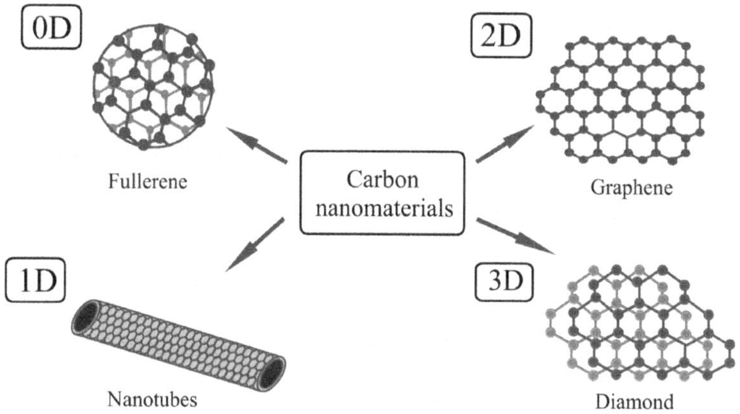

Fig. 10.1 Structural diversity of CNMs: fullerenes (0D), graphene (2D), nanotubes (1D), and diamond (3D)

are specific to an application. Graphene oxide, characterized by a random distribution of oxygen-containing functional groups on a primarily sp^2 hybridized carbon lattice, can undergo reduction processes to yield graphene, thus exemplifying the adaptability and potential of carbon-based materials in technological advancements. Experimental investigations have revealed this thermal conductivity is altered by more than 50% the degree of stratification in graphene. The thermal conductivity thus declines from about 2800 W/mK as the number of atomic graphene layers grows and to ~ 1300 W/mK when expanding a two-layered (bi-layer) structure into one made up of four atomic layers [8, 9]. This observation highlights the profound impact of atomic layering on the material properties of graphene, underscoring the intricate relationship between structural composition and functional characteristics in nanoscale materials [10].

Green synthesis is found to be a clean, efficient, and environmentally sustainable method for generating NMs, with plant extracts playing a pivotal role in this process. These botanical extracts serve as biogenic mediators, facilitating the formation of NMs through a series of biochemical reactions. The shape, size, and overall morphology of the nanoparticles are intricately influenced by the active compounds and precursors present within the plant extracts, such as various metal salts. Additionally, the use of plant extracts in green synthesis imbues the resulting NMs with a range of beneficial properties, including antimicrobial [11], reducing, and stabilizing effects. These advantageous characteristics are attributed to the active constituents within the plant extracts, a revelation that underscores the potential of botanicals in the sustainable production of NMs. This approach not only aligns with green chemistry principles but also opens up new avenues for leveraging the vast diversity of plant-based compounds in nanotechnology [12]. One of the most intriguing and environmentally sustainable methodologies in green synthesis involves the utilization of plant or food waste to fabricate NMs. This process typically entails the extraction of specific chemical compounds from the organic waste material. The procedure involves drying the plant or food remnants, followed by mechanical processing, such as grinding or chopping. Subsequently, the processed material is immersed in hot water for a designated period, after which it undergoes filtration. The resulting extract is then preserved at 4 °C for further use. This approach not only exemplifies an eco-friendly strategy for nanomaterial synthesis but also promotes the recycling of biological waste, aligning with principles of sustainability and environmental stewardship [13].

10.2 Synthesis of CNMs from Agricultural Waste

Agriculture stands as a pivotal and consistent sector, primarily as it furnishes raw materials essential for the food and feed industries. The cultivation and harvesting processes yield agricultural by-products, which, despite their potential utility, often become environmental burdens due to voluminous accumulation and suboptimal disposal practices. Consequently, there is a burgeoning interest in repurposing such agricultural residues—encompassing sugarcane bagasse, banana peels, watermelon rinds, orange peels, and pineapple leaves—into CNMs. Carbon, ranking as the 15th most prevalent element within the Earth's crust, exhibits significant and adjustable luminescence, with its quantum effects holding substantial fundamental and technological significance. Notably, CNMs, including carbon nanotubes, fullerenes, and graphene, are lauded for their unparalleled physical and chemical attributes, such as superior tensile strength, remarkable corrosion resistance, and outstanding electrical and thermal conductivity and stability. Furthermore, carbon quantum dots (CQDs) represent an emergent category within CNMs, distinguished by their potent and modifiable fluorescence emission characteristics. In the context of green synthesis, biomass sources, such as wood, foliage, and low-value biomaterials like plant husks, serve as precursors for creating carbon-based materials. This review acknowledges the criteria for green synthesis, which includes the use of renewable feedstocks, process attributes, or their combinations. Notably, employing biomass as the initial material in green synthesis can obviate the need for metal catalysts in the production of CNMs. The specific reaction conditions and starting materials dictate the resultant morphologies of CNMs, which can range from nanofibers and nanotubes to nanoporous structures. For example, graphitic carbon nanostructures in various allotropic forms, exhibiting a coiled morphology, have been successfully synthesized through a precipitation process at 900 °C, involving cellulose hydrothermal treatment [6]. Moreover, the synthesis of carbon nanosheets can be done from coconut coir-derived lignocellulosic biomass through hydrothermal carbonization and then pyrolysis [14]. The influence of clay mineral particles has been thoroughly examined in numerous studies, where their role in various chemical and physical processes has been investigated to understand their properties and potential applications [15, 16]. Cellulose derived from husk has been utilized to fabricate carbon hollow nanostructures through a triphasic acid digestion method, subsequently undergoing carbonization and high-temperature pyrolysis under CO_2 laser at 2200 °C [17].

10.2.1 Green Synthesis of CNTs

To enhance the quality and yield desired for CNTs production, novel chemical vapor deposition (CVD) techniques have been exploited. Such advancements are radio frequency (RF), enhanced CVD, plasma-enhanced CVD (PECVD), and microwave-enhanced CVD methods [18, 19]. Moreover, spray pyrolysis has been reported as a valid alternative method to produce CNTs on a large scale [20]. The conversion process is very attractive considering its relatively simple processing technology, low-cost apparatus required, and lack of dependence on reducing agents or high vacuum. This approach is particularly well suited for large-scale commercial manufacturing. The technology becomes more interesting when combined with renewable biomass as the initial material. One example of this is the use of wood sawdust, which has been successfully integrated into a spray pyrolysis process, producing large yields and high-quality CNMs. This advancement has made it possible to produce both carbon fibers and nanotubes [21]. Hydrogen was produced as a by-product, along with a small amount of hazardous gases, such as NO_2 and CO. Significantly, the synthesis of CNMs with a diameter of 50 nm was made possible by operating at a relatively low temperature of 750 °C, which is notably lower than the 1200 °C necessary for CVD in the formation of carbon nanotubes (CNT/CE) [22]. Oils are used as precursors in the synthesis of both multi-walled CNTs (MWCNTs) and single-walled CNTs (SWCNTs), which have been successfully synthesized by spray pyrolysis. The pyrolysis of neem oil has been employed to prepare MWCNTs with vertically aligned and regular morphology [23].

Many CNTs have been synthesized using camphor ($C_{10}H_{16}O$) from a crystalline latex of the tree, *Cinnamomum camphora* [20]. *Cinnamomum camphora* is primarily distributed in subtropical countries, such as Japan, India, China, and Indonesia. Camphor was carbonized at 875 °C in an argon atmosphere to yield multiwall tubes, of diverse types. These nanotubes are a mixture of aligned CNTs, MWCNTs, and SWCNTs [24]. CNTs that are arranged in a vertical orientation are known as vertically aligned carbon nanotubes (VACNTs). These are 1D carbon-based materials anchored onto a solid substrate. VACNTs are not randomly oriented like the rest of CNTs-based materials and have thus a well-defined geometric orientation. By using camphor as a starting material, just a small amount of catalyst is needed. This again prevents the formation of undesired amorphous carbon, avoiding a subsequent heat treatment step after deposition as typically in thermal decomposition methods [25]. A precursor formed by a mix of chicken fat and ferrocene, similar to the mixture with vegetable oil, is also adequate for the production CNTs [26]. Further, plant extracts like walnut (*Juglans regia*), neem (*Azadirachta indica*), garden grass (*Cynodon dactylon*), and rose (*Rosa*) have been employed for the production of CNTs. A large number of active carbons in these extracts find a dual role as carbon source and catalyst. The CVD of nanocarbon using walnut extract remarkably performs better than those made by the plant-extracted methods which added hot filament and iron substrates with pyrolytic process. Moreover, elevating the temperature to 800 °C resulted in a more pronounced formation of a specific kind of carbon structure, CNBs. CNBs were confirmed to be present by multiple techniques, such as scanning electron microscopy, infrared (IR) spectroscopy, and Raman spectroscopy [27]. Subsequently, the carbon powders, using a low-temperature pyrolysis method, were employed for synthesizing 27–31 nm diameter type of CNTs. $NiCl_2$ was being used in this process as the catalyst and olive (*Olea europaea*), and coconut oils were being utilized as bio-based carbon sources to make sustainable materials. The technique successfully produced uniform SWCNTs with olive oil, while carbon rods were formed using coconut oil. This can be ascribed to the presence of saturated fats in a large part (82.5%) of coconut oil, which are not usually good solvents for SWCNTs. These saturated fats are less reactive than unsaturated hydrocarbons [28]. Simply spray pyrolysis was used to synthesize 0.79–1.71 nm diameters SWCNTs in high-quality preparations. The method used was a combination of iron/cobalt (Fe/Co) catalyst, which was impregnated in high-silica-zeolite-based support. After that, eucalyptus oil was decomposed catalytically at 850 °C [28].

10.2.2 Green Synthesis of Carbon Nanodiamonds

Nanodiamonds (NDs) are a unique type of CNMs that have a tetrahedral arrangement of carbon atoms at the nanoscale. The accidental discovery of NDs occurred in 1963 during an experiment involving the shock compression modification of carbon in a blast chamber, leading to the formation of this distinct allotrope [29]. During the large-scale production of NDs, the detonation method is used in a closed chamber that is either filled with an inert gas or water (ice) to create a sudden shift in pressure caused by an explosion. The synthesis process is categorized as dry or wet depending on whether a gas or liquid coolant is used, respectively. The sudden increase in pressure results in the formation of a by-product that contains diamond particles of around 4–5 nm in size, as well as other carbon allotropes and impurities [30]. NDs produced by the detonation process have oxygen-containing functional groups attached to their surface, such as –OH and –COOH. The surface chemistries of NDs enable further functionalization procedures, which are crucial in enhancing the biological, physical, and chemical properties of NDs, hence expanding their range of industrial applications [31, 32]. In a study, *Moringa oleifera* leaves were used for the bioreduction of chemically prepared NDs. Initially, NDs were dispersed in distilled water and mixed with an extract obtained from ground Moringa leaves, followed by overnight incubation, which led to the reduction of NDs, as indicated by a change in solution color. This method leverages the plant's biochemical constituents, offering an eco-friendly and efficient alternative to conventional synthetic processes [33].

10.2.3 Green Synthesis of Carbon Quantum Dots

The quantum dot is one of the important allotropes in CNMs [34]. QDs are typically around 1–5 nm in diameter and host a wide range of properties that make them valuable. It has advantages, such as better photostability, high luminescent brightness, and broad absorption band compared with other carbon-based nanostructures. In addition to the high quantum yield, long fluorescence lifetime and surface functionalization using biological entities with photostability have enabled QDs for a myriad of luminescence applications [35, 36]. The chemical methodology for the synthesis of fluorescent CQDs includes an environment-friendly route using biogenic resources like pomegranate extract, D-glucose, and aqueous beetroot extract [37, 38]. A by-product from the pyrolysis of coconut milk waste was used as a precursor for synthesizing CQDs. Coconut oil, known for its high levels of saturated fat with lauric acid forming around 50%, is produced by thermal pyrolysis of coconut milk. During the thermal pyrolysis process, a black residue typically considered waste can serve as a precursor for CQD production through additional pyrolysis. Water-dispersible CQDs are then prepared from this process without the need for surface passivation or acidic reagents. Tunable blue-white light-emitting CQDs to selectively detect Fe^{3+} ions in coconut milk can be achieved via varied pH and thermal conditions [39]. Furthermore, *Azadirachta indica* (neem) leaf extract was also used for the synthesis of CQDs in a one-step hydrothermal treatment process [40]. This method is very eco-friendly and low cost, having CQDs showing huge fluorescent quantum yield up to 27.2% [41].

10.2.4 Green Synthesis of Graphene

A graphene lattice consists of a honeycomb defined by the simple structure through a sp^2 hybridization neutral carbon–carbon network [42]. Graphene synthesis is conducted using both top-down and bottom-up methods. Mechanical and chemical techniques are noteworthy among the various methods. More specifically, chemical methods involve the exfoliation of graphite with silicon carbide (SiC) (SiC) to exfoliate graphite [43], and titanium carbide (TiC) is used to synthesize epitaxial graphene [44] and tantalum carbide (TaC). Moreover, the unzipping of CNTs [45], alongside the use of metals, such as cobalt (Co), nickel (Ni), platinum (Pt), copper (Cu), iridium (Ir), and ruthenium (Ru) [46, 47], has been explored. Additional approaches include solvothermal synthesis [48], CVD, and organic synthesis methods [49]. Furthermore, the reduction of graphene oxide (GO) is a well-established route to obtain graphene [50]. Ruan et al. described the use of low-priced carbon feedstocks, such as food, agro-waste, and insects, for creating high-quality single-layer graphene sheets. It is directly manufactured on a copper foil surface under an argon atmosphere, and no further purification is required to achieve the necessary graphene quality [51]. Besides, the use of an ammonia solution as a reducing agent allows glucose to be converted into a graphene nanosheet utilizing glucose oxide reduction [52]. Kim et al. used dextran to reduce the GO into graphene. This method is particularly eco-friendly as dextran does possess biocompatible properties [53]. In addition, triethylamine and Ginkgo biloba extract are effective reagents for reducing graphene in green chemical synthesis. In addition, the basic amino acid glycine has been used to reduce GO [54].

10.2.5 Green Synthesis of Nanobiochar

In the framework of nanotechnology, due to the environmentally friendly synthesis techniques and energy-saving modes, a new type of innovative material, namely nanobiochar, has been obtained. In a recent study, a special graveness has been given to physical and chemical properties, which, according to the research, are peculiar to nanobiochar particles that are much smaller than macrochar. These properties include a great surface area, a large volume of pores, a relatively smaller hydrodynamic radius, a more negative zeta-potential, as well as much more oxygen-containing functional groups and carbon defect. The presence of nanobiochar achieves a great amount of ROS; hence, it activates the ability of the material to be adsorbed. It will be demonstrated further how the presented properties make their properties substantially similar to the black carbon, which is still eligible for the included properties [55, 56]. Various types of agricultural waste, wood and forest residuals, municipal solid waste, animal manure, and sewage sludge are the most frequently used feedstock for biochar preparation [57]. Moreover, the agriculture leftovers and residues, such as sugar cane bagasse, rice husk, and peanut shell, as well as rice straw, bamboo, grass, oil palm empty fruit bunch, soybean stover, and the wood of forest and associated industries are used for the production of nanobiochar. The properties of nanobiochar, similar to macrobiochar conditions, are determined by the applied feedstock. It was noted that over 98% of the nanobiochar variants are obtained due to mechanical grinding. Genovese et al. and Li et al. have represented the usage of more cost-effective agricultural waste, such as natural layered nanosheets obtained from corn cob and soybean stover [58, 59]. Their innovative synthesis technique was based on pre-hydrothermally treating the biomass and then, substring to the thermal-chemical flash exfoliation, which abstains the grinding [59–61]. There are also available novel preparation techniques, such as microwave pyrolysis. Besides, the softwood biomasses have been sufficient for the efficient obtainment of nanobiochar, as softwood chips and hemp stalks have been cited as being possible for the nanobiochar powder preparation [62]. Moreover, the hemicellulose, which is present in agricultural waste, is the most frequent in the woody biomass, and it is responsible for the lignin percentage. Biomass that has been torrefied and has a high hemicellulose concentration exhibits superior grindability in comparison with biomass with a higher lignin percentage. Torrefaction of biomass takes place under oxygen-depleted conditions at temperatures below 300 °C, resulting in the production of bio-coal. On the other side, as the temperature is up 300 °C, the biochar is produced [63]. The agricultural waste feedstocks with high hemicellulose concentration may introduce favorable grindability and the wild range of particle sizes [64].

10.3 Characterization Techniques for CNMs

In general, it should be noted that NMs possess diverse properties including magnetic, optical, electrical, mechanical, as well as chemical and physical, which differ significantly from atoms and bulk materials. Hence, it can be defined that the factors can be attributed to the differences, and these factors involve nanosize, surface area and composition, the surface/charge interaction as well as the crystallography [65–68]. Most important, it should be noted that the purity and performance of the nanoparticle are determined by its elemental or chemical composition. It is worth mentioning that there is a vast variety of characterization techniques involved in the study of nanotechnology [66].

One of the primary methods is the particle size measurement using the dynamic light scattering technique. Apart from this, electron microscopy (scanning and transmission) also looks into size measurement as well as the distribution pattern [66]. Furthermore, the relationship between the surface area-to-volume ratio and the characteristics and performance of a nanomaterial was found to have a substantial influence, often assessed through the application of the Brunauer–Emmett–Teller (BET) technique [68]. Furthermore, the relationship between the surface area-to-volume ratio and the characteristics and performance of a nanomaterial was found to have a substantial influence, often assessed by the utilization of the BET approach [66]. CNMs encompass fullerenes, graphene, carbon black, carbon nanohorns, NDs, carbon nanofibers, CNTs, QDs, as well as CDs. A review by Mansuriya and Atlintas [69] discussed the commonly studied characterization techniques, with a focus on CDs. The techniques specified are listed in Fig. 10.2.

The following Table 10.1 summarizes the common techniques and their principles of action that have been applied in the evaluation of CNM properties.

Table 10.2 presents the available studies on various CNMs synthesized using agricultural waste materials, focusing on the characterization aspects of these NMs.

10.4 CNMs in Tissue Engineering and Regeneration

The disciplines of regenerative medicine and tissue engineering integrate concepts from engineering, chemistry, biology, and materials science to create functional tissues. These tissues aim to repair or substitute damaged organs and tissues in the human body. Tissue engineering, a subset of regenerative medicine, encompasses the utilization of specific cell types, engineered biomaterials, and biochemical factors with three primary objectives: (1) enhancing cell proliferation to reinstate

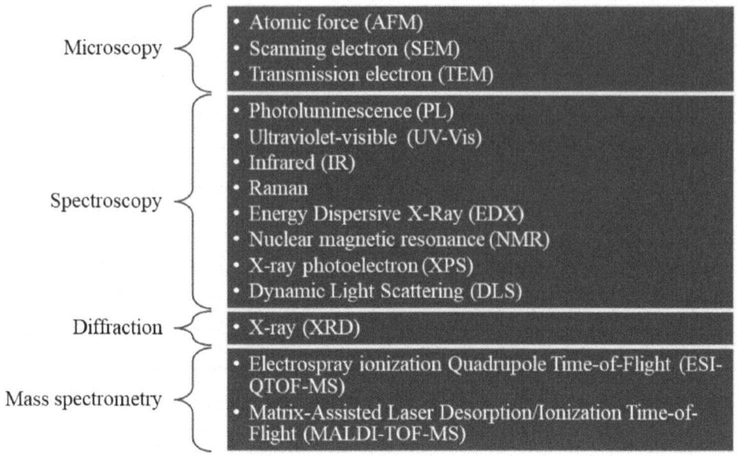

Fig. 10.2 Overview of characterization techniques of CNMs/CDs. Adapted from [69]

normal physiological functions, (2) encouraging the regeneration of organs previously deemed irreparable, and (3) cultivating organs or tissues outside the body for subsequent surgical implantation. In essence, the primary aim of tissue engineering extends beyond merely generating tissue to fill defects; it involves creating fully functional biological replacements that can seamlessly integrate with the body's systems and fulfill essential biological functions [102]. In order to successfully develop a tissue, the process often makes use of scaffolds, cells, and bioactive chemicals to promote cell attachment, proliferation, and differentiation. The goal is to imitate the natural extracellular matrix. However, hurdles like reproducing the complex structure of the natural extracellular matrix, guaranteeing biocompatibility, and obtaining adequate vascularization in the engineered tissue are major obstacles. Addressing these challenges, the potential of CNMs has emerged as a focal point of recent research.

CNMs represent a transformative advancement in the field of nanotechnology, with their impact spanning across a multitude of disciplines, including biomedicine. These materials, defined by their nanometer-scale dimensions, exhibit exceptional properties that diverge significantly from those of their bulk counterparts, capturing the interest of researchers and clinicians alike. In the fields of tissue engineering and regenerative medicine, CNMs are recognized as crucial elements. They hold the potential to tackle significant biomedical issues, providing novel approaches for the repair and regeneration of impaired tissues [103]. The extensive range of CNMs includes diverse structures, each with distinct features: 0D particles display quantum mechanical properties, 1D fibers boast unmatched tensile strength, 2D sheets demonstrate superior electrical conductivity and surface area, and 3D networks provide strong mechanical strength and high porosity. This structural diversity underpins the versatility of CNMs, allowing for their tailored application in different aspects of tissue engineering [104, 105]. A critical factor driving the integration of CNMs

Table 10.1 Common techniques and their principles of action have been applied in evaluation of CNM properties

Technique	Principles/uses	References
AFM	• High-resolution imaging with scanning probe approach • It provides the topographic view of the surface sample • It comes from the tip-sample interface interactions	[70, 71]
SEM	• Provides a spatially resolved capture of images • The retrieved image provides morphology, topography, chemical composition, granular orientation, crystallographic information, etc.	[70]
TEM	• Obtains chemical information and images at atomic-level spatial resolution • Quantifies particle size exactly, and morphology or composition information as well • The lattice and surface imperfections were studied in high-resolution TEM	[70, 72]
PL, UV–Vis	• Examines the optical properties • Used mostly to analyze photoluminescence lifetime	[70, 73, 74]
UV–Vis	• Analyzes the optical properties	[70, 73, 74]
IR	• Describes the chemical functionalities • Probes the carbonyl and hydroxyl groups of CNM • Also determines the heteroatoms doped on their surface • Fourier-transform IR characterizes CNMs with hydroxyl, carboxyl, carboxylic acid, ether, or epoxy functionalities	[39, 72, 75–78]
Raman	• The types of carbon states are most commonly determined by non-destructive and therefore non-invasive ways	[41, 79–81]
EDX	• Analyzes the elemental composition and purity	[82–84]
NMR	• Furnishes additional and valuable structural data • The ability to detect the type of chemical bonds that are taking place on a surface, elemental composition as well as various functional groups present	[69, 70, 85, 86]
XPS	• Original application in the analysis of surface electronic states, elemental composition, and other aspects related to surfaces • Spatial resolution is too low to resolve single nanoparticles	[70]
DLS	• Used for determination of particle size	[70]
Zeta potential	• Determines surface charge and particle size • Characterization of chemical functionalization on the surface of CDs	[84, 87]
XRD	• Based on the diffraction pattern, it represents and determines crystalline structure, phase purity, and particle size • Inefficient for characterizing amorphous CDs	[70, 88–96]
BET	• Determination of surface area	[97]
MS	• Chemical structure clarification	[70]

Table 10.2 Characterization of CNMs synthesized using agricultural waste materials

Waste materials	CNM	Technique	Findings	References
Papaya peels	Reduced graphene oxide–carbon dots nanocomposites (RGO-CDs-NCs), Carbon dots (CDs)	XRD	• The merger of the components formed a structure with graphitic properties • The narrower interlayer spacing present in rGO than GO is evidence of GO reduction to rGO • The barely visible carbon plane-related CD peak in the composites should be attributed to the low diffraction intensity and minor presence of CDs in rGO	[98]
		FT-IR	• The spectrum of the CDs indicated that the CDs' surface contained a high number of different functional groups • The O–H stretching vibrations are responsible for the most prominent absorbance peak at 3370 cm^{-1} • C–H stretching vibrations can be seen at 2962 cm^{-1}, C=C stretching vibration at 1656 cm^{-1}, and N–H bending vibration at 1386 cm^{-1} • This absorption band at 1068 cm^{-1} is associated with C–N stretching vibrations • The spectrum ranges from 1465 to 1050 cm^{-1} revealed the existence of epoxy C–O groups and the symmetric stretching vibrations of COO–. These different functional groups made CDs more soluble in water • It was proved that the reduction or establishment of oxygen-containing functional groups peak intensity in the GO, such as C=O, COO–, C–O–C, and –OH, in the rGO in the composite was accurate	

(continued)

Table 10.2 (continued)

Waste materials	CNM	Technique	Findings	References
		TEM	• The carbon dots were uniformly distributed on the graphene sheets indicating an efficient synthesis of graphene and CDs • The size of the CDs varied from 1 to 8 nm, and its average diameter was 4.5 nm, indicating that it has a consistent and narrow-distribution	
		SEM	• This image displays near-spherical CDs accumulated on the rGO folds	
		UV–Vis	• Two clear absorption peaks at 221 and 332 nm indicate $\pi \rightarrow \pi^*$ and n$\rightarrow$ π^* transition in the aromatic sp2 regions, e.g., –C=C– and –C–C–, of the graphitic core and sp3 hybridized regions, e.g., C=O, C–N, or –C–OH from the carboxyl/amine group of the CDs • There was no strong peak observed in the case of rGO, and instead, a new robust peak appeared at 282 nm	
		PL	• A strong peak at 470 nm, known as the recombination of photogenerated excitons, clearly dominated the CDs' emission spectrum • Nanocomposites of rGO and CD had weaker PL emission peaks is a possible explanation to the charge transport at the CD/rGO interface which was relatively efficient, given the quenching effect indicated • PL intensity of the composites' emission spectrum upon excitation by a wave length; varying from 290 to 410 nm appeared to increase the peak intensity with increasing wavelength	

(continued)

Table 10.2 (continued)

Waste materials	CNM	Technique	Findings	References
Orange peels	Carbon quantum dots silver nanocomposites (CQDs-Aq-NCs)	UV–Vis	• It was observed that the CQD absorption peaks at 270 nm, but the CQDs-Ag NCs displayed a peak at 430 nm	[99]
		FT-IR	• Absorption peaks were observed at 3329 cm^{-1} for O–H stretching, 1639 cm^{-1} for C=O and aromatic C=C stretching, and 1017 cm^{-1} for C–O stretching • The study listed many compounds like polyphenolic acids, coumarins, flavonoids, alkaloids, and alcohols, which made the bioreduction of Ag^{+} to Ag^{0} possible	
		XRD	• The XRD pattern was noted with distinctive, sharp, and intense diffraction, further suggested Ag nanoparticles/CQDs-Ag NCs with a crystalline nature • Bragg's diffraction peaks were in agreement with face-centered cubic (FCC) crystalline structure of Ag NPs. This can be consistently proved by the broadening of Bragg's peaks in all the samples after CQDs-Ag NCs formation	
		SEM	• The CQDs-Ag NCs synthesized showed the homogenous and spherical shape. The average size of CQDs is about 5–8 nm, and that of the nanoparticles (CQDs-Ag NCs) are in range between 15 and 20 nm	

(continued)

Table 10.2 (continued)

Waste materials	CNM	Technique	Findings	References
Fruit peels (banana, pomegranate, orange) and date stones	Activated carbon	TEM	• The image showed a rough surface of the activated carbon with numerous cavities in different shapes and sizes	[97]
		EDX	• Carbon, oxygen, silicon, and potassium were found in the analysis of activated charcoal from banana peel • The components of the activated carbon made from pomegranate peels were also very similar, with aluminum being added • Only carbon and oxygen were identified for activated carbon obtained from orange peels and date stones	

(continued)

Table 10.2 (continued)

Waste materials	CNM	Technique	Findings	References
		FT-IR	• The activated carbon (AC) developed from banana peels has shown the positions of bands at discrete wavenumbers. These bands can be assigned to the different molecular vibrations at 2843.85, 1940.31, 1582.17, 1453.45, 1118.41, 757 and 530 cm^{-1}, respectively • The band at 2843.85 cm^{-1} is attributed to O–H bond stretching, and the bands at 1940.42 and 1582.14 cm^{-1} assigned forward evidence of COOH in carboxylic acids, while those centered on 1454 cm^{-1} are due to C=C (aliphatic hydrocarbons) groups, respectively. The band at 1118.41 cm^{-1} corresponds to C–H bending, the band at 757.36 cm^{-1} corresponds to C–O stretching, and the band at 530.02 cm^{-1} corresponds to C–Cl stretching. The bands seen at 2350.40 and 2325.00 cm^{-1} were associated with the presence of the O=C=O functional group • The pomegranate peels were found to exhibit absorption bands at 2712.2, 1690.75, 1120.32, and 665.14 cm^{-1}, which were identified as corresponding to O–H stretching, carboxylic acid, C=O stretching, and C-O stretching, respectively, along with C=C bending. The bands seen at 2360.4, 2349.47, and 2337.64 cm cm^{-1} were attributed to the presence of O=C=O • The orange peels' AC exhibited distinct bands at 3681.19, 2794.19, 2383.40, 2140.59, 1572.92, 1453.10, and 1128.73 cm^{-1}, signifying the presence of O–H stretching, O–H stretching, O=C=O, N=C=N stretching, C=C stretching, and C–H bending. Additionally, there were bands at 1999.62 and 1951 cm^{-1}. The value of 55 cm^{-1} corresponds to the stretching of the C=C=C bond • The date stone AC exhibits bands at specific wavenumbers, namely 3665.45, 2817.79, 2124.28, 1905.70, 1444.61, 1123.54, and 529.99 cm^{-1}. These bands have been identified as corresponding to the stretching of O–H bonds, the stretching of O–H bonds, the stretching of N=C=N bonds, the bending of C–H bonds, the stretching of C–O bonds, and the stretching of C–Cl bonds, respectively	

(continued)

Table 10.2 (continued)

Waste materials	CNM	Technique	Findings	References
		BET	• The SBET and VT of the prepared AC correlated with the type of agricultural waste utilized • The AC derived from banana peel exhibited the highest specific surface area (SBET) of 289.86 m^2/g, whereas the AC derived from pomegranate peel demonstrated the largest total pore volume (VT) of 0.462 cm^3/g and the highest surface area determined by the BJH method (125.11 m^2/g) • Orange peel AC exhibited a lower BET surface area (94.26 m^2/g), and date stone AC had a smaller VT (0.132 cm^3/g) • The ratios of SBJH/SBET and VBJH/VBET indicated an increase in mesoporosity in AC derived from pomegranate and orange peels	
Avocado peels	Carbon quantum dots (CQDs)	UV–Vis	• The primary characteristics are absorption peaks at approximately 285 nm, which are likely the result of the π–π^* transition from the carbon core, and a lesser peak at 343 nm, which suggests the n–π^* transitions in the sp2 hybridized carbon • The band gap energy was determined to be 3.4 eV	[100]

(continued)

Table 10.2 (continued)

Waste materials	CNM	Technique	Findings	References
		XRD	• The diffraction pattern of the synthesized QDs showed distinct peaks at $2\theta = 24.5°$ and $42.8°$, which correspond to the (002) and (100) crystallographic planes of carbonaceous materials, respectively	
		FT-IR	• The prominent peak observed at around 1635 cm^{-1} is most likely attributed to the stretching of the C=O bond in a secondary amide • The in-plane bending vibration of the N–H bond, which is a unique characteristic of the secondary amide, is represented by the peak at a wavenumber of 1534 cm^{-1} • The NH stretching vibrations, which are internally bonded to other molecules, can be detected at a frequency of 3278 cm^{-1}. Following these vibrations, the –C–H stretching vibrations occur at a frequency of 2924 cm^{-1} • The peak at 1382 cm^{-1} corresponds to the C–N stretching vibrations commonly observed in amide compounds	
		PL	• The highest level of fluorescence emission was detected at a wavelength of 404 nm, while an excitation wavelength of 350 nm was used • A noticeable decrease in fluorescence intensity was seen when the excitation wavelength was raised from 350 to 500 nm • The size-dependent PL features of carbon dots are a characteristic feature of quantum confinement, which is seen in their excitation-dependent PL behavior • The PL property of the brown-yellow carbon quantum dots was confirmed by their emission of blue light when exposed to UV light • An increase in the excitation wavelength (λ_ex) from 440 to 500 nm led to the PL emission peaking at approximately 440 nm, indicating a red shift from 404 nm	

(continued)

Table 10.2 (continued)

Waste materials	CNM	Technique	Findings	References
		TEM	• The results indicate that the particles have a spherical shape, are evenly distributed, and do not aggregate together • The QDs exhibited a limited size distribution, with an estimated particle size of approximately 2.9 nm • The HRTEM image exhibited distinct lattice fringes with a lattice spacing of 2.1 nm, corresponding to the (100) lattice planes of CQDs, indicating a high degree of crystallinity in the QDs • The QDs exhibited a lattice spacing of approximately 0.23 nm, attributed to the (100) plane characteristic of graphene	
		SEM	• The images display nanoparticles exhibiting a spherical morphology and distinct boundaries, hence validating the findings from the TEM examination	
		EDX	• The examination indicated the existence of carbon, oxygen, and nitrogen, with these constituents being evenly dispersed	
Watermelon rind	Graphene quantum dots (Graphene QDs)	TEM	• The investigation revealed the existence of spherical particles with an average diameter of 6.46 nm and a corresponding crystal structure characterized by a lattice spacing of around 0.22 nm, which matches the lattice fringe of graphene	[101]

(continued)

Table 10.2 (continued)

Waste materials	CNM	Technique	Findings	References
		FT-IR	• The appearance of a wide absorption band ranging from 3150 to 3400 cm^{-1} indicates the creation of N–H bonds, which can be attributed to the inclusion of nitrogen atoms in the graphene QDs and the existence of OH groups • The peak seen between 1600 and 1675 cm^{-1} corresponds to the presence of C=C aromatic and C=O carboxylic groups	
		Raman spectroscopy	• The study revealed the distinctive characteristics of graphene, as evidenced by the peaks observed at 1344 and 1597 cm^{-1}, which correspond to the D and G bands, respectively • Second-order features were noted at 2633 and 2927 cm^{-1}, associated with the 2D and D+G bands, respectively	
		XPS	• Compounds with strong similarity were detected at 286.08, 400.08, and 532.08 eV, corresponding to the C 1 s, N 1 s, and O 1 s peaks, respectively • This indicates that the primary components of the graphene QDs are carbon, nitrogen, and oxygen and that they are free of contaminants	
		UV–Vis, PL	• In the UV–vis absorption spectra, two prominent peaks were identified at 468 and 334 nm, respectively. These are attributed to the Π–Π* transition of the nitrogen group and the Π–Π* transition of the sp2 hybridized nanocarbon • A prominent PL emission band was observed at 446.5 nm under UV illumination, with an excitation at 365 nm	

into biomedical applications is their biocompatibility. Essential for any material intended for implantation, biocompatibility ensures that CNMs can interface with biological tissues without eliciting adverse immune responses. Numerous research efforts have emphasized the significance of surface modifications in improving the biocompatibility of CNMs, thereby making them more appropriate for in vivo applications. Additionally, the electrical conductivity of CNMs like carbon nanotubes and graphene matches the needs for developing electrically active tissues, such as neural and cardiac tissues. Their ability to conduct electrical signals can support or even stimulate the electrical functionality of regenerating tissues, an attribute that is crucial for the successful integration of engineered tissues with native systems. The mechanical strength of CNMs is crucial in scaffold development for tissue engineering. Integrating CNMs into scaffold materials markedly improves their strength and mechanical properties, enabling the scaffolds to sustain tissue growth in physiological environments. This is especially important for load-bearing tissues like bone and cartilage, where mechanical robustness is essential. Furthermore, CNMs' large surface area and chemical adaptability allow for surface functionalization with different biomolecules, enhancing cell attachment, growth, and differentiation. Such interactions are vital for the effective integration and functioning of engineered tissues in the host organism [104, 105].

In this section, we explore the role of CNMs in the field of tissue engineering, with a focus on the potential for applying these advanced materials to foster innovations in regenerative medicine. While the ideal scope of this discussion aims to center around CNMs derived from agricultural waste, current research in this specific area is still nascent. Consequently, substantial examples and applications of waste-derived CNMs within tissue engineering remain limited.

10.4.1 0D-CNMs

0D-CNMs, including carbon dots (CDs), fullerenes, and nanodiamonds, represent the smallest class of CNMs characterized by their particle-like behavior and quantum mechanical properties. Unlike their one- or two-dimensional counterparts, 0D NMs are defined by their discrete, point-like structure, offering unique optical, electronic, and chemical properties that have captured the interest of researchers across various fields, especially in nanomedicine and tissue engineering.

Because of their unique qualities, CDs are a flexible class of zero-dimensional carbon-based materials that have garnered significant interest in tissue engineering. CDs have evolved into a wide range of materials, including graphene quantum dots, carbon quantum dots, and carbon nanodots, illustrating their diversity in uses and progression beyond their first identification as fluorescent fragments from oxidized carbon nanotubes [106]. These nanoscale materials show excitation-wavelength-dependent photoluminescence, which makes them appropriate for a variety of uses, including drug administration and bioimaging. In tissue engineering, CDs are widely used because of their capacity to promote osteogenic differentiation in mesenchymal stem cells (MSCs). Increased alkaline phosphatase (ALP) activity and the activation of osteogenic gene markers are signs that research has demonstrated the significant contribution of CDs to osteogenic development [107, 108]. By adding CDs to hydroxyapatite and polymers, this osteoinductive capacity is further increased. This increases the mechanical qualities, osteoconductivity, and osteoinductivity of the composite materials [109, 110]. An instance of this is the discovery that when nitrogen-doped carbon dots are mixed with hydroxyapatite nanoparticles, they have a beneficial effect on the process of osteogenic differentiation and calcium mineralization in MC3T3-E1 cells. This effect is most likely achieved through the activation of the bone morphogenetic protein signaling pathway [111]. CDs provide exceptional biocompatibility, photoluminescent properties, and the ability to be customized for specific applications, making them a highly promising material for advancements in tissue engineering and regenerative medicine.

Fullerene (C_{60}), discovered in 1985 by Kroto's team, has been extensively studied due to its unique structural, electrical, and chemical properties [113]. The production of fullerene can be conceptualized as the process of enclosing a single layer of graphene within a spherical structure. Fullerene is distinct from graphene in that it is composed of 60 carbon atoms organized in a closed spherical configuration. Fullerene and its derivatives, such as fullerene films and nanowhiskers, have demonstrated potential applications in tissue engineering. Recent studies have highlighted its utility in bone tissue engineering (BTE) despite the challenges associated with fullerene in forming 3D porous scaffolds, adhering to cells, and interacting with bone tissues due to its nonpolar carbon–carbon bonds and potential cytotoxic effects through reactive oxygen species (ROS) generation [112]. Aged fullerene films have shown improved physicochemical properties, making them more suitable for cell cultivation [113]. Nanowhiskers, which are pre-aligned or patterned fullerene structures, have demonstrated efficacy in directing the adherence and proliferation of osteoblast cells. For instance, the orientation of osteoblast cell growth has been directed along the axis of fullerene C_{60} nanowhiskers, which are produced via vortex rotation and subsequently transferred onto glass substrates. This demonstrates the capacity of fullerene structures to promote and guide the proliferation and maturation of osteoblasts [114, 115]. Moreover, studies have demonstrated that polyhydroxylated fullerene promotes the growth of human adipose-derived stem cells (hADSCs) and improves the formation of bone tissue and mineralization. This highlights the potential of modified fullerene derivatives in enhancing the process of bone regeneration [115].

NDs have become an important type of carbon-based nanomaterial in the realm of tissue engineering due to their outstanding characteristics like extraordinary hardness, top-notch mechanical strength, chemical durability, excellent thermal conductivity, reactive surfaces, and superb biocompatibility. Their chemical inertness as well as high surface-to-volume ratio make NDs ideal for easy surface functionalization, enhancing their integration into biomaterial scaffolds [116]. Research has shown that nanostructured nanocrystalline diamond (NCD) films greatly enhance the attachment of cells, metabolic functions, and bone-forming capabilities compared to traditional polystyrene culture dishes, highlighting the potential of NDs in aiding bone regeneration [117]. For example, integrating octadecylamine-functionalized nanodiamond

(ND-ODA) into poly (l-lactic acid) (PLLA) scaffolds has led to significant enhancements in mechanical properties, such as Young's modulus and tensile strength. Additionally, it has promoted the formation of bone-like apatite, showcasing superior mineralization capabilities that are advantageous for bone regeneration [118]. Moreover, incorporating NDs into poly (lactic-co-glycolic acid) (PLGA) nanofibrous scaffolds has significantly improved their mechanical attributes and exhibited excellent biocompatibility by facilitating cell adhesion, spreading, and proliferation, all while not causing substantial inflammatory responses [119]. The intrinsic fluorescence of NDs under certain light wavelengths presents a valuable tool for in vivo monitoring of implant devices, offering insights into the healing process. These results highlight the extensive potential of ND-based scaffolds as versatile biomaterials in bone regenerative engineering. They emphasize the ability of these scaffolds to promote cell proliferation and differentiation, improve mineralization, and contribute mechanical strength to scaffolds. This positions ND-based systems as a promising avenue for developing advanced tissue engineering solutions.

Continuing research is focused on comprehensively understanding the interplay between NMs and biological systems to ensure their safe and effective use in therapeutic settings. The primary objective is to create zero-dimensional CNMs with customized characteristics to meet the requirements of various biomedical applications. This emphasizes the interdisciplinary nature of tissue engineering and underscores the pivotal role of materials science in promoting regenerative therapies.

10.4.2 1D-CNMs

1D-CNMs, which include CNTs and carbon nanofibers (CNFs), have unique features that make them extremely valuable in the field of nanotechnology. CNTs are widely used in tissue engineering and regenerative medicine due to their exceptional mechanical strength, electrical conductivity, and chemical stability. CNTs, which are composed of coiled graphene sheets in a cylindrical shape, have emerged as important contributors in the field of tissue engineering. Due to their unique characteristics, such as strong tensile strength, electrical conductivity, and a high ratio of surface area to volume, these materials are highly adaptable [120]. The integration of CNTs into scaffolds used in tissue engineering shows great potential for the regeneration of intricate tissues, including those found in the nervous system, heart, and bones. CNTs have demonstrated favorable utility in the field of bone tissue engineering since they possess distinctive characteristics that augment the process of bone regeneration and repair. The addition of MWCNTs to a scaffold made of polyurethane/zein/chitosan, by modifying the carboxyl groups, greatly enhances the growth, multiplication, penetration, and on-site specialization of MC3T3-E1 cells. This leads to improved bone formation and transmission [121]. In addition, it has been demonstrated that MWCNTs coated with polycaprolactone (PCL) nanofibers, which have been modified with hydroxyl/carboxyl groups, can improve the adherence of MSCs, promote their differentiation into bone cells, and boost vascularization. In addition,

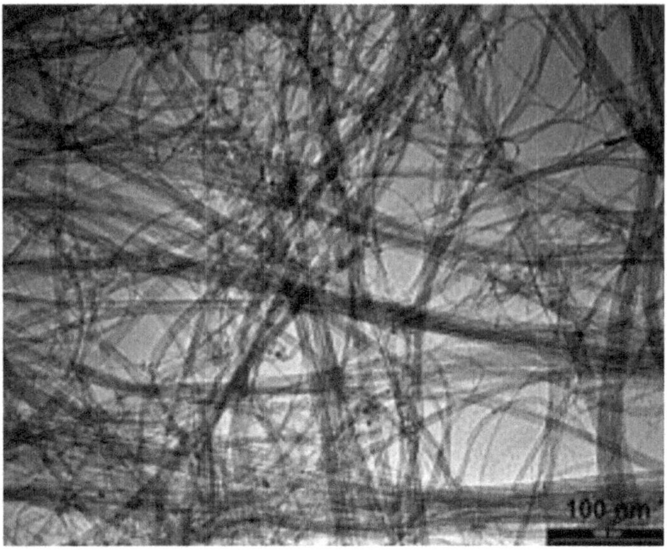

Fig. 10.3 TEM image depicting structural attributes of CNTs. Adapted from Wang et al. [123]. © 2023 Mohamed Abdelmonem. Used under license

they have been found to decrease inflammation in both laboratory and live animal experiments, which promotes the recovery of skull injuries in rats and enhances the density of bone minerals [122]. In addition, it has been shown that MWCNT-Poly (methyl methacrylate) bone cement, which has been modified with carboxyl groups, increases the expression of genes that are crucial for bone formation in rat bone marrow mesenchymal stem cells. Furthermore, it enhances the mineral density in scaffolds, hence promoting the integration of bone in adult male rabbits. Figure 10.3 displays a TEM image that showcases the structural makeup of CNTs, as illustrated by Wang et al. (2019). This graphic exemplifies the structural characteristics of CNTs [123].

These studies highlight the important contribution of CNTs in enhancing bone tissue engineering through their ability to stimulate bone formation and integrate with scaffolds. 1D-CNMs also have significant advantages in brain and cardiac tissue engineering due to their high electrical conductivity. Scaffolds using CNTs have been shown to enhance the growth of nerve fibers and facilitate the development of neural networks, providing potential opportunities for repairing nerve damage and treating neurodegenerative disorders. For example, CNTs have demonstrated their usefulness in promoting the formation of neurons and improving the connections between brain cells. CNT-polyethylene glycol (PEG) hydrogels have been used to assist adult neural stem cells, leading to a greater proportion of neurons to astrocytes and enhanced synaptic connection, which is crucial for the formation of functional neural networks [124]. In addition, transparent substrates composed of MWCNT/Silica have proven to be successful in supporting organotypic cultures of the entorhinal-hippocampal complex. They greatly enhance signal synchronization and fiber growth, hence improving communication between the cortex and hippocampus [125]. There is no text provided. Incorporating CNTs into serine conductive catheters has been demonstrated to enhance the attachment and growth of Schwann cells. Significant structural repair and functional recovery were demonstrated when these catheters, in conjunction with electrical stimulation, were surgically inserted into rat models with sciatic nerve lesions [126]. The conductive properties of these nanoparticles in cardiac tissue engineering can imitate the electrical conditions of the heart, which has the potential to improve the integration and performance of synthetic cardiac tissues. CNTs have demonstrated significant potential in facilitating the restoration and regrowth of heart tissue. Research on the incorporation of electrically aligned CNTs specifically carboxylated MWCNTs, into silk electrospun scaffolds, has shown that it enhances the growth of rat cardiomyocytes, promotes the production of heart-specific proteins, and facilitates the development of sarcomeres and gap junctions, which are crucial for the proper functioning of cardiac muscle [127]. In addition, the creation of biohybrid conductive CNT-hydrogels, which include solubilized pericardial matrix and hydrazide-functionalized multi-walled carbon nanotubes, has played a crucial role in promoting the growth of hMSCs and improving their natural capacity to heal cardiac tissue. This improvement is partially ascribed to enhanced calcium handling, increased contraction amplitudes, and accelerated contraction rates in cardiomyocytes produced from human-induced pluripotent stem cells (hiPSCs). Furthermore, there is a rise in the expression of connexin-43, which plays a crucial role in the connection and elongation of heart muscle cells [128]. CNTs play a crucial role in the development of sophisticated tissue-engineered blood arteries in the field of vascular tissue engineering. The utilization of wet electrospinning techniques involving PCL/Gelatin/CNT yarns has been implemented to produce scaffolds that facilitate the attachment and growth of vascular endothelial cells. The presence of CNTs significantly enhances the alignment of cells, mimicking the inherent orientation observed in arterial cells. The alignment is essential for accurately imitating the functionality of natural vessels [129].

Various natural substances have been employed in the fabrication of sustainable nanofibers. For example, cellulose, derived from natural sources like wood and cotton pulp through chemical and physical methods, is characterized by its biodegradability, biocompatibility, hydrophilicity, and high mechanical toughness [130]. Owing to these properties, cellulose finds extensive application in the field of tissue engineering. Cellulose acetate (CA), a cellulose derivative, possesses the capability to adsorb organic constituents and certain toxins. Despite its inferior mechanical characteristics, these can be improved through amalgamation with additional polymers [131–133]. Ahmed et al. utilized a CA/PLA composite to augment the chemical and biological attributes of the material. Enhancements in the properties of elastic elongation, cell adhesion, and biocompatibility were observed in the CA/PLA nanofibers [134]. Alginate, a natural polysaccharide sourced from brown seaweeds, serves as another natural polymer. Its derivative, sodium alginate (SA), is favored in biomedical fields due to its well-established properties like high solubility, biodegradability, biocompatibility, affordability, hydrophilicity, nontoxicity, and inherent structural integrity [135]. Alginate has been utilized in the fabrication of nanofibers through the process of electrospinning. Nevertheless, to improve the strength and stability of their structure, various cross-linking agents have been utilized in different research investigations. The stability and solubility of alginate fibers in biological fluids and water have been enhanced through cross-linking with divalent ions like calcium and barium, as well as through dual cross-linking with epichlorohydrin, adipic acid (ADA), and hexamethylene diisocyanate (HMDI) [136, 137]. As advancements in materials science continue to unfold, the role of 1D-CNMs in tissue engineering is poised to expand further. Their unique ability to meld mechanical reinforcement with electrical conductivity and biochemical compatibility offers a comprehensive toolkit for addressing the complex challenges of tissue regeneration. This multidisciplinary approach underscores the evolving nature of regenerative medicine, highlighting the potential of nanotechnology to drive forward the development of sophisticated therapeutic solutions for repairing and regenerating damaged tissues.

10.4.3 2D-CNMs

Graphene and its derivatives, such as GO and rGO, are characterized by their two-dimensional honeycomb structure and have emerged as revolutionary NMs in tissue engineering. Their large surface area and ability to be modified for specific purposes make them well suited for creating a wide range of NMs with unique features, which is why they are highly suitable for biological applications. This includes the synthesis of biocompatible scaffolds in tissue engineering that promote cell growth and distribute bioactive molecules. Graphene's exceptional mechanical strength, electrical conductivity, and biocompatibility render it a preferred material for scaffold development in tissue engineering.

The use of graphene-based NMs in tissue engineering is extensive and varied, as illustrated in Fig. 10.4. In the context of bone tissue engineering, graphene and its derivatives promote the osteogenic differentiation and proliferation of human mesenchymal stem cells. Research has shown that graphene scaffolds can expedite cellular differentiation without external growth factors, possibly due to their ability to enhance local dexamethasone concentration. Additionally, GO nanosheets have been effective in encouraging cell proliferation and osteogenic differentiation at lower doses, demonstrating their potential for bone regeneration. Ricci et al. have documented that graphene acts as a potent substrate for bone tissue engineering, boosting the proliferation of hMSCs and aiding their differentiation into the osteogenic lineage, an effect linked to the osteoinductive and osteoconductive properties of graphene [139]. Graphene has a wide range of applications, including bone regeneration as well as the engineering of heart, cutaneous, and neurological tissues. Graphene-based materials enhance the attachment, growth, and development of skeletal muscle cells in the field of muscle tissue engineering [140].

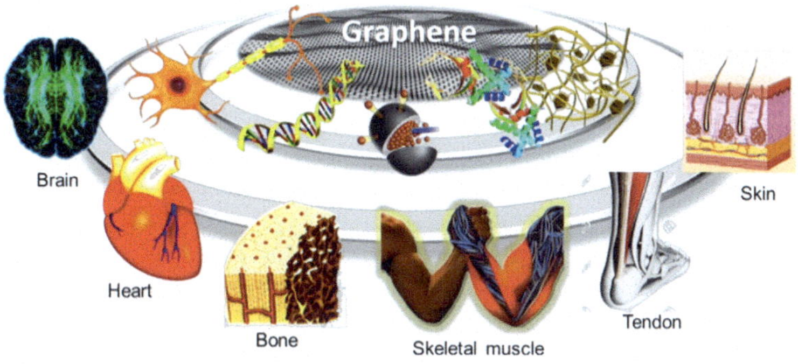

Fig. 10.4 Utilizing graphene-based nanoparticles in tissue engineering for regenerative medicine. Used with authorization from [138]. Copyright © 2024, Elsevier Ltd

Tissue engineering offers a practical remedy for cardiovascular ailments, such as myocardial infarction and cardiomyopathy. This is achieved by fabricating scaffolds and cardiac patches that imitate the extracellular matrix (ECM) of heart tissue. These artificially created tissues display properties that resemble those of natural cardiac tissue, with inherent electrical conductivities ranging from 0.005 S/m (transverse) to 0.1 S/m (longitudinal), which is comparable to the electrical conductivity of the heart's Purkinje fibers, known for their electrical conductivity [141]. Traditional scaffolds used in tissue engineering are typically non-conductive and lack electrical influence on the targeted site. Additionally, they often do not possess the fibrous microstructure critical for mimicking and supporting cardiac tissue functions. To address these difficulties and enhance the efficacy of cardiac tissue engineering, CNMs, specifically those based on graphene, are employed. Integrating graphene-based NMs into scaffolds for cardiac tissue regeneration has led to significant enhancements in cellular functionality and the overall manipulation of heart tissue. NMs have a crucial role in improving cellular function and are essential for boosting cellular conductivity, which is a critical factor in cardiac tissue engineering. In addition, the integration of graphene nanoparticles into scaffolds improves the durability of the materials, strengthening their structural function in heart regeneration projects [142–147].

Traditional methods for skin tissue regeneration, like autologous and allogeneic skin grafts, encounter limitations due to possible donor site injury, immune rejection, and high surgical costs. Challenges also arise from insufficient vascularization and the absence of growth factor incorporation. However, developments in nanotechnology, stem cell biology, and vascularization techniques present promising paths for developing engineered skin substitutes that reduce the risks of immune rejection and pathogen contamination. However, the application of these substitutes is hindered by their inferior mechanical properties and non-porous structure compared to natural extracellular matrix scaffolds. Graphene-based NMs, renowned for their exceptional mechanical durability, expansive surface area, and capacity to adsorb proteins, present a practical answer for enhancing artificial skin replacements. Nyambat and colleagues demonstrated that a composite sponge made from genipin-crosslinked adipose stem cell-derived extracellular matrix and nano-GO has the potential as a highly efficient material for skin tissue creation. The integration of GO into the scaffold resulted in improvements in the adhesion, proliferation, and motility of L929 cells. Follow-up investigations performed on rats using these scaffolds augmented with GO demonstrated no local reactions or immunological responses, suggesting its potential for use in regenerating skin tissue [145]. Research has shown that using decellularized scaffolds that are modified with graphene can enhance the growth, reproduction, and movement of skin cells [146]. Jian and colleagues have recently created a substance called polyurethane-modified graphene oxide (PMGO) that is specifically designed for promoting wound healing. This material has the unique ability to maintain antibacterial qualities for an extended period. The PMGO composite scaffold exhibited remarkable biocompatibility, enhanced water vapor transport rates, and maintained its wide-ranging antibacterial characteristics for 30 days [147].

Using graphene-based materials as a tool in tissue engineering provides an easy approach to accelerate the neural differentiation process by promoting embryonic stem cells toward mature functioning neuronal cell lines. Various in vivo studies have also highlighted the ability of graphene-related compounds to regenerate brain tissues. For example, the team led by Gao et al. showed that the incorporation of graphene oxide nanosheets (GON) into poly(D, L-lactide-co-caprolactone) (PL-PCL) conduits (GON PL-PCL) greatly enhances the regeneration of peripheral nerves in animal models [148]. In addition, Serrano and his colleagues developed a specialized framework composed of reduced graphene oxide microfibers (rGO MFs) designed exclusively for repairing injured cervical spinal cords [149]. To sum up, the chemical characteristics of 2D-CNMs particularly graphene and its variations have firmly established them as vital elements, in tissue engineering. Their versatility across different tissue types, along with their capability to improve cellular functions and facilitate regeneration, highlights their significant potential in advancing regenerative medicine.

10.4.4 3D-CNMs

Ever-more innovative solutions to difficult tissue engineering and regenerative medicine problems, such as those emerging from 3D-CNMs, are at the frontier of materials science. These materials, such as graphene foam or carbon aerogels, and others of carbon-based architectures have network structures not with straight lines but a connected three-dimensional similar to the architecture of extracellular matrix. This mimicry is essential because it creates a framework for cells to stick and grow, as well as the necessary porosity that allows nutrients in and waste out—critical operations of tissue growth and regeneration. The distinctive attributes of 3D-CNMs, including their lightweight characteristics, remarkable mechanical strength, and extensive surface area, have shown to be advantageous in creating scaffolds for tissue engineering. These scaffolds facilitate essential processes, such as cell attachment, growth, and differentiation, which are crucial for generating functional tissues. The use of these materials covers a broad spectrum of regenerative medicine, encompassing the repair of bone and cartilage, as well as neural tissue engineering, demonstrating their adaptability and capacity to improve tissue regeneration.

3D graphene foams are becoming prominent materials for tissue engineering because of their porous and interconnected scaffold structure, which enhances cell proliferation and differentiation by facilitating efficient nutrition supply. These 3D scaffolds outperform their 2D counterparts in mimicking the extracellular matrix in vivo, closely matching its physical and chemical characteristics. 3D graphene foam has been acknowledged in the field of bone tissue engineering for its ability to enhance the growth and specialization of osteoblasts. Studies have demonstrated that these foams promote the development of bone cells without relying on substances that induce bone formation. Additionally, the osteoconductive attributes of graphene are known to boost both cell growth and differentiation [150]. A separate study examined the application of 3D porous scaffolds composed of rGO and nano-hydroxyapatite composites (nHA@RGO) in the field of bone tissue engineering [151].

Neural tissue engineering also benefits from the conductive properties of 3D carbon materials. Their ability to support neural cell growth and function offers promising avenues for repairing damaged neural pathways and treating neurodegenerative disorders. The conductive scaffolds can enhance signal transmission between neurons, a critical aspect of neural tissue functionality [152]. In recent research, Guo et al. demonstrated that a unique scaffold combining bacterial cellulose and graphene in a three-dimensional structure (3D-BC/G) has shown promising results, in regulating the growth and specialization of neural stem cells. The initial findings suggest that these innovative scaffolds not support the development and attachment of stem cells but also preserve their original characteristics and boost their ability to proliferate [153]. Utilizing 3D graphene foam in conjunction with bone marrow-derived MSCs has demonstrated favorable results in promoting wound healing within the realm of skin tissue regeneration. This composite material exhibited biocompatibility and promoted the proliferation and migration of MSCs. The injected MSCs enhanced the synthesis of molecules that stimulate angiogenesis, such as vascular endothelial growth factor (VEGF) and basic fibroblast growth factor (bFGF). Furthermore, there was a reduction in scar formation as indicated by decreased levels of transforming growth factor-beta 1 (TGF β1) and alpha-smooth muscle actin (α SMA), along with an increase in TGF β3 levels [154].

Current research and development efforts in the field of 3D-CNMs are primarily aimed at improving their interaction with biological systems, increasing their compatibility with living organisms, and customizing their physical and chemical characteristics for specific uses. This effort aims to improve the integration of these materials into host organisms, ensuring that they can effectively support tissue regeneration without eliciting adverse responses. As tissue engineering evolves, ongoing advancements in the synthesis, functionalization, and manufacturing of 3D-CNMs are expected to expand their use in regenerative medicine, signaling the advent of a new era in sophisticated therapeutic approaches.

10.5 Conclusion

The exploration of CNMs derived from agricultural waste presents a compelling narrative of sustainable innovation, where environmental challenges are transformed into opportunities for advancing biomedical research. The synthesis and characterization of these NMs are critical, as they directly influence their properties and efficacy in biomedical applications. Recent advancements in synthesis methodologies have demonstrated the potential to convert agricultural waste into high-quality CNMs efficiently. Tailored synthesis techniques are paramount for controlling the NMs' size, morphology, and surface characteristics, which in turn determine their interaction with biological systems. The detailed characterization of these CNMs sheds light on their structural and functional attributes, which are essential for their applicability in the biomedical field. These NMs present advantageous properties for tissue engineering, where their support for tissue growth and regeneration is highly valued. By creating structures conducive to cell growth, CNMs derived from agricultural waste could significantly contribute to the progress of tissue engineering and regenerative medicine. Their application in these fields shows promise, but future investigations need to tackle issues like the scalability of production and long-term biocompatibility. Additionally, there is a notable scarcity of research on CNMs from agricultural waste, a gap this chapter highlights, urging further exploration into agricultural by-products to enhance sustainability in materials science. Upcoming research should delve into this underexplored avenue, offering sustainable and environmentally friendly progress in the discipline. Recommended actions include the development of cost-efficient, scalable synthesis methods and thorough in vivo evaluations to confirm their safety and effectiveness. Improvements in functionalization techniques are essential to customize these materials for precise medical uses. Overcoming these obstacles is crucial for fully realizing the potential of AWD-CNMs in medical innovation.

References

1. Iijima S (1991) Helical microtubules of graphitic carbon. Nature 354(6348):56–58. https://doi.org/10.1038/354056a0
2. Malik S, Muhammad K, Waheed Y (2023) Nanotechnology: a revolution in modern industry. Molecules 28(2):661. https://doi.org/10.3390/molecules28020661
3. Onugwu AL et al (2023) Nanotechnology based drug delivery systems for the treatment of anterior segment eye diseases. J Control Release 354:465–488. https://doi.org/10.1016/j.jconrel.2023.01.018
4. Chakraborty M, Hashmi MSJ (2018) Graphene as a material—an overview of its properties and characteristics and development potential for practical applications. In: Reference module in materials science and materials engineering. Elsevier. https://doi.org/10.1016/B978-0-12-803581-8.10319-4
5. Geim AK (2009) Graphene: status and prospects. Science (1979) 324(5934):1530–1534. https://doi.org/10.1126/science.1158877

6. Egbedina AO, Bolade OP, Ewuzie U, Lima EC (2022) Emerging trends in the application of carbon-based materials: a review. J Environ Chem Eng 10(2):107260. https://doi.org/10.1016/j.jece.2022.107260
7. Nguyen TD, Nguyen MTN, Lee JS (2023) Carbon-based materials and their applications in sensing by electrochemical voltammetry. Inorganics (Basel) 11(2):81. https://doi.org/10.3390/inorganics11020081
8. Mbambo MC et al (2020) Remarkable thermal conductivity enhancement in Ag—decorated graphene nanocomposites based nanofluid by laser liquid solid interaction in ethylene glycol. Sci Rep 10(1):10982. https://doi.org/10.1038/s41598-020-67418-3
9. Nakagawa K, Satoh K, Murakami S, Takei K, Akita S, Arie T (2021) Controlling the thermal conductivity of multilayer graphene by strain. Sci Rep 11(1):19533. https://doi.org/10.1038/s41598-021-98974-x
10. Makhsin SR (2023) Physical properties of graphene. J Mech Eng SI 12(1):225–267. https://doi.org/10.24191/jmeche.v12i1.24649
11. Sivaraj A, Kumar V, Sunder R, Parthasarathy K, Kasivelu G (2020) Commercial yeast extracts mediated green synthesis of silver chloride nanoparticles and their anti-mycobacterial activity. J Clust Sci 31(1):287–291. https://doi.org/10.1007/s10876-019-01626-4
12. Osman AI et al (2024) Synthesis of green nanoparticles for energy, biomedical, environmental, agricultural, and food applications: a review. Environ Chem Lett 22(2):841–887. https://doi.org/10.1007/s10311-023-01682-3
13. Yap YH et al (2020) Green synthesis of silver nanoparticle using water extract of onion peel and application in the acetylation reaction. Arab J Sci Eng 45(6):4797–4807. https://doi.org/10.1007/s13369-020-04595-3
14. Omoriyekomwan JE, Tahmasebi A, Dou J, Wang R, Yu J (2021) A review on the recent advances in the production of carbon nanotubes and carbon nanofibers via microwave-assisted pyrolysis of biomass. Fuel Process Technol 214:106686. https://doi.org/10.1016/j.fuproc.2020.106686
15. Kumari N, Mohan C (2021) Basics of clay minerals and their characteristic properties. In: Clay and clay minerals. IntechOpen.https://doi.org/10.5772/intechopen.97672
16. Iranfar S, Karbala MM, Shakiba M, Shahsavari MH (2023) Effects of type and distribution of clay minerals on the physico-chemical and geomechanical properties of engineered porous rocks. Sci Rep 13(1):5837. https://doi.org/10.1038/s41598-023-33103-4
17. Barin GB, de Fátima Gimenez I, da Costa LP, Filho AGS, Barreto LS (2014) Hollow carbon nanostructures obtained from hydrothermal carbonization of lignocellulosic biomass. J Mater Sci 49(2):665–672. https://doi.org/10.1007/s10853-013-7747-1
18. Snyders R, Hegemann D, Thiry D, Zabeida O, Klemberg-Sapieha J, Martinu L (2023) Foundations of plasma enhanced chemical vapor deposition of functional coatings. Plasma Sources Sci Technol 32(7):074001. https://doi.org/10.1088/1361-6595/acdabc
19. Bertran-Serra E et al (2023) Advancements in plasma-enhanced chemical vapor deposition for producing vertical graphene nanowalls. Nanomaterials 13(18):2533. https://doi.org/10.3390/nano13182533
20. Qasim M, Clarkson AN, Hinkley SFR (2023) Green synthesis of carbon nanoparticles (CNPs) from biomass for biomedical applications. Int J Mol Sci 24(2):1023. https://doi.org/10.3390/ijms24021023
21. Kumar K et al (2023) Biomass waste-derived carbon materials for sustainable remediation of polluted environment: a comprehensive review. Chemosphere 345:140419. https://doi.org/10.1016/j.chemosphere.2023.140419
22. Bernd MGS, Bragança SR, Heck N, da Silva Filho LCP (2017) Synthesis of carbon nanostructures by the pyrolysis of wood sawdust in a tubular reactor. J Mater Res Technol 6(2):171–177. https://doi.org/10.1016/j.jmrt.2016.11.003
23. Kumar R, Tiwari RS, Srivastava ON (2011) Scalable synthesis of aligned carbon nanotubes bundles using green natural precursor: neem oil. Nanoscale Res Lett 6(1):92. https://doi.org/10.1186/1556-276X-6-92
24. Das R, Das Tuhi S (2018) Carbon nanotubes synthesis. Carbon Nanotubes Clean Water 27–84

25. Huang S, Du X, Ma M, Xiong L (2021) Recent progress in the synthesis and applications of vertically aligned carbon nanotube materials. Nanotechnol Rev 10(1):1592–1623
26. Suriani AB et al (2016) Parametric study of waste chicken fat catalytic chemical vapour deposition for controlled synthesis of vertically aligned carbon nanotubes. Cogent Phys 3(1). https://doi.org/10.1080/23311940.2016.1247486
27. Tripathi N, Pavelyev V, Islam SS (2017) Synthesis of carbon nanotubes using green plant extract as catalyst: unconventional concept and its realization. Appl Nanosci 7(8):557–566. https://doi.org/10.1007/s13204-017-0598-3
28. Battiston LL, Oliveira KD, Lião LM, Oliveira GAR, Kolicheski MB, Ávila-Neto CN (2023) Towards a correlation between iron/cobalt content, support pore size and metal particle size in supported catalysts. ChemNanoMat 9(10):e202300264
29. Borchardt R, Fromm T, Haase A, Fecher J, Rosiwal SM (2020) World's first thermoelectric generator made of tailored carbon allotropes. Adv Eng Mater 22(9):2000108
30. Dolmatov VY (2001) Detonation synthesis ultradispersed diamonds: properties and applications. Russ Chem Rev 70(7):607–626. https://doi.org/10.1070/RC2001v070n07ABEH000665
31. Mahdavi M, Mahmoudi N, Rezaie Anaran F, Simchi A (2016) Electrospinning of nanodiamond-modified polysaccharide nanofibers with physico-mechanical properties close to natural skins. Mar Drugs 14(7):128. https://doi.org/10.3390/md14070128
32. Chipaux M., van der Laan KJ, Hemelaar SR, Hasani M, Zheng T, Schirhagl R (2018) Nanodiamonds and their applications in cells. Small 14(24). https://doi.org/10.1002/smll.201704263
33. Anand A et al (2021) A novel approach towards green synthesis of nanodiamonds as biocompatible agents. J Nano Electron Phys 13(3):03040-1–03040-6. https://doi.org/10.21272/jnep.13(3).03040
34. Gaur M et al (2021) Biomedical applications of carbon nanomaterials: fullerenes, quantum dots, nanotubes, nanofibers, and graphene. Materials 14(20):5978. https://doi.org/10.3390/ma14205978
35. Wagner AM, Knipe JM, Orive G, Peppas NA (2019) Quantum dots in biomedical applications. Acta Biomater 94:44–63
36. Sobhanan J, Rival JV, Anas A, Shibu ES, Takano Y, Biju V (2023) Luminescent quantum dots: synthesis, optical properties, bioimaging and toxicity. Adv Drug Deliv Rev 114830
37. Rawat P et al (2023) An overview of synthetic methods and applications of photoluminescence properties of carbon quantum dots. Luminescence 38(7):845–866
38. Manikandan V, Lee NY (2022) Green synthesis of carbon quantum dots and their environmental applications. Environ Res 212:113283
39. Preethi M, Viswanathan C, Ponpandian N (2021) A green path to extract carbon quantum dots by coconut water: another fluorescent probe towards Fe^{3+} ions. Particuology 58:251–258. https://doi.org/10.1016/j.partic.2021.03.019
40. Singh A, Banerjee SL, Gantait A, Kumari K, Kundu PP (2023) Metal-based nanoparticles: synthesis and biomedical applications. In: Nanoparticles reinforced metal nanocomposites: mechanical performance and durability. Springer, pp 365–408
41. Yadav PK et al (2019) Green synthesis of fluorescent carbon quantum dots from *Azadirachta indica* leaves and their peroxidase-mimetic activity for the detection of H_2O_2 and ascorbic acid in common fresh fruits. ACS Biomater Sci Eng 5(2):623–632. https://doi.org/10.1021/acsbiomaterials.8b01528
42. Saba N, Jawaid M, Fouad H, Alothman OY (2019) Nanocarbon: preparation, properties, and applications. In: Nanocarbon and its composites. Elsevier, pp 327–354. https://doi.org/10.1016/B978-0-08-102509-3.00009-2
43. Vourgidi C, Giannopoulou I, Kourtis A, Magganiari M, Xenidis A (2024) Utilization of waste graphite for the sustainable production of silicon carbide. Mater Proc 15(1):82
44. Mhadhbi M, Driss M (2021) Titanium carbide: synthesis, properties and applications. J Brill Eng 2:1–11

45. Mao H et al (2021) Microstructure and mechanical properties of carbide reinforced TiC-based ultra-high temperature ceramics: a review. Coatings 11(12):1444
46. Voiry D et al (2016) High-quality graphene via microwave reduction of solution-exfoliated graphene oxide. Science (1979) 353(6306):1413–1416. https://doi.org/10.1126/science.aah3398
47. Hughes AE, Haque N, Northey SA, Giddey S (2021) Platinum group metals: a review of resources, production and usage with a focus on catalysts. Resources 10(9):93
48. Sha J et al (2017) Three-dimensional printed graphene foams. ACS Nano 11(7):6860–6867. https://doi.org/10.1021/acsnano.7b01987
49. Chen L et al (2017) Oriented graphene nanoribbons embedded in hexagonal boron nitride trenches. Nat Commun 8(1):14703. https://doi.org/10.1038/ncomms14703
50. Phiri J, Gane P, Maloney TC (2017) General overview of graphene: Production, properties and application in polymer composites. Mater Sci Eng B 215:9–28. https://doi.org/10.1016/j.mseb.2016.10.004
51. Abbas Q et al (2022) Graphene synthesis techniques and environmental applications. Materials 15(21):7804
52. Hu Z, Huang Y, He X, Guo W, Yan K (2022) Solution-phase conversion of glucose into semiconductive carbonaceous nanosheet photocatalysts for enhanced environmental applications. Chem Eng J 427:131464
53. Kim Y-K, Kim M-H, Min D-H (2011) Biocompatible reduced graphene oxide prepared by using dextran as a multifunctional reducing agent. Chem Commun 47(11):3195. https://doi.org/10.1039/c0cc05005a
54. Palomba M, Carotenuto G, Longo A (2022) A brief review: the use of L-ascorbic acid as a green reducing agent of graphene oxide. Materials 15(18):6456
55. Qin Y, Li G, Gao Y, Zhang L, Ok YS, An T (2018) Persistent free radicals in carbon-based materials on transformation of refractory organic contaminants (ROCs) in water: a critical review. Water Res 137:130–143. https://doi.org/10.1016/j.watres.2018.03.012
56. Lyu H et al (2018) Experimental and modeling investigations of ball-milled biochar for the removal of aqueous methylene blue. Chem Eng J 335:110–119. https://doi.org/10.1016/j.cej.2017.10.130
57. Oleszczuk P, Ćwikła-Bundyra W, Bogusz A, Skwarek E, Ok YS (2016) Characterization of nanoparticles of biochars from different biomass. J Anal Appl Pyrolysis 121:165–172. https://doi.org/10.1016/j.jaap.2016.07.017
58. Li L et al (2017) Mass preparation of micro/nano-powders of biochar with water-dispersibility and their potential application. New J Chem 41(18):9649–9657. https://doi.org/10.1039/C7NJ00742F
59. Genovese M, Jiang J, Lian K, Holm N (2015) High capacitive performance of exfoliated biochar nanosheets from biomass waste corn cob. J Mater Chem A Mater 3(6):2903–2913. https://doi.org/10.1039/C4TA06110A
60. Luong DX et al (2020) Gram-scale bottom-up flash graphene synthesis. Nature 577(7792):647–651. https://doi.org/10.1038/s41586-020-1938-0
61. Li X et al (2017) Flexible and self-healing aqueous supercapacitors for low temperature applications: polyampholyte gel electrolytes with biochar electrodes. Sci Rep 7(1):1685. https://doi.org/10.1038/s41598-017-01873-3
62. Wallace CA, Afzal MT, Saha GC (2019) Effect of feedstock and microwave pyrolysis temperature on physio-chemical and nano-scale mechanical properties of biochar. Bioresour Bioprocess 6(1):33. https://doi.org/10.1186/s40643-019-0268-2
63. Ahmad KA, Ahmad E, Al Mesfer MK, Nigam KDP (2023) Bio-coal and bio-coke production from agro residues. Chem Eng J 145340
64. Weber K, Quicker P (2018) Properties of biochar. Fuel 217:240–261. https://doi.org/10.1016/j.fuel.2017.12.054
65. Baig N, Kammakakam I, Falath W (2021) Nanomaterials: a review of synthesis methods, properties, recent progress, and challenges. Mater Adv 2(6):1821–1871

66. Mekuye B, Abera B (2023) Nanomaterials: an overview of synthesis, classification, characterization, and applications. Nano Select
67. Rai M, Nguyen TA (2021) Nanomaterials recycling. Elsevier
68. Yaqoob AA et al (2020) Cellulose derived graphene/polyaniline nanocomposite anode for energy generation and bioremediation of toxic metals via benthic microbial fuel cells. Polymers (Basel) 13(1):135
69. Mansuriya BD, Altintas Z (2021) Carbon dots: classification, properties, synthesis, characterization, and applications in health care—an updated review (2018–2021). Nanomaterials 11(10):2525
70. Hu Q, Gong X, Liu L, Choi MMF (2017) Characterization and analytical separation of fluorescent carbon nanodots. J Nanomater 2017
71. Li Q, Zhang T, Pan Y, Ciacchi LC, Xu B, Wei G (2016) AFM-based force spectroscopy for bioimaging and biosensing. RSC Adv 6(16):12893–12912
72. Tajik S et al (2020) Carbon and graphene quantum dots: a review on syntheses, characterization, biological and sensing applications for neurotransmitter determination. RSC Adv 10(26):15406–15429
73. Dager A, Uchida T, Maekawa T, Tachibana M (2019) Synthesis and characterization of monodisperse carbon quantum dots from fennel seeds: photoluminescence analysis using machine learning. Sci Rep 9(1):14004
74. Jorns M, Pappas D (2021) A review of fluorescent carbon dots, their synthesis, physical and chemical characteristics, and applications. Nanomaterials 11(6):1448
75. Demirci S, McNally AB, Ayyala RS, Lawson LB, Sahiner N (2020) Synthesis and characterization of nitrogen-doped carbon dots as fluorescent nanoprobes with antimicrobial properties and skin permeability. J Drug Deliv Sci Technol 59:101889
76. Kazemifard N, Ensafi AA, Rezaei B (2020) Green synthesized carbon dots embedded in silica molecularly imprinted polymers, characterization and application as a rapid and selective fluorimetric sensor for determination of thiabendazole in juices. Food Chem 310:125812
77. Mourdikoudis S, Pallares RM, Thanh NTK (2018) Characterization techniques for nanoparticles: comparison and complementarity upon studying nanoparticle properties. Nanoscale 10(27):12871–12934
78. Zhu Q, Zhou R, Liu J, Sun J, Wang Q (2021) Recent progress on the characterization of cellulose nanomaterials by nanoscale infrared spectroscopy. Nanomaterials 11(5):1353
79. Khairol Anuar NK, Tan HL, Lim YP, So'aib MS, Abu Bakar NF (2021) A review on multifunctional carbon-dots synthesized from biomass waste: design/fabrication, characterization and applications. Front Energy Res 9:67
80. Saraswat V, Yadav M (2020) Carbon dots as green corrosion inhibitor for mild steel in HCl solution. ChemistrySelect 5(25):7347–7357
81. Zaib M, Akhtar A, Maqsood F, Shahzadi T (2021) Green synthesis of carbon dots and their application as photocatalyst in dye degradation studies. Arab J Sci Eng 46(1):437–446
82. Atchudan R, Edison TNJI, Perumal S, Vinodh R, Lee YR (2018) In-situ green synthesis of nitrogen-doped carbon dots for bioimaging and TiO_2 nanoparticles@ nitrogen-doped carbon composite for photocatalytic degradation of organic pollutants. J Alloys Compd 766:12–24
83. Chandrasekaran P, Arul V, Sethuraman MG (2020) Ecofriendly synthesis of fluorescent nitrogen-doped carbon dots from coccinia grandis and its efficient catalytic application in the reduction of methyl orange. J Fluoresc 30:103–112
84. Yahaya Pudza M, Zainal Abidin Z, Abdul Rashid S, Md Yasin F, Noor ASM, Issa MA (2020) Eco-friendly sustainable fluorescent carbon dots for the adsorption of heavy metal ions in aqueous environment. Nanomaterials 10(2):315
85. Duan P, Zhi B, Coburn L, Haynes CL, Schmidt-Rohr K (2020) A molecular fluorophore in citric acid/ethylenediamine carbon dots identified and quantified by multinuclear solid-state nuclear magnetic resonance. Magn Reson Chem 58(11):1130–1138
86. Talib AB, Mohammed MH (2023) Preparation, characterization and preliminary cytotoxic evaluation of 6-mercaptopurine-coated biotinylated carbon dots nanoparticles as a drug delivery system. Mater Today Proc 80:2327–2333

87. Esfandiari N, Bagheri Z, Ehtesabi H, Fatahi Z, Tavana H, Latifi H (2019) Effect of carbonization degree of carbon dots on cytotoxicity and photo-induced toxicity to cells. Heliyon 5(12)
88. Bajpai SK, D'souza A, Suhail B (2019) Blue light-emitting carbon dots (CDs) from a milk protein and their interaction with *Spinacia oleracea* leaf cells. Int Nano Lett 9:203–212
89. He Y, He J, Zhang H, Liu Y, Lei B (2017) Luminescent carbon dots assembled into mesoporous aluminas for oxygen sensing. Opt Mater Express 7(3):945–954
90. Jlassi K, Eid K, Sliem MH, Abdullah AM, Chehimi MM, Krupa I (2020) Rational synthesis, characterization, and application of environmentally friendly (polymer–carbon dot) hybrid composite film for fast and efficient UV-assisted Cd^{2+} removal from water. Environ Sci Eur 32(1):1–13
91. Lin H, Ding L, Zhang B, Huang J (2018) Detection of nitrite based on fluorescent carbon dots by the hydrothermal method with folic acid. R Soc Open Sci 5(5):172149
92. Lou Y et al (2021) Recent advances of biomass carbon dots on syntheses, characterization, luminescence mechanism, and sensing applications. Nano Select 2(6):1117–1145
93. Pal A, Sk MP, Chattopadhyay A (2020) Recent advances in crystalline carbon dots for superior application potential. Mater Adv 1(4):525–553
94. Ramachandran S, Sathishkumar M, Kothurkar NK, Senthilkumar R (2018) Synthesis and characterization of graphene quantum dots/cobalt ferrite nanocomposite. In: IOP conference series: materials science and engineering. IOP Publishing, p 12139
95. Rojas-Valencia OG, Regules-Carrasco M, Hernández-Fuentes J, Reza-San Germán CM, Estrada-Flores M, Villagarcía-Chávez E (2021) Synthesis of blue emissive carbon quantum dots from Hibiscus Sabdariffa flower: surface functionalization analysis by FT-IR spectroscopy. Materialia (Oxf) 19:101182
96. Shaikh AF, Tamboli MS, Patil RH, Bhan A, Ambekar JD, Kale BB (2019) Bioinspired carbon quantum dots: an antibiofilm agents. J Nanosci Nanotechnol 19(4):2339–2345
97. Hussain OA, Hathout AS, Abdel-Mobdy YE, Rashed MM, Rahim EAA, Fouzy ASM (2023) Preparation and characterization of activated carbon from agricultural wastes and their ability to remove chlorpyrifos from water. Toxicol Rep 10:146–154
98. Shahraki HS, Bushra R, Shakeel N, Ahmad A, Ahmad M, Ritzoulis C (2023) Papaya peel waste carbon dots/reduced graphene oxide nanocomposite: from photocatalytic decomposition of methylene blue to antimicrobial activity. J Bioresources Bioprod 8(2):162–175
99. Chugh R, Kaur G (2023) Citrus sinensis agro-waste peels mediated CQDs-Ag nanocomposite potentiates numerous applications
100. Ferjani H, Abdalla S, Oyewo OA, Onwudiwe DC (2024) Facile synthesis of carbon dots by the hydrothermal carbonization of avocado peels and evaluation of the photocatalytic property. Inorg Chem Commun 160:111866
101. Rodwihok C et al (2022) Preparation and characterization of photoluminescent graphene quantum dots from watermelon rind waste for the detection of ferric ions and cellular bio-imaging applications. Nanomaterials 12(4):702
102. Esdaille CJ, Washington KS, Laurencin CT (2021) Regenerative engineering: a review of recent advances and future directions. Future Medicine Ltd. https://doi.org/10.2217/rme-2021-0016
103. Akhtar A et al (2021) Neurodegenerative diseases and effective drug delivery: a review of challenges and novel therapeutics. J Control Release 330:1152–1167
104. Fang H et al (2022) Emerging zero-dimensional to four-dimensional biomaterials for bone regeneration. BioMed Central Ltd. https://doi.org/10.1186/s12951-021-01228-1
105. Peng Z, Zhao T, Zhou Y, Li S, Li J, Leblanc RM (2020) Bone tissue engineering via carbon-based nanomaterials. Wiley. https://doi.org/10.1002/adhm.201901495
106. He C, Xu P, Zhang X, Long W (2022) The synthetic strategies, photoluminescence mechanisms and promising applications of carbon dots: current state and future perspective. Carbon N Y 186:91–127
107. Shao D et al (2017) Carbon dots for tracking and promoting the osteogenic differentiation of mesenchymal stem cells. Biomater Sci 5(9):1820–1827. https://doi.org/10.1039/c7bm00358g

108. Han Y et al (2019) Bioactive carbon dots direct the osteogenic differentiation of human bone marrow mesenchymal stem cells. Colloids Surf B Biointerfaces 179:1–8. https://doi.org/10.1016/j.colsurfb.2019.03.035
109. Lu Y et al (2018) Zero-dimensional carbon dots enhance bone regeneration, osteosarcoma ablation, and clinical bacterial eradication. Bioconjug Chem 29(9):2982–2993. https://doi.org/10.1021/acs.bioconjchem.8b00400
110. Shafiei S et al (2019) Egg shell-derived calcium phosphate/carbon dot nanofibrous scaffolds for bone tissue engineering: fabrication and characterization. Mater Sci Eng C 100:564–575. https://doi.org/10.1016/j.msec.2019.03.003
111. Khajuria DK, Kumar VB, Gigi D, Gedanken A, Karasik D (2018) Accelerated bone regeneration by nitrogen-doped carbon dots functionalized with hydroxyapatite nanoparticles. ACS Appl Mater Interfaces 10(23):19373–19385. https://doi.org/10.1021/acsami.8b02792
112. Sierakowska A, Jasiewicz B, Piosik Ł, Mrówczyńska L (2023) New C8-substituted caffeine derivatives as promising antioxidants and cytoprotective agents in human erythrocytes. Sci Rep 13(1):1785
113. Gudkov SV, Simakin AV, Sarimov RM, Kurilov AD, Chausov DN (2021) Novel biocompatible with animal cells composite material based on organosilicon polymers and fullerenes with light-induced bacteriostatic properties. Nanomaterials 11(11):2804
114. Krishnan V et al (2015) Vortex-Aligned fullerene nanowhiskers as a scaffold for orienting cell growth. ACS Appl Mater Interfaces 7(28):15667–15673. https://doi.org/10.1021/acsami.5b04811
115. Chen Y, Li X (2022) The utilization of carbon-based nanomaterials in bone tissue regeneration and engineering: respective featured applications and future prospects. Med Nov Technol Devices 16:100168
116. Nunn N, Torelli M, McGuire G, Shenderova O (2017) Nanodiamond: a high impact nanomaterial. Elsevier Ltd. https://doi.org/10.1016/j.cossms.2016.06.008
117. Bauso LV, La Fauci V, Longo C, Calabrese G (2024) Bone tissue engineering and nanotechnology: a promising combination for bone regeneration. Biology (Basel) 13(4):237
118. Mallik AK, Chisty AH, Nuruzzaman Khan M, Shahruzzaman M, Haque P, Mizanur Rahman M (2022) Poly (lactic acid)(PLA)-based nanosystems in biomedical applications. Nanoeng Biomater 63–89
119. Siafaka PI, Özcan Bülbül E, Dilsiz P, Karantas ID, Okur ME, Üstündağ Okur N (2021) Detecting and targeting neurodegenerative disorders using electrospun nanofibrous matrices: current status and applications. J Drug Target 29(5):476–490
120. Mane PV, Rego RM, Yap PL, Losic D, Kurkuri MD (2024) Unveiling cutting-edge advances in high surface area porous materials for the efficient removal of toxic metal ions from water. Prog Mater Sci 101314
121. Shrestha S, Shrestha BK, Ko SW, Kandel R, Park CH, Kim CS (2021) Engineered cellular microenvironments from functionalized multiwalled carbon nanotubes integrating Zein/Chitosan @Polyurethane for bone cell regeneration. Carbohydr Polym 251. https://doi.org/10.1016/j.carbpol.2020.117035
122. Patel KD et al (2020) Coating biopolymer nanofibers with carbon nanotubes accelerates tissue healing and bone regeneration through orchestrated cell- and tissue-regulatory responses. Acta Biomater 108:97–110. https://doi.org/10.1016/j.actbio.2020.03.012
123. Wang C et al (2019) Incorporation of multi-walled carbon nanotubes to PMMA bone cement improves cytocompatibility and osseointegration. Mater Sci Eng C 103. https://doi.org/10.1016/j.msec.2019.109823
124. Ye L et al (2021) Carbon nanotube–hydrogel composites facilitate neuronal differentiation while maintaining homeostasis of network activity. Adv Mater 33(41). https://doi.org/10.1002/adma.202102981
125. Pampaloni NP et al (2020) Transparent carbon nanotubes promote the outgrowth of enthorino-dentate projections in lesioned organ slice cultures. Dev Neurobiol 80(9–10):316–331. https://doi.org/10.1002/dneu.22711

126. Park J et al (2020) Electrically conductive hydrogel nerve guidance conduits for peripheral nerve regeneration. Adv Funct Mater 30(3). https://doi.org/10.1002/adfm.202003759
127. Zhao G, Zhang X, Li B, Huang G, Xu F, Zhang X (2020) Solvent-free fabrication of carbon nanotube/silk fibroin electrospun matrices for enhancing cardiomyocyte functionalities. ACS Biomater Sci Eng 6(3):1630–1640. https://doi.org/10.1021/acsbiomaterials.9b01682
128. Roshanbinfar K et al (2019) Carbon nanotube doped pericardial matrix derived electroconductive biohybrid hydrogel for cardiac tissue engineering. Biomater Sci 7(9):3906–3917. https://doi.org/10.1039/c9bm00434c
129. Jiang C, Wang K, Liu Y, Zhang C, Wang B (2021) Using wet electrospun PCL/Gelatin/CNT yarns to fabricate textile-based scaffolds for vascular tissue engineering. ACS Biomater Sci Eng 7(6):2627–2637. https://doi.org/10.1021/acsbiomaterials.1c00097
130. Nagarajan KJ, Balaji AN, Ramanujam NR (2019) Extraction of cellulose nanofibers from cocos nucifera var aurantiaca peduncle by ball milling combined with chemical treatment. Carbohydr Polym 212:312–322. https://doi.org/10.1016/j.carbpol.2019.02.063
131. Kumar M et al (2019) Use of cellulose acetate/polyphenylsulfone derivatives to fabricate ultrafiltration hollow fiber membranes for the removal of arsenic from drinking water. Int J Biol Macromol 129:715–727. https://doi.org/10.1016/j.ijbiomac.2019.02.017
132. Ghaseminezhad SM, Barikani M, Salehirad M (2019) Development of graphene oxide-cellulose acetate nanocomposite reverse osmosis membrane for seawater desalination. Compos B Eng 161:320–327. https://doi.org/10.1016/j.compositesb.2018.10.079
133. Aboamera NM, Mohamed A, Salama A, Osman TA, Khattab A (2019) Characterization and mechanical properties of electrospun cellulose acetate/graphene oxide composite nanofibers. Mech Adv Mater Struct 26(9):765–769. https://doi.org/10.1080/15376494.2017.1410914
134. Ahmed MK, Menazea AA, Abdelghany AM (2020) Blend biopolymeric nanofibrous scaffolds of cellulose acetate/ε-polycaprolactone containing metallic nanoparticles prepared by laser ablation for wound disinfection applications. Int J Biol Macromol 155:636–644. https://doi.org/10.1016/j.ijbiomac.2020.03.257
135. Ye S, Xie C, Agar OT, Barrow CJ, Dunshea FR, Suleria HAR (2024) Alginates from brown seaweeds as a promising natural source: a review of its properties and health benefits. Food Rev Int 1–29
136. Yang JM et al (2016) Cell proliferation on PVA/sodium alginate and PVA/poly(γ-glutamic acid) electrospun fiber. Mater Sci Eng, C 66:170–177. https://doi.org/10.1016/j.msec.2016.04.068
137. Malik S et al (2021) Sustainable nanofibers in tissue engineering and biomedical applications. Mater Des Process Commun 3(6):e202
138. Shin SR et al (2016) Graphene-based materials for tissue engineering. Adv Drug Deliv Rev 105:255–274. https://doi.org/10.1016/j.addr.2016.03.007
139. Ricci A, Cataldi A, Zara S, Gallorini M (2022) Graphene-oxide-enriched biomaterials: a focus on Osteo and chondroinductive properties and immunomodulation. Materials 15(6):2229
140. Kandhola G et al (2023) Nanomaterial-based scaffolds for tissue engineering applications: a review on graphene, carbon nanotubes and nanocellulose. Tissue Eng Regen Med 20(3):411–433
141. Shin SR et al (2013) Carbon-nanotube-embedded hydrogel sheets for engineering cardiac constructs and bioactuators. ACS Nano 7(3):2369–2380. https://doi.org/10.1021/nn305559j
142. Norahan MH, Amroon M, Ghahremanzadeh R, Mahmoodi M, Baheiraei N (2019) Electroactive graphene oxide-incorporated collagen assisting vascularization for cardiac tissue engineering. J Biomed Mater Res A 107(1):204–219. https://doi.org/10.1002/jbm.a.36555
143. Ovcharenko EA et al (2020) A new nanocomposite copolymer based on functionalised graphene oxide for development of heart valves. Sci Rep 10(1). https://doi.org/10.1038/s41598-020-62122-8
144. Zhao G et al (2018) Reduced graphene oxide functionalized nanofibrous silk fibroin matrices for engineering excitable tissues. NPG Asia Mater 10(10):982–994. https://doi.org/10.1038/s41427-018-0092-8

145. Nyambat B, Chen CH, Wong PC, Chiang CW, Satapathy MK, Chuang EY (2018) Genipin-crosslinked adipose stem cell derived extracellular matrix-nano graphene oxide composite sponge for skin tissue engineering. J Mater Chem B 6(6):979–990. https://doi.org/10.1039/c7tb02480k
146. Jafarkhani M et al (2020) Graphene functionalized decellularized scaffold promotes skin cell proliferation. Can J Chem Eng 98(1):62–68. https://doi.org/10.1002/cjce.23588
147. Jian Z et al (2020) Polyurethane-modified graphene oxide composite bilayer wound dressing with long-lasting antibacterial effect. Mater Sci Eng C 111. https://doi.org/10.1016/j.msec.2020.110833
148. Zhang D et al (2020) Surface-Anchored graphene oxide nanosheets on cell-scale micropatterned poly(d, l-lactide-co-caprolactone) conduits promote peripheral nerve regeneration. ACS Appl Mater Interfaces 12(7):7915–7930. https://doi.org/10.1021/acsami.9b20321
149. Domínguez-Bajo A, González-Mayorga A, López-Dolado E, Munuera C, García-Hernández M, Serrano MC (2020) Graphene oxide microfibers promote regenerative responses after chronic implantation in the cervical injured spinal cord. ACS Biomater Sci Eng 6(4):2401–2414. https://doi.org/10.1021/acsbiomaterials.0c00345
150. Xie W et al (2018) Mechanically robust 3D graphene–hydroxyapatite hybrid bioscaffolds with enhanced osteoconductive and biocompatible performance. Crystals (Basel) 8(2). https://doi.org/10.3390/cryst8020105
151. Nie W et al (2017) Three-dimensional porous scaffold by self-assembly of reduced graphene oxide and nano-hydroxyapatite composites for bone tissue engineering. Carbon N Y 116:325–337. https://doi.org/10.1016/j.carbon.2017.02.013
152. Qian Y, Zhao X, Han Q, Chen W, Li H, Yuan W (2018) An integrated multi-layer 3D-fabrication of PDA/RGD coated graphene loaded PCL nanoscaffold for peripheral nerve restoration. Nat Commun 9(1). https://doi.org/10.1038/s41467-017-02598-7
153. Guo R et al (2021) Biomimetic 3D bacterial cellulose-graphene foam hybrid scaffold regulates neural stem cell proliferation and differentiation. Colloids Surf B Biointerfaces 200. https://doi.org/10.1016/j.colsurfb.2021.111590
154. Lingzhi Z, Meirong L, Xiaobing F (2020) Biological approaches for hypertrophic scars. Int Wound J 17(2):405–418

Chapter 11
Waste-Driven Carbon Nanomaterials for Drug Delivery Application

Sunita Barik and Rahul Kumar Sharma

Abstract In the present decade, urbanization, industrial expansion, overpopulation, etc. have generated huge volumes of waste. It is becoming a universal challenge that needs an imperative solution. In addition, significant progress in the biomedical field has influenced the complete spectrum of medicine and healthcare. This has broadened the further discovery of different biomedical strategies for the detection and healing of different diseases. A wide variety of nanomaterials have been developed for diverse biomedical applications like bio-imaging, bio-sensors, vaccinations, drug delivery, etc. The exploitation of waste to generate precious products has unexpected compensation so far as climate and economy are concerned. Carbon-based nanomaterials comprising graphene, fullerenes, carbon dots, carbon nanofibers, carbon nanotubes, etc. have broad biomedical applications on account of their distinct thermal, optical, electrical, mechanical, and superior biocompatible nature. Although several methods have been reported highlighting the synthesis of high-value carbon nanomaterials, the production of carbon nanomaterials recycled or prepared using wastes as precursors is noteworthy. These not only evade harmful effects on the environment but also execute circular economy practices. In the present chapter, a general discussion on different carbonous waste materials, along with waste-driven carbon nanomaterials in particular for drug delivery applications is highlighted.

Keywords Biomass · Graphene · Carbon nanomaterials · Carbon quantum dots · Drug delivery

S. Barik (✉)
Dibakar Pattanaik Institute of Advanced Studies in Education, Berhampur, Odisha 761008, India
e-mail: bariksunita2010@gmail.com; sunitabarik2010@gmail.com

R. K. Sharma
Govt. Shyam Sundar Agrawal PG College Sihora, Jabalpur, Madhya Pradesh 483225, India

N. Talreja et al. (eds.), *Waste-Derived Carbon Nanostructures*, Nanostructure Science and Technology, https://doi.org/10.1007/978-3-031-75247-6_11

11.1 Introduction

Globally waste management and its harmful outcomes on the ecosystem and living organisms are some of the foremost topics that are threatening our present and future. Population explosion and extraordinary rise in the global economy in consort with global productivity are the main causes of the huge production of solid waste materials [1]. Waste materials starting from domestic homes to industry significantly influence our environment and the biological system of humans as well as animals. Every day discarding a huge portion of this waste to the earth's surface and into the water body is becoming the greatest hazard to the survival of living organisms, particularly human beings and aquatic as well as terrestrial animals. This dumping is leading to a decrease in our productive agricultural terrestrial, in addition to a potable water crisis, food intoxication, genetic mutation, respirational issues, new diseases, and many more. At the same time, waste management of solid waste is deprived owing to the speedy rise in global population [2]. To have a healthier solution to waste management, the finest way is an invent advanced scientific approaches that will convert waste into valuable value-added products [3]. In the present century, waste materials are chosen as precursors for the creation of useful products that are of prime importance in different fields of science and technology [4, 5].

11.2 Carbon-Based Nanomaterials

Owing to the growing demand for carbon-based nanomaterials (CNMs), researchers have put forth unbelievable efforts in the production of CNMs using waste materials as precursors. This not only helps in reducing the adverse consequences of the waste on the environment but is also economical. A variety of restrained carbon-based nanostructures like graphene, fullerene, carbon nanofibers, carbon dots, carbon nanotubes (CNTs), etc. as shown in Fig. 11.1, have attracted considerable worldwide attention and have raised as a fascinating material in recent technology time owing to their characteristic physicochemical, electrical, mechanical, optical, electronic, thermal along with improved biocompatibility nature [6]. CNMs have been used commonly in catalysis, fuel cells, supercapacitors, batteries, and water purifiers, as nanofillers in the polymer industry, bio-imaging, drug delivery systems, and many more [1]. Though there are several conventional methods adopted for the production of CNMs, the production of CNMs using waste as the starting material is likely to decrease the volume of waste along with lesser production costs. More significantly, the use of waste as a resource is an inspiring step for recycling technology. Generally, high carbon content wastes particularly waste plastics along with several biowastes are marked as precursors for making CNMs [7].

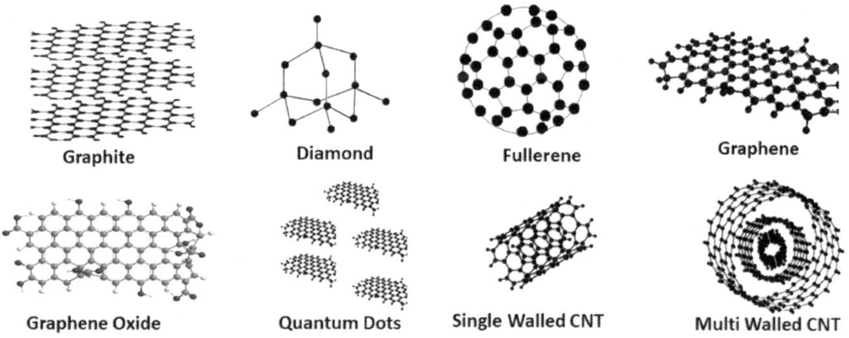

Fig. 11.1 Different types of CNMs

11.2.1 Graphene-Based Nanomaterials (GBNS)

Graphene is an allotrope of carbon comprising a monolayer of atoms organized in a hexagonal lattice structure. The carbon atom in graphene is sp^2 hybridized and is linked to its adjacent carbon by σ-bonds and a delocalized π-bond, that contributes to a valence band, which ranges over the whole sheet. The graphene layers stacked on top of each other result in graphite, in which the distinct layers are held together by van der Waals forces with an interplanar space of 0.335 nm. It is a very good conductor of heat and electricity along its plane, and it sturdily absorbs visible light. Owing to its exciting thinness, a graphene sheet is nearly transparent, and it is the toughest material ever measured. Owing to its transparency, electrical conductivity, high tensile strength, etc., it has drawn worldwide attention among researchers and found wide applications in bioengineering, biotechnology, biomedicine, and several interdisciplinary sciences [8].

GBNs comprise only carbon. Still, it is a matter of discussion to realize how derivatives of carbon like reduced graphene oxide (rGO) and graphene oxide (GO) perform in a biotic system and the time taken by these materials to get evacuated from living organisms [9]. During fabrication, GBNs generally undertake several chemical treatment processes for doping, oxidation, and functionalization, which leads to the introduction of functional groups. This reflects that graphene derivatives found suitable for biomedical applications comprise metals and/or impurities in addition to carbon. To be a suitable biomaterial, graphene being hydrophobic needs functional modification. It may be non-covalent or covalent functionalization [10]. Covalent functionalization progresses processability and solubility whereas non-covalent functionalization advances biocompatibility, sensing or binding capacity, dispersibility, and reactivity [11]. In comparison to other carbon-based materials, GBNs possess a large surface area, superior solubility, and easily modifiable surface by various functional groups, which makes them a very good choice for biomedical use. There are numerous structural differences among the extensive family members of GBNs i.e., Graphene, chemically modified graphene, GO, and rGO that govern their physicochemical behavior.

Graphene is a novel nanocarrier for drug delivery application because of its inherent optical properties, small size, cost-effectiveness, vast specific surface area along suitable non-covalent interactions with aromatic drug molecules [8]. GO possesses certain special characteristics, for example, purity, layer number, surface area, surface chemistry, lateral dimension, etc. which are appropriate for drug delivery applications. Due to the abundant oxygen functional groups (epoxy, hydroxyl, and carboxyl) on the surface, it becomes easily soluble and gets conjugated to different molecules effortlessly, which makes GO an outstanding multifunctional nanomaterial for drug delivery applications [8, 10]. rGO is reflected as a halfway structure in between extremely oxidized GO and graphene sheet. Very good biocompatibility nature, high water dispersibility, ease of surface functionalization, and ample oxygen functional groups (hydroxyl, carboxyl, epoxy, carbonyl) at the edges of the plane surface, etc. makes rGO an outstanding nanomaterial for drug delivery applications [8].

11.2.2 Carbon Quantum Dots (CQDS)

Carbon quantum dots (CQDs) are carbon nanoparticles with a size below 10 nm. Worldwide widely these materials are drawing much more consideration owing to their unique features of photoluminescence, high chemical stability, biocompatibility, degradability, less toxicity, easy synthesis, etc. in comparison to conventional semiconductor quantum dots. Owing to their exceptionally gifted fluorescence behavior, they have vast applications in different fields of science and technology like optoelectronic devices, energy conversion, catalysis, chemical sensing, bio-imaging, drug delivery, and many more [7, 8].

11.2.3 CNTs

Among the different allotropic forms of carbon, CNTs are one-dimensional allotropical forms with diameters and lengths in the order of nanometers and micrometers, correspondingly. They are described as extended hollow hexagonal structures of graphene, rolled up into cylindrical tubular form. Depending on the number of walls present, further, they are classified as single-walled carbon nanotubes (SWCNTs) and multi-walled carbon nanotubes (MWCNTs). These materials display excellent features like large surface area, high aspect ratio, good electrical conductivity, outstanding mechanical strength, and stiffness, etc., for which they find wide applications in diverse fields. Owing to their extraordinary features like nanosized hollow huge structure, outstanding chemical resistance, distinct surface properties, effortless attachment of required functional groups on the outer layers, etc., CNTs are used as nanocarriers for drug delivery application to targeted cells [12].

11.3 Synthesis of CNMs from Waste

Synthesis of nanomaterials using wastes is considered a greener method in which wastes are converted into different types of CNMs in a controlled way [13, 14]. To date, several preparative methods (i.e., thermal/combustion methods, microwave pyrolysis, template method, solvothermal/hydrothermal, laser ablation, arc discharge, ultrasonic passivation, electrochemical exfoliation, etc.) have been developed for the preparation of carbon-based nanomaterials from waste products [15, 16]. Some of the carbonaceous wastes and diverse methods employed for the synthesis of CNMs are shown in Fig. 11.2.

Based on the nature of the precursor, these methods are broadly discussed under two categories "top-down" and "bottom-up" approaches. Transforming the bulk materials into nanoscale materials is considered as top-down approach, however, the formation of nanomaterials from molecular precursors is termed as bottom-up

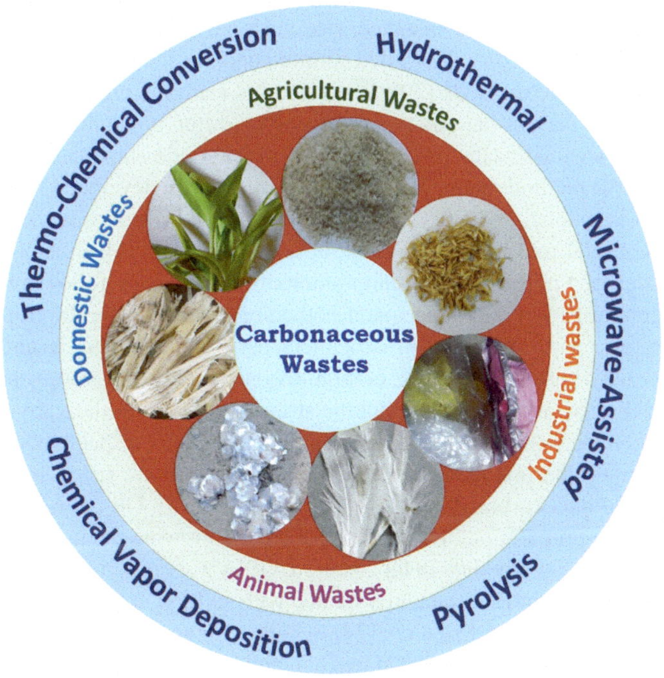

Fig. 11.2 Some of the carbonaceous wastes

approach [17]. In different synthesis methods, judicious selection of different physical parameters like temperature, pressure, nature of precursor, environmental conditions (inert or air), catalyst, etc., are the prerequisites to obtain desired carbon-based nano products. Consequently, numerous carbon-based nanomaterials like graphene and its derivatives, quantum dots, CNTs, etc. have been prepared to date. Next, a brief discussion on different waste sources for the preparation of different carbon-based nanomaterials is presented. Depending on the generated sources, wastes released to the environment can be broadly discussed under the following headings. These are:

1. Agricultural wastes
2. Domestic wastes
3. Animal wastes
4. Industrial wastes

11.3.1 Agricultural Wastes

In agricultural wastes, the composition of carbon is usually high which can be utilized for the synthesis of carbon-based nanomaterials. Numerous agricultural wastes like rice husk, wheat straw, coconut shell, tea tree plants, oil palm leaves, sugarcane bagasse, etc. are extensively employed for the preparation of various kinds of carbon nanomaterials [4, 18, 19]. For instance, rice husk has been used for the synthesis of graphene, single-walled/multi-walled carbon nanotubes, and carbon nano-onions by various preparative methods [20]. Wang and his team members have developed different types of carbon nanomaterials following different routes using different sources. Using rice husk as the precursors following a single-step microwave plasma irradiation (MPI) method, under H_2 and Ar flow, they have developed nano carbons starting from graphene, and CNTs to graphene-CNT hybrid. In this method, pressure plays a crucial role in the formation of nanomaterials [21]. In another report, Jin et al. developed porous carbon nano-onions from rice husk following a combination of graphitization and subsequent activation process. The developed material exhibits an extraordinary electrochemical performance. This is attributed to the tremendously high surface area, carbon–oxygen functional groups, along with 3Dimensional micro-open porous graphitic structures [22]. Similarly, by using rice residue and glycine as the precursor for carbon and nitrogen, respectively, following a single-step hydrothermal method, Qi and his co-workers have successfully synthesized nitrogen-doped carbon quantum dots (N-CQDs) [23]. Oil palm leaves (OPL) were used for the preparation of nanoporous carbon following the pyrolysis method by Yallappa and his co-workers. In this method, OPLs were initially heated at 600 °C in an N_2 environment and subsequently treated with sulfuric and nitric acids mixture (3:1 vol/vol), in order to oxidize them [24]. Another cheap and easily available agricultural waste i.e., wheat bran, is used by John and his associates to develop carbon dots by following the hydrothermal method, where the suspension of wheat barn was heated at 180 °C for 3 h [25]. Fahmi et al. prepared the CDs by pyrolyzing the cellulose of bamboo leaves at 300 °C [26]. Similarly, by pyrolysis Manaf et al. have also prepared carbon nanosphere from the sago bark of Metroxylon palm [27]. Sugarcane bagasse char is also employed as a precursor for synthesizing CDs. Using this as starting material, following hydrothermal method, Chung et al. have synthesized CDs at different temperatures such as 150, 170, 190, and 210 °C. However, CDs synthesized at 190 °C exhibit the greatest fluorescence behavior and it was merged with the hydroxyapatite [19]. Vikarm et al. prepared MWCNTs from an agricultural waste of chickpea peels by pyrolyzing it at 400 °C in air condition. On heating at such a high temperature, in addition to carbon and hydrogen, oxygen is also produced which played a crucial role not only in the initiation of the CNT precipitation but also assisted in the formation of an embryonic cap resulting in the formation of MWCNT [28]. Furthermore, waste of walnut (Juglans regia) shells is also used to form CNTs through pyrolysis. In this process, walnut shell waste is first converted into charcoal, and subsequently on further heating in the presence of $ZnCl_2$, it results activated carbon. Thereafter, activated carbon along with benzene and ferrocene is pyrolyzed at 700 °C for 4 h resulting in the formation of CNTs [29].

11.3.2 Domestic Wastes

Domestic wastes like fruit peels, vegetable peels, waste oil, and plastic bags, are mainly composed of hydrocarbon. That means they can be used as a source of carbon to produce numerous carbon-based nanomaterials [30–32]. In this context, currently, extensive work has been carried out in which domestic wastes are transformed into nanoscale materials by employing different preparative methods. For example, CNTs are synthesized by using shopping plastic bags (i.e., linear low-density polyethylene (LLDPE)) through the pyrolysis of small pieces of plastic bags in the CVD reactor. Herein the LLDPE is used as a precursor of carbon [33]. Waste oil is also an important waste that can be used as a free source of carbon [31, 32, 34]. Cooking palm oil is used in household frying, so its waste can also be used to prepare carbon-based nanomaterials. Jung et al. utilized waste frying oil to synthesize onion-like carbon nanoparticles (CNOs) by following a simple and catalyst-free pyrolysis technique. These as-prepared CNO particles form a chain-like interconnected network in which nanoparticles are linked to each other [34]. Vertically aligned carbon nanotubes (VACNT) were synthesized by Rusop and co-workers by using a green source of carbon i.e., waste cooking palm oil. To prepare the VACNT, refined palm oil was used to fry three folds at 180 °C and then blackened oil was filtered. Subsequently, the waste palm oil was mixed with 5.33 wt% ferrocene (served as a catalyst) and heated in a furnace to obtain the VACNT [31]. Azam et al. have used palm-based waste cooking oil as a carbon source for preparing graphene nanoplatelets (GNPs). These GNPs were prepared by catalytic chemical vapor deposition (CVD) technique in which nickel oxide powder was used as the catalyst [32]. The conversion of fruit peels to carbon nanomaterials is another economical, and environmentally benign method. Different fruit peels have been used to produce carbon nanomaterials [30]. Fruit peels like watermelon rind waste and orange peels have been used for the synthesis of Graphene quantum dots and carbon dots through green and simple one-pot hydrothermal technique at moderate temperature [35–38]. Groundnut shell waste is a natural biowaste that can be used as a source of carbon. Nano carbons have been synthesized from groundnut shells via a single-step pyrolysis technique in which the sample is heated in an inert environment (N_2 atmosphere) at 550, 750, and 950 °C temperatures [39]. On the other hand, the Peanut shell waste was used to prepare the carbon quantum dots (CQDs) at moderate temperature by using a simple one-step hydrothermal method [40].

11.3.3 Animal Wastes

It has been noticed that animal wastes like waste crab shells, fish scales, chicken feathers, and so on can also be feedstock of carbon to synthesize the myriads of nanoscale carbon materials [41–45]. Yao et al. used the Chitin of crab shell waste to prepare the carbon quantum dots (CQDs) via a one-pot microwave-assisted pyrolysis method. During synthesis, chitin is utilized as a source of carbon as well as a chelating ligand leading to the formation of Gd@CQDs, Mn@CQDs, and Eu@CQDs with Gd^{3+}, Mn^{2+}, and Eu^{3+}, respectively [41]. Fish scale is also mainly made up of chitin which is a highly abundant natural biopolymer, and collagen. Therefore, in the fish scale, ample carbon, hydrogen, oxygen, and nitrogen are present that can be used as a source for the synthesis of carbon nanomaterial. The scales of Sea Bass (Dicentrarchus labrax) were used for the synthesis of nitrogen-doped carbon dots by hydrothermal method, without the requirement of any other reagent [42]. Pajarito et al. used the waste chicken feathers (WCFs) to synthesize GNPs where WCFs were first graphitized and then by using high-speed homogenization and sonication method, the graphitic carbon was exfoliated [43]. Hastuti and his co-workers have used chicken feathers for the preparation of carbon dots (CDs) following the hydrothermal method. They have studied the effect of different reaction conditions such as reaction time (5 and 7 h) and temperature (160 °C, 180 °C, and 200 °C) on the yield on CDs. It was reported that the highest yield of CDs was obtained when reaction condition was maintained at 200 °C for 7 h [45].

11.3.4 Industrial Wastes

Industries produce some carbonaceous products, especially plastics waste, Petroleum refining waste, chemical waste, and many more. Owing to being rich in carbon, industrial wastes can be used as precursors to prepare carbon nanomaterials like CNTs, MWCNTs, etc. To synthesize these nanomaterials, various synthesis methods have been explored Arc discharge, CVD pyrolysis, CVD (plasma), laser deposition in the presence and absence of catalyst, and others [46]. For instance, Wu and his co-workers have reported the synthesis of CNTs from waste plastics following the thermos-chemical conversion method. To scale up the yield of CNTs, the effect of different concentrations of Mn (from 0 to 10 wt%), and Fe-based catalyst (from 23.4 to 32.9 wt%) was investigated. They noticed that on increasing the concentration of Mn and catalyst, the production of CNTs also significantly gets increased [47]. Singh et al. synthesized CNTs from the composite of fly ash/ red mud/ rock sample and PVA (polyvinyl alcohol). In this method, by alkali activation, the fly ash/red mud/ rock sample was converted into an activated sample. These fly ash/red mud/ rock sample was used as a filler and assisted in the fabrication of the bio-composite film of PVA. When bio-composite film was pyrolyzed at 500 °C, CNTs were formed [48].

11.4 Waste-Derived Carbon Nanomaterials for Drug Delivery Application

Drug delivery is the process of managing a therapeutic compound to achieve a beneficial effect in both humans and animals. In the present decade, nanoparticles are highly employed as a carrier of drugs and deliver them to a specified area in the body. The use of nanotechnology in the biomedical field, in particular for drug delivery has several advantages in comparison to traditional drug intake. It increases the drug potency, lowers toxicity to the targeted cells, helps in precise targeting, reduces side effects, and many more [49]. Targeted delivery, otherwise known as smart delivery, is believed to improve efficacy by reducing side effects. In traditional drug delivery for example intravascular injection or oral ingestion, the medicine is circulated throughout the whole body through blood circulation. As a result in most cases, a very small amount of the medicine reaches the affected organ, whereas in targeted drug delivery, the medicine gets concentrated in the desired tissues. Nanoparticles loaded with drugs release the drug in a quantified manner as a result uniform effect of the drug is noticed along with a decrease in the frequency of dosages. On the contrary, the high cost of the nanocarriers makes efficiency more tough [50, 51]. Among the different carbon nanomaterials, to date, CQDs have been much explored for targeted drug delivery. An overall mechanism of drug delivery is shown in Fig. 11.3 [14].

Some of the waste-driven CNMs, particularly for drug delivery applications are discussed below and are summarized in Table. 11.1.

11.4.1 CQD for Drug Delivery Application

Although a wide variety of synthesis routes are available for CQD preparation using different chemicals, still a good number of green carbon precursors like plant leaves, vegetables, waste papers, waste kitchen materials, fruit peels, and juices, and many more have been optimized for producing CQDs. Biowaste sources are drawing much

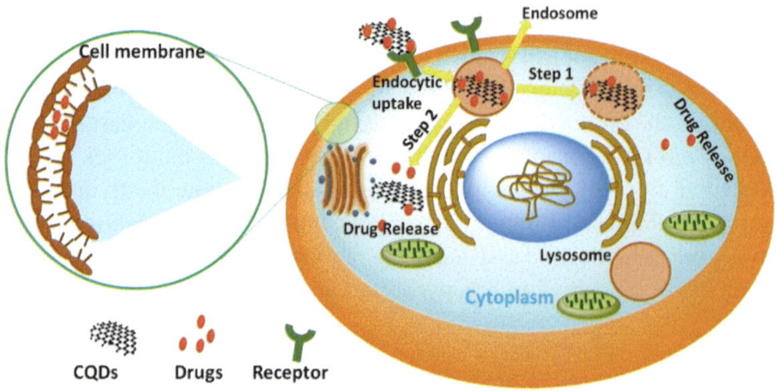

Fig. 11.3 An overall mechanism of drug delivery using nanocarriers

more attention as these are cheap and renewable and they possess very good biocompatibility and biodegradability. The synthesis strategies and optimization conditions of CQD not only affect its geomorphology and properties but also play a crucial role in understanding the quantum phenomena that control their biomedical applications. Presently, some of the conventional drug delivery methods like transdermal administration, parenteral route, oral administration, pharmaceutical drugs inhalation, etc. face disadvantages as there is a probability of drug dosage mislaid and drug level fluctuation leading to drip in healing effect.

In accordance with green chemistry principles, Himaja et al. have synthesized extremely fluorescent water-soluble C-dots from kitchen waste (fresh pineapple and cucumber peel) in a small-time duration at a temperature just above 100 °C. The overall process was less time-consuming, simple, and cost-effective. A comparative study reveals better results for cucumber peel-extracted C-dots. Different test results, biocompatibility along with non-toxic nature, suggest the suitability of the materials for biosensing, bioimaging, and drug delivery application [52]. Following a microwave-assisted synthesis method, novel water-soluble magnetic fluorescent CQDs have been developed from waste crab shells by Yao and his team members. Diverse heteroatoms like Mn $^{2+}$, Eu $^{3+}$ & Gd $^{3+}$ transition metal ions having five, six, and seven no. of unpaired electrons respectively were chosen as dopants and were incorporated into the carbon matrix. Due to the presence of unpaired electrons, the material possesses fairly high electronic spin that produces a strong local magnetic field in response to the applied external field. The magneto fluorescent CQDs (MFCQDs) on conjugation with folic acid (FA) empowered precise targeting to folate receptor-positive HeLa cells. They have selected Doxorubicin (DOX) as a model drug for conjugation with FAGd@CQDs. The as-prepared nanocomposites showed significantly higher cytotoxicity toward HeLa cells than free DOX. In vivo and in vitro studies reveal promising biocompatibility and superior cytotoxicity by as-prepared FA-Gd@CQDs/DOX in comparison to free DOX, signifying it a hopeful material for drug delivery application [41]. *Escherichia coli (E Coli)* is a rod-shaped,

Table. 11.1 Carbon nanomaterials along with sources, method of preparation, and application

S. No.	CNMs	Carbon source	Process	Application	Year	References
1	Carbon dots	Fresh pineapple and cucumber peel	Green synthesis	Drug delivery, bioimaging, biosensing	2014	[52]
2	Carbon dots	Waste crab shell	Microwave-assisted pyrolysis	Drug delivery, bioimaging	2017	[41]
3	Carbon dots	Genomic DNA of *Escherichia coli*	Hydrothermal synthesis	Drug delivery, bioimaging	2015	[53]
4	Carbon dots	Bamboo leaves cellulose	Pyrolysis	Drug delivery	2018	[26]
5	Carbon dots	Sugarcane bagasse char	Hydrothermal method	Drug delivery	2020	[19]
6	Carbon dots	Wheat bran	Hydrothermal method	Drug delivery	2020	[25]
7	Meso-SiO_2/C-dots complex	Rice husks	Hot injection method	Drug delivery, bioimaging	2014	[54]
8	Carbon dots@SiO_2	Rice husks	Hot injection method	Biomedical application	2020	[55]
9	Carbon dots	Cocoa pod husk	Microwave irradiation method	Drug delivery, Bioimaging	2022	[57]
10	Nitrogen-doped carbon dots	Water hyacinth	Hydrothermal method	Anti-cancer activity	2020	[58]
11	Carbon nanotube	Plastic waste	CVD synthesis	Drug delivery	2017	[33]
12	Nanoscale carbon	Groundnut shell	Pyrolysis	Drug delivery, bioimaging, growth inhibitor	2017	[39]
13	Nanoporous carbon	Oil palm leaves	Pyrolysis	Drug delivery, cellular imaging	2018	[24]

(continued)

Table. 11.1 (continued)

S. No.	CNMs	Carbon source	Process	Application	Year	References
14	Carbon nanoscale systems	Sago bark	Pyrolysis	Drug delivery, Bioimaging	2017	[27]
15	CRGO-MC composite	Disposed battery cell	Electrolysis	Drug delivery	2020	[59]

gram-negative, coliform bacteria most commonly found in the lower intestine of humans and animals. Its presence in water gives a strong indication of animal or sewage waste contamination. DNA is hardly taken as a starting material seeing its probable for synthesis of carbon-based material, in particular with CDs. Ding and his co-workers have reported the development of biocompatible, fluorescent CDs using *E. Coli* bacteria genomic DNA as the precursor following an oxidant-free hydrothermal method, that can be used as a delivery vehicle to deliver drugs into eukaryotic cells [53]. Fahmi and his associates have developed carbon dots from bamboo tree waste. By the pyrolysis of bamboo leaf cellulose, they have synthesized C-dots. To make the material more suitable for multifunctional applications, the developed C-dots were modified with a targeting agent i.e., 4-carboxybenzylboronic acid that stabilizes it. Additionally, doxorubicin a tumor drug was attached to C-dots to study its drug delivery ability to HeLa tumor cells [26].

Drugs like acetaminophen are safe at therapeutic dosage but its over-dosage leads to serious liver failure in humans. However, controlled drug delivery technology like the use of polymeric carriers, liposomes, nanoparticles, etc. can aid in sustaining drug dosage at an effective level with supreme therapeutic effectiveness [19]. Hydroxyapatite (HAP) is a gifted biomaterial with a wide range of applications. It can be used for cell culture, antibody purification on an engineering scale, as a mock blood vessel, as a scaffold in bone tissue engineering, for drug delivery, and many more. However, targeting ability, and precise drug release are the challenges associated with pure HAP which can be overcome by use of its composite [18]. Owing to exceptional biocompatibility along with decent hydrophilicity nature and other characteristics, CDs are active medium for drug delivery. However, longtime quenching leads to its destruction and it gets diffused during the imaging and tracking process. By combining these two nanoparticles, the drawbacks can be overcome. HAP nanoparticles act as a shell and shield the CDs from demolition. Simultaneously, due to the fluorescence actions of CDs, the drug gets traced efficiently. Following the hydrothermal method, using sugarcane bagasse Chung and his associates synthesized carbon dots which were further used for the preparation of CD-HAP nanocomposite. The developed material was found suitable to be used as a drug carrier of acetaminophen in humans [19].

Amoxicillin (AMX) a semisynthetic drug, closely allied to ampicillin belongs to the penicillin family. Alike other members, AMX is bactericidal with very good activity against certain gram-negative bacilli and some gram-positive cocci. For ample annihilation of bacterial infections, penetration of the drug through the mucus layer along with its presence at a fixed concentration for an appropriate period at the infection site is desired. If AMX is administered in high doses, it can seriously damage the stomach cell lining along with it can also develop drug resistance in the bacteria. Thus, a drug delivery system that will release the drug at a reliable rate for a longer duration even in an acidic medium is desired. Through hydrothermal treatment, John et al. have prepared photoluminescent CDs from wheat bran, an agricultural by-product, and subsequently conjugated it with AMX. The drug release characteristics of AMX–CD conjugate were studied at diverse pH to find out its probable use as a drug delivery agent for AMX. The result reveals the suitability of AMX-loaded CDs as an antibacterial drug delivery agent with minimum cytotoxic effect [25].

Agricultural wastes are inexpensive and plentily available in our surroundings. They are good reservoirs of cellulose, hemicellulose, and lignin with high carbon content for which they are extensively used as carbon precursors for carbon nanomaterials production. Rice husks and agricultural waste are rich in carbon and silica. Taking it as precursors, using the hot injection method Pandey et al. have synthesized mesoporous silica oxide micro-particles/C-dot complex. The extremely biocompatible nature of the developed complex for Vero cells was noticed. Although independently both mesoporous silica and C-dots can act as drug delivery vehicles, however, a high drug loading proficiency can be achieved for their composite along with bio-imaging application [54]. Similarly, by using rice husks as precursors, Sun and his associates have developed a simple cost-effective method for preparing carbon dots in situ at room temperature with a synchronized lengthy lifetime, superior stability along high phosphorescence quantum efficiency. Different analysis results reflect the suitability of the material for biomedical applications [55]. In the cocoa industry after the exclusion of cocoa beans from the fruit, cocoa pod husk (CPH) is obtained. It is a waste by-product which is about 52–76% of the cocoa fruit weight. In order to support the global production of cocoa beans, presently quite a lot of tropical countries are encouraging the farmers to plant more and more cocoa trees. Along with the upsurge in the production of cocoa beans, CPH generation is also increasing [56]. The chemical analysis reflects a high percentage of cellulose, hemicellulose, and lignin in it, for which CPH is a suitable precursor for the production of CDs. Following the microwave irradiation method, Norman and his co-workers have developed CPH-CDs. In vitro, cytotoxicity study of the developed material was carried out via the MTT assay using human colon adenocarcinoma cells which reflects the suitability of the material for drug delivery application [57]. In addition to the above, some waste-driven carbon dots exhibit anti-cancerous activity. Following the hydrothermal method, Paul and Kurian have reported the development of spherically water-soluble fluorescent nitrogen-doped carbon dots (N-CDs) in a single step using water hyacinth as the precursor. The as-prepared carbon dots display anti-cancerous activity against DLA tumor cells [58].

11.4.2 CNT with Drug Delivery Application

Non-degradable plastic bags were commercially converted into CNTs by Mezni et al. following an eco-friendly chemical vapor deposition method. The developed CNTs were coated with doxorubicin followed by functionalization with chitosan, a biodegradable polymer for advanced cancer treatment. The functionalized chitosan-coated CNTs were subsequently tested for their confined and slow drug release property inside the cellular vicinity of the human breast cancer cell line. Different test results suggest the suitability of the material for innovative therapeutic applications [33].

11.4.3 Miscellaneous CNMs with Drug Delivery Application

Nano-sized allotropes of carbon are wisely known as nanocarbon (NC). They are in huge demand because of their biocompatibility, high conductivity and stability, lower toxicity, fluorescence emission properties, eco-friendliness, etc. Additionally, and most importantly, they can be easily functionalized with bioactive particles. Yallappa and his associates have carried out the synthesis of NCs using groundnut shells (biowaste) as a carbon source following the pyrolysis technique. The developed NCs exhibited remarkable antimicrobial activity against human pathogens, owing to their even morphology and compactness of nanoparticles. This suggests the suitability of the developed NCs for bio-imaging and drug delivery applications [39].

Similarly, Yallappa et al. have developed nanoporous carbon by pyrolysis of oil palm leaves and subsequently conjugated it with a fluorescent dye, coumarin-6. The developed conjugated system possesses high dispersion stability in a water medium. Different test result reflects the highly biocompatible nature of the material for normal cells and the toxic nature of human cancer cells. This suggests conjugated nanoporous carbon to be a promising material for drug delivery applications along with cellular imaging [24]. Manaf and his team members have synthesized carbon nano-scale systems (CNS) using Sago bark (biowaste) following the pyrolysis technique. Further functionalization of the developed CNS was carried out to broaden its applications in various biomedical fields. Different analysis result confirms the suitability of the material for bio-imaging along with anti-cancer drug delivery [27].

In our environment, predisposed battery is a vast source of graphite. Ghosh and his team members have successfully recovered graphite from disposed battery cells through electrolysis. Subsequently following Hummer's method from the recovered graphite, graphene oxide (GO) is synthesized. Further reduction of the GO results in chemically reduced graphene oxide (CRGO). Subsequently, the developed CRGO was used as filler material in ketorolac tromethamine (a non-steroidal anti-inflammatory drug) loaded methylcellulose (MC) matrix to develop nanocomposite films. The introduction of CRGO as filler material increases the environmental stability of the developed composite. Additionally, it also enhances the thermal stability of the material. The in vitro drug release studies reflect the suitability of the CRGO-MC composite film for precise transdermal drug delivery [59].

11.5 Conclusions

In the present chapter, the different carbon sources based on waste materials are discussed. Further brief descriptions of different procedures adopted for the synthesis of CNMs using these waste resources are presented. These CNMs have versatile applications in different fields of science and technology like gas sensors, carbon fixation, fuel cells, catalysis, and biomedical fields. Out of the different biomedical fields, many studies have been reported on waste-driven CNMs in bio-imaging applications whereas the drug delivery field has not been much explored using these CNMs. The available drug delivery findings have been mentioned in the present chapter still vast scope lies ahead for further research in this area.

References

1. Tiwari SK, Bystrzejewski M, De Adhikari A, Huczko A, Wang N (2022) Methods for the conversion of biomass waste into value-added carbon nanomaterials: Recent progress and applications. Prog Energy Combust Sci 92. https://doi.org/10.1016/j.pecs.2022.101023
2. Samaddar P, Ok YS, Kim KH, Kwon EE, Tsang DCW (2018) Synthesis of nanomaterials from various wastes and their new age applications. J Clean Prod 197:1190–1209
3. Deng J, You Y, Sahajwalla V, Joshi RK (2016) Transforming waste into carbon-based nanomaterials. Carbon 96:105–115. https://doi.org/10.1016/j.carbon.2015.09.033
4. Article R, Ikram R, Jan BM, Szabo T (2023) Recycling waste sources into nanocomposites of graphene materials: overview from an energy—focused perspective
5. Kumar K et al (2023) Biomass waste-derived carbon materials for sustainable remediation of polluted environment: a comprehensive review. Chemosphere 345:140419
6. Tatrari G et al (2021) Solid waste-derived carbon nanomaterials for supercapacitor applications: a recent overview. Mater Adv 2:1454–1484
7. Omar RA, Talreja N, Chuhan D, Ashfaq M (2024) Waste-derived carbon nanostructures (WD-CNs): an innovative step toward waste to treasury. Environ Res 246:118096
8. Dasari TP, Danielle S, Asok M, Paul KD (2018) A review on graphene-based nanomaterials in biomedical applications and risks in environment and health. Nano-Micro Lett 10:1–34
9. Zhu Y et al (2010) Graphene and graphene oxide: synthesis, properties, and applications. Adv Mater 22:3906–3924
10. Liu J, Cui L, Losic D (2013) Acta biomaterialia graphene and graphene oxide as new nanocarriers for drug delivery applications. Acta Biomater 9:9243–9257
11. Georgakilas V et al (2016) Noncovalent functionalization of graphene and graphene oxide for energy materials, biosensing, catalytic, and biomedical applications. Chem Rev 116:5464–5519
12. Modekwe HU, Daramola MO, Mamo MA, Moothi K (2024) Heliyon Recent advancements in the use of plastics as a carbon source for carbon nanotubes synthesis—a review. Heliyon 10:e24679
13. Ikram R, Jan BM, Ahmad W (2020) Advances in synthesis of graphene derivatives using industrial wastes precursors; prospects and challenges. J Mater Res Technol 9:15924–15951

14. Sultana A, Zare M, Thomas V, Kumar TSS, Ramakrishna S (2022) Nano-based drug delivery systems: conventional drug delivery routes, recent developments and future prospects. Med Drug Discov 15
15. Saha JK, Dutta A (2022) A review of graphene: material synthesis from biomass sources. Waste Biomass Valorization 13
16. Kumar R, Singh RK, Singh DP (2016) Natural and waste hydrocarbon precursors for the synthesis of carbon based nanomaterials: graphene and CNTs. Renew Sustain Energy Rev 58:976–1006
17. Kang C, Huang Y, Yang H, Yan XF, Chen ZP (2020) A review of carbon dots produced from biomass wastes. Nanomaterials 10:1–24
18. Kattimani VS, Kondaka S, Lingamaneni KP (2016) Hydroxyapatite—past, present, and future in bone regeneration. Bone Tissue Regen Insights **7**:BTRI.S36138
19. Chung HK, Wongso V, Sambudi NS, Isnaeni (2020) Biowaste-derived carbon dots/ hydroxyapatite nanocomposite as drug delivery vehicle for acetaminophen. J Sol-Gel Sci Technol 93 214–223
20. Ali SH, Emran MY, Gomaa H (2021) Rice husk-derived nanomaterials for potential applications. Topics in mining, metallurgy and materials engineering. Springer International Publishing. https://doi.org/10.1007/978-3-030-68031-2_19
21. Wang Z et al (2015) Nanocarbons from rice husk by microwave plasma irradiation: from graphene and carbon nanotubes to graphenated carbon nanotube hybrids. Carbon N. Y. 94:479–484
22. Jin H et al (2019) Synthesis of porous carbon nano-onions derived from rice husk for high-performance supercapacitors. Appl Surf Sci 488:593–599
23. Qi H et al (2019) Biomass-derived nitrogen-doped carbon quantum dots: highly selective fluorescent probe for detecting Fe^{3+} ions and tetracyclines. J Colloid Interface Sci 539:332–341
24. Yallappa S, Aruni S, Manaf A, Hegde G (2018) Synthesis of a biocompatible nanoporous carbon and its conjugation with florescent dye for cellular imaging and targeted drug delivery to cancer cells. 33
25. John TS, Yadav PK, Kumar D, Singh SK, Hasan SH (2020) Highly fluorescent carbon dots from wheat bran as a novel drug delivery system for bacterial inhibition. Luminescence 35:913–923
26. Fahmi MZ et al (2018) Bamboo leaf-based carbon dots for efficient tumor imaging and therapy. RSC Adv 8:38376–38383
27. Aruni S, Manaf A, Hegde G, Mandal UK, Wui W (2018) Functionalized carbon nano-scale drug delivery systems from biowaste sago bark for cancer cell Imaging. 1071–1077. https://doi.org/10.2174/1567201813666161017130612
28. Singh V et al (2021) Chickpea peel waste as sustainable precursor for synthesis of fluorescent carbon nanotubes for bioimaging application. Carbon Lett. 31:117–123
29. Roquia A, Khalfan Hamed Alhashmi A, Hamed Abdullah Alhasmi B (2021) Synthesis and characterisation of carbon nanotubes from waste of *Juglans regia* (walnut) shells. Fullerenes Nanotub Carbon Nanostruct 29:860–867
30. Perumal S, Atchudan R, Edison TNJI, Lee YR (2021) Sustainable synthesis of multifunctional carbon dots using biomass and their applications: a mini-review. J Environ Chem Eng 9. https://doi.org/10.1016/j.jece.2021.105802
31. Suriani AB, Md Nor R, Rusop M (2010) Vertically aligned carbon nanotubes synthesized from waste cooking palm oil. J Ceram Soc Jpn 118:963–968
32. Azam MA, Mudtalib NESAA, Seman RNAR (2018) Synthesis of graphene nanoplatelets from palm-based waste chicken frying oil carbon feedstock by using catalytic chemical vapour deposition. Mater Today Commun 15:81–87
33. Mezni A et al (2017) Highly biocompatible carbon nanocapsules derived from plastic waste for advanced cancer therapy. J Drug Deliv Sci Technol 41:351–358
34. Jung S et al (2020) Carbon nano-onions from waste oil for application in energy storage devices. New J Chem 44:7369–7375
35. Zhou J, Sheng Z, Han H, Zou M, Li C (2012) Facile synthesis of fluorescent carbon dots using watermelon peel as a carbon source. Mater Lett 66:222–224

36. Li Y, Bai G, Zeng S, Hao J (2019) Theranostic carbon dots with innovative NIR-II emission for in vivo renal-excreted optical imaging and photothermal therapy. ACS Appl Mater Interfaces 11:4737–4744
37. Hu Y et al (2021) Simple and eco-friendly synthesis of crude orange-peel-derived carbon nanoparticles for detection of Fe3+ and ascorbic acid. Luminescence 36:1385–1394
38. Prasannan A, Imae T (2013) One-pot synthesis of fluorescent carbon dots from orange waste peels. Ind Eng Chem Res 52:15673–15678
39. Yallappa S, Deepthi DR, Yashaswini S, Hamsanandini R, Chandraprasad M (2017) Nanostructures and nano-objects natural biowaste of groundnut shell derived nano carbons : synthesis, characterization and its in vitro antibacterial activity. 12:84–90
40. Zhu J et al (2019) Waste utilization of synthetic carbon quantum dots based on tea and peanut shell. J Nanomater **2019**
41. Yao Y et al (2017) Magneto fl uorescent carbon dots derived from crab shell for targeted dual-modality bioimaging and drug delivery. https://doi.org/10.1021/acsami.7b01599
42. Campalani C et al (2021) Biobased carbon dots: From fish scales to photocatalysis. Nanomaterials 11:1–14
43. Pajarito B, Belarmino AJ, Calimbas RM, Gonzales JR (2020) Graphite nanoplatelets from waste chicken feathers. Materials (Basel) 13
44. Gao L et al (2014) Conversion of chicken feather waste to N-doped carbon nanotubes.pdf. Environ Sci Technol 48:10191–10197
45. Hastuti E, Salsadilla C, Sari AY, Hikmah U (2024) The hydrothermal effect of time and temperature on the synthesis of carbon dots (CDs) from chicken feathers. IOP Conf Ser Earth Environ Sci 1312:3–9
46. Kerdnawee K et al (2017) Present advancement in production of carbon nanotubes and their derivatives from industrial waste with promising applications. KONA Powder Part J 2017:24–43
47. He S, Xu Y, Zhang Y, Bell S, Wu C (2021) Waste plastics recycling for producing high-value carbon nanotubes: investigation of the influence of Manganese content in Fe-based catalysts. J Hazard Mater 402
48. Singh R, Volli V, Lohani L, Purkait MK (2020) Synthesis of carbon nanotubes from industrial wastes following alkali activation and film casting method. Waste Biomass Valorization 11:4957–4966
49. Sanopoulou M, Papadokostaki KG (2017) Controlled drug release systems: Mechanisms and kinetics. Biomed Membr (Bio)artificial Organs 1–33. https://doi.org/10.1142/9789813223974_0001
50. Ghosh TK et al (2020) Chemically reduced graphene oxide (CRGO) from waste batteries and morphological assessment of CRGO/methyl cellulose transdermal film. Nano-struct Nano-objects 22:100454
51. Sharma G (2024) Biogenic carbon nanostructured materials for detection of cancer and medical applications: a mini review. Hybrid Adv. 5:100166
52. Himaja AL, Karthik PS, Sreedhar B, Singh SP (2014) Synthesis of carbon dots from kitchen waste: conversion of waste to value added product. J Fluoresc 24:1767–1773
53. Ding H, Du F, Liu P, Chen Z, Shen J (2015) DNA-carbon dots function as fluorescent vehicles for drug delivery. ACS Appl Mater Interfaces 7:6889–6897
54. Pandey S et al (2014) Synthesis of mesoporous silica oxide/C-dot complex (meso-SiO_2/C-dots) using pyrolysed rice husk and its application in bioimaging. RSC Adv 4:1174–1179
55. Sun Y et al (2020) Ultralong lifetime and efficient room temperature phosphorescent carbon dots through multi-confinement structure design. Nat Commun 11:1–11
56. Yaya Ouattara L et al (2021) Cocoa pod husk: a review. BioResources 16:1988–2020
57. Norman A et al (2022) Cocoa pod husks as precursors for biosynthesis of carbon dots as potential bioimaging tool. Int J Nanosci Nanotechnol 18:79–92

58. Paul A, Kurian M (2020) Materials today : proceedings N-doped photoluminescent carbon dots from water hyacinth for tumour detection. 1–5
59. Ghosh TK et al (2020) Chemically reduced graphene oxide (CRGO) from waste batteries and morphological assessment of CRGO/methyl cellulose transdermal film. Nano-struct Nano-objects 22

Chapter 12
Waste-Derived Carbon Nanotubes (CNTs): A Revolutionary Product Toward Energy Applications

Mehdi Abedi, Vahid alimardani, Younes Ghasemi, and Radmehr Kazemi

Abstract The management of waste globally is a cause of concern for the environment, nature, and people's health. This poses a significant issue for the future. With the world's population expanding and their consumption levels increasing, the quantity of waste being produced each day is also growing. Waste material management can be improved by converting the carbon-containing waste material into valuable materials, such as carbon nanomaterials (CNMs). These materials have the potential to be employed in different industrial devices with a variety of applications such as energy storage devices. This book chapter's comprehensive review discusses the latest advancements and investigation of global solid waste that is converted into CNMs. The modifications to the design of supercapacitors and their influence on operational dynamics have been addressed. The chapter also discusses the techniques utilized by scientists to assess the energy storage capabilities of different supercapacitors.

Keywords Carbon nanotubes · Waste material · Capacitor · Battery · Energy storage · Nanomaterial

M. Abedi (✉) · Y. Ghasemi · R. Kazemi
Pharmaceutical Sciences Research Center, Shiraz University of Medical Sciences, Shiraz, Iran
e-mail: mehdi.abedi.a@gmail.com

M. Abedi
Research and Development Department, Danesh Salamat Kowsar Co, Shiraz, Iran

V. alimardani
Student Research Committee, Shiraz University of Medical Sciences, Shiraz, Iran

Y. Ghasemi
Department of Pharmaceutical Biotechnology, School of Pharmacy, Shiraz University of Medical Sciences, Shiraz, Iran

N. Talreja et al. (eds.), *Waste-Derived Carbon Nanostructures*, Nanostructure Science and Technology, https://doi.org/10.1007/978-3-031-75247-6_12

12.1 Introduction

Carbon-based nanomaterials are an important part of material science. Due to their ability to easily change structure, their safety for living organisms, and their ease of modification, nanomaterials have a wide range of potential applications in energy, catalysis, adsorption, and biology [1, 2]. Numerous advances in technology and materials have been made in the last two decades to address the challenges of energy production and storage. Carbon-derived materials are receiving increased attention for their special and changeable attributes. In the past decade, scientists have come across carbon nanotubes (CNTs) and a variety of carbon-based materials such as graphene. This has prompted extensive research into the utilization of carbon in creating novel materials [3–5]. Unfortunately, even though CNTs and graphene have very interesting qualities, their wide practical uses are limited because they are expensive to make, not produced in large quantities, require expensive equipment, and are difficult to make. Additionally, their manufacturing process is harmful to the environment due to the use of hazardous chemicals and the generation of toxic waste. This causes more problems for the environment. So, scientists are trying to create new materials from carbon that are cheap and easy to make in large amounts. This is essential for the research at this time [6].

Scientists have recently turned their attention to various types of waste that contain carbon as a potential source for creating valuable materials. These waste materials are valuable because of their abundance of carbon, low cost, and widespread availability. Numerous studies have achieved the conversion of trash into valuable carbon products. The design of these materials consists of nano- and microstructure and can be modified to exhibit diverse surface properties. Waste-derived products have shown success in various applications, such as energy storage. The waste utilized for making inexpensive carbon-based raw materials can be divided into two categories. The initial category of biowaste is comprised of carbohydrates and can decompose naturally within the carbon cycle. This kind of waste typically has a small influence on the earth. Some examples are dead plants, agricultural leftovers, and household waste [7]. Another type of waste is items that are not easily biodegradable and have a negative impact on the environment. This kind of waste harms the environment and includes things like plastic bags and bottles, industrial waste, and city garbage [6].

Waste materials like plastic, agricultural waste, garbage, paper, and industrial waste have carbon-based structures. These carbon materials can be converted to carbon nanomaterial by different procedures including calcination, hydrothermal, pyrolysis, physical vapor deposition (PVD), chemical oxidation, chemical vapor deposition (CVD), and thermal dissociation, etc. Figure 12.1 demonstrates the diverse approaches to producing CNMs from carbon waste. Depending on the method used, the activation can modify the carbon structure's form [8]. The method for producing carbon nanomaterials (CNMs) varies depending on the surface reactions and the substances utilized. CVD appears to be the most efficient method for producing large quantities of items. Nonetheless, researchers are continuing their efforts to discover more efficient and cost-effective methods for producing large quantities of CNMs.

CNMs have multiple uses, allowing us to utilize residual waste to produce additional CNMs using solid waste materials. CNMs, CNTs, CQDs, and graphene have special characteristics like excellent heat-conducting electrical properties and mechanical stability. The temperature, pressure, and heating rate of pyrolysis are crucial and can impact the formation of layers and the properties of CNMs [8]. Recently, many scientists have been studying how to turn waste into certain materials called CNMs. Kukovitskii and his team utilized Ni-plates as highly effective catalysts in producing 10–40 nm CNTs from waste polyethylene. We made CNTs by heating up waste material at 420–450 °C in helium. The nanotubes were 10–40 nm wide [9]. By heating polypropylene (PP) in a special reactor, Zhang et al. produced small CNTs with 7 nm diameter. The reactor containing the combination of PP/maleated PP/ferrocene was sealed and maintained at temperatures ranging from 600 to 700 °C for 12 h. After cooling down to room temperature, the dark products were gathered, processed, sanitized, and dried in a vacuum for 6 h. At a temperature of 600 °C, microspheres composed of CNTs exhibit rough walls and range in size from 35 to 55 nm. However, when heated to 700 °C, many well-organized and smooth nanotubes with outer diameters of 40–70 nm, inner diameters of 7–30 nm, and lengths up to tens of micrometers are produced [10]. In the same way, Liu et al. used a two-reaction process consisting of catalytic pyrolysis and then the decomposition of PP in a special container to break it down into smaller components and convert waste PP into CNT. They heated plastic to 1023 °C with a catalyst in a special reactor to decompose. Multi-walled carbon nanotubes (MWCNTs) can be synthesized at various temperatures using nickel oxide as a catalyst. MWCNTs demonstrate their highest yield at 973 °C. The selection and quantity of catalysts are crucial when creating CNTs from plastic waste [11]. Some studies have found that Ni is good at helping to make CNTs. By modifying the ratio of organic material to inorganic catalysts and the temperature, the highest quality MWCNTs were produced.

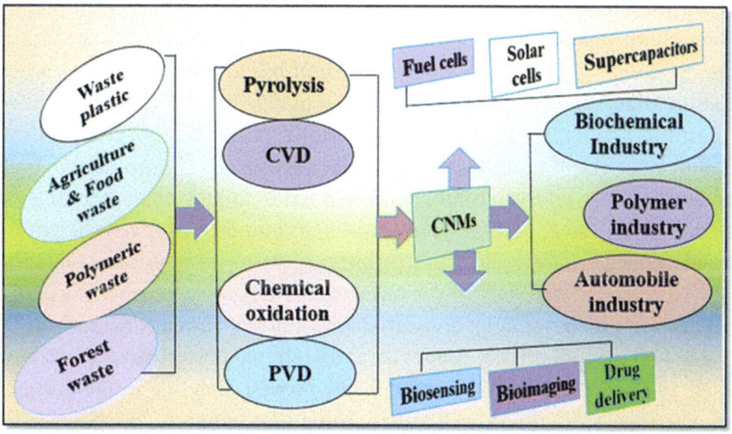

Fig. 12.1 Schematic diagram showing the synthesis pathways and industrial applications of various waste-derived CNMs. The figure was reproduced with permission [8]

12.2 Energy Storage

Finding more sustainable ways to produce energy without damaging the environment is crucial. The global climate is shifting, and our fossil fuel supply is diminishing. Fortunately, we are making progress in utilizing clean and sustainable energy such as solar power, wind power, geothermal energy, and biomass [12, 13]. Fortuitously, there has been a significant advancement in the utilization of renewable and sustainable energy sources such as geothermal, wind solar, and biomass. In addition, there is widespread usage of portable devices and hybrid vehicles that run on both electricity and fuel in daily life [14, 15]. Carbon-based materials are highly effective in energy storage devices such as batteries, fuel cells, and supercapacitors. There is a high amount of carbon in non-biodegradable waste such as food scraps and other non-decomposable materials. Therefore, making new materials from waste to use as good and cheap electrodes for storing energy has become a popular and interesting topic.

12.2.1 Battery

The demand for efficient and affordable batteries is increasing as the popularity of portable electronic devices and electric cars grows. Right now, the Li-ion battery is the most popular electric storage device because it can hold a lot of energy, has excellent performance, and is very stable. The basic Li-ion battery is made with a special kind of carbon and a lithium-based substance, which are put into a liquid that conducts electricity. The expensive price of commercial graphite electrodes stops them from being used a lot. Plus, using too much graphite minerals will make the natural supply run out fast [16, 17]. Several different forms of carbon, such as graphite, CNTs, and graphene, have been shown to have good electrochemical properties. On the other hand, the difficult manufacturing techniques and the expensive production expenses present them with significant difficulties. This is making it difficult to produce them on a large scale [18]. Therefore, making carbon-based materials from renewable resources like plants is very important because plants are found all over the world, are not harmful, and are inexpensive. Additionally, it can be converted into porous material with a variety of functionalities, which are essential for producing batteries. Carbon materials derived from biomass often possess unique characteristics such as a significant surface area, diverse pore structures, numerous functional groups, the ability to adsorb water, and conductive properties. As a result of these features, they are suitable for use as materials in the electrode of a battery. Furthermore, biomass carbon materials can be readily converted into porous formations and utilized in batteries due to their exceptional longevity and high rate performance [19]. Utilizing plant-based and other renewable materials has proven effective in the composition of Li-ion batteries. Recent research has revealed that biowaste can be transformed into a uniform type of carbon. This carbon can be used in Li-ion batteries as the part that stores electricity. Tian and his team combined red phosphorus and hard carbon from a coconut shell by filling the holes of the hard carbon with small particles of red phosphorus. The hard carbon was comprised of phosphorus that accounted for nearly half of its weight. Tian and his team mixed red phosphorus with hard carbon extracted from coconut shells by filling the pores of the hard carbon with tiny particles of red phosphorus. When this composite was used as an anode in a sodium-ion battery, it was found to contain nearly half of its weight in phosphorus. Even after being used many times, it can still store a good amount of energy and use it efficiently [20]. Jiang and his team suggested a process for repurposing used bamboo chopsticks into carbon fibers suitable for use in batteries. First, they extract cellulose fibers from the used bamboo chopsticks by treating them with hot water and a strong base solution. Subsequently, the cellulose fibers were degraded and changed into carbon fibers by being heated in a specific atmosphere. This resulted in uniform carbon fibers with microscopic pores throughout [21]. Due to their long and thin nature, carbon fibers allow for fast movement of electricity. Simultaneously, the small pores in the carbon fibers allowed for deep penetration of Li + ions. When the prepared carbon fibers were used as the positive electrode in Li-ion batteries, they performed very well over many cycles and at different rates. The fibers had a starting capacity of about 500 mA h/g and a charge capacity of about 283 mA h/g. This shows that the fibers were about 56.6% efficient in their energy storage, which is better than bulky bamboo carbons and similar to transition metal oxides [21]. The study indicates that the behavior of carbon materials is greatly influenced by the manner in which they are produced, particularly the temperature utilized during the process. The performance of carbon materials as the negative electrode in batteries is heavily influenced by these characteristics [22, 23]. Factors such as the type of plants, burning method, chemicals used, and surrounding air can all impact the visual characteristics of the resulting carbon product [24]. When the temperature rises in the production of carbon material, the surface reactivity and oxygen concentration decrease. The release of oxygen and hydrogen atoms into the air is a result of the biomass compounds heating up and breaking down [24]. Introducing varied atoms is the frequently utilized approach for enhancing the capabilities of carbon derived from biomass. Introducing different atoms to carbon causes a shift in electron distribution, resulting in a more appealing surface for molecules or ions. This enables the carbon surface to serve as a catalyst in a variety of reactions, such as those occurring in batteries [25]. Prompun and his team synthesize CNTs from waste cooking oil by the chemical decomposition process. The results indicate that as the reaction temperature increased, the pores in the material became smaller. At the same time, the specific surface area of the CNT also increased. The ability of CNT samples to function as an energy storage component in batteries was examined by researchers. The study showed that CNTs had the best storage capacity for lithium-ion batteries when they were made at 900 °C (474.34 mAh/g for the first cycle). This was because they had the right size holes and a large surface area for storing the lithium ions [26]. Christopher and colleagues researched the possibility of reusing single-wall CNTs from old lithium-ion battery electrodes. Simply, the SWCNT-Li cells were used a lot and then treated with acid and heat to recycle them, using a process originally made for purifying SWCNT material. The recycling process effectively removed old battery byproducts and restored the SWCNT material to its original state. The energy requirements for different recycling methods were examined and contrasted with the energy consumption for producing SWCNT. The recycled-SWCNT material was utilized in Li+ battery coin cells and demonstrated equal performance to the pure-SWCNT electrodes, with insertion and extraction capacities of 650 mA h g^{-1}. The amount of energy needed to reuse SWCNTs is less than half of the energy needed to make new material. Finding ways to manage SWCNTs when they are no longer needed while minimizing their negative impact presents new opportunities [27]. These materials are being considered by some scientists for the development of supercapacitors, solar cells, fuel cells, and batteries. Furthermore, waste with high carbon content that cannot decompose was converted into electrodes for batteries. Zhang and colleagues successfully fabricated carbon-fiber monolith through using filamentous fungi. The fungi cell wall is highly durable and effectively binds the fungi fibers together. This plays a crucial role in the construction of carbon fiber structures. With a combination of N and O with carbon-fiber monolith as dopant, the carbon block exhibits superb electrical conductivity and a significant surface area. The potential of using carbon monolith as a conductive layer to improve the performance of Li–S batteries is emphasized. When Li–S batteries are being charged and discharged, they perform significantly better. The batteries can still maintain a capacity of 650 mA h/g even after undergoing 100 cycles at 5 °C [28].

12.2.2 Capacitor and Supercapacitor

Energy storage can be achieved with supercapacitors, also referred to as electrochemical double-layer capacitors. They have a power level that is intermediate between regular batteries and capacitors but with lower energy [29]. Supercapacitors excel at storing energy and are a preferred option for energy storage technology [30]. Electrochemical double-layer capacitors store energy by collecting charges and ions at the interface between the electrodes and electrolyte using electrostatic forces [31]. Therefore, the ability of a material to store energy is expected to increase as its specific surface area. High-performing supercapacitor electrodes require materials that are both porous and conductive [32, 33]. Porous carbon materials are commonly employed in supercapacitor electrodes because of their porous nature, large surface area, effective electrical conductivity, and low cost [34]. However, when the pores of porous carbon are clogged, it can result in a reduction in the number of ions it can store during high levels of electrical current. Carbon-based supercapacitor electrodes are highly effective as they have a large power storage capacity and are durable enough to be used repeatedly without degradation. This is due to their excellent conductive properties and their high strength. The primary goal in creating carbon-based supercapacitors is to increase their energy storage capacity without compromising their overall performance. Scientists have researched various carbon materials obtained from waste to determine their potential to improve energy storage performance without requiring significant investment [35]. The various micro and nanostructures in these materials can enhance the energy storage capacity. Activated carbon is the most widely used carbon electrode material due to its large surface area, electrical conductivity, and cost. The method of producing activated carbon from various waste materials can impact its performance in a supercapacitor. The number of locations for storing electric charge is determined by the surface area of the activated carbon structure. However, the reported capacities do not always show a direct correlation with the surface area. It is widely understood that activated carbon generally possesses a large number of minuscule pores. Despite the fact that adding micro holes to activated carbon can increase its surface area, many of these minuscule openings are too small for electrolyte ions to penetrate. Therefore, the capacity for storing energy is significantly reduced [36, 37]. Therefore, it is important to design activated carbon-based supercapacitors with tiny pores that are the same size as electrolyte ions. This will help improve their energy storage capacity and operational lifespan. Qian and his team prepared carbon porous material by pyrolyzing hair fibers and KOH at 800 °C for 2 h. The porous structure of carbon retained a high electric charge and maintained its stability after 20,000 cycles of usage. The specific capacitance of 126 F/g is also confirmed when using a 1 M LiPF6 ethylene carbonate/diethyl carbonate organic electrolyte at a current density of 1 A/g1. Due to its numerous micro/mesoporous and extensive surface area, the supercapacitor functions effectively, especially after being enhanced with specific chemical reagents for improved energy storage [38]. Moreover, scientists have invented various carbon materials from waste for the purpose of using them in supercapacitors. For example, Feng and his team made a special 3D carbon structure using waste bagasse. They applied the hydrothermal and chemical activation methods for carbon material preparation. The porous 3D structure exhibits excellent porosity and a specific surface area of 2296 m^2 g^{-1}. At 5 A, the fabricated device possessed a capacitance of 320 F $g^{-1}$and was still at 71% capacitance at 50 A g^{-1} [39]. Liu and his colleagues discovered that by heating sawdust with FeCl3, they could produce lengthy and slender carbon fibers. The combination of Fe and Cl was determined to play a complementary role in the development of carbon nanofibers on mesoporous carbon. The carbon nanofibers remained separate, potentially enhancing their capacity to store energy [40]. The nanofibers may not have as high of a specific capacitance as other carbon-based materials, but they are still a viable option for supercapacitors due to their low cost, ability to maintain performance, and long-term stability.

Two-dimensional (2D) carbon nanosheets are also prepared from waste material for high-performance supercapacitors. 2D carbon sheets with holes were produced from old coconut shells by Sun and his team through heating and treating them simultaneously. It was shown that the iron catalyst helped create a 2D graphene-like structure when heated. The significant presence of graphitic carbon within the carbon nanosheets that were formed enables them to have a large surface area and efficient electrical conduction. The porous carbon structure facilitated rapid and efficient ion movement, making it ideal for storing energy. Prepared carbon nanosheets exhibited a specific capacitance of up to 268 F/g at 1 A/g when utilized for supercapacitors. Also, it lasts a very long time and maintains a high level of efficiency (99.5%) even after being used 5000 times in KOH [41]. Li and his team discovered a simple method to convert willow catkin into carbon nanosheets with porous structures. Additionally, they included nitrogen and sulfur to enhance their quality. They discovered that by adjusting the quantity of alkali used in the raw material, they were able to manipulate the pore structure of the carbon they created, as willow catkin fiber possesses a unique hollow and layered arrangement. The porous carbon nanosheets containing nitrogen and sulfur have demonstrated the capability to store energy, with 298 F/g at 0.5 A/g and 233 at F/g 50 A/g respectively, highlighting their versatility across different speeds. Also, the electrode can maintain its performance over a long time, with only a 2% decrease in capacity after 10,000 cycles. Moreover, when all the cell parts are combined in a symmetric cell, it can handle a diverse range of 1.8 V and store a large amount of energy, specifically 21 Wh/kg at a power output of 180 W/kg. Their work has resulted in the production of an inexpensive and sustainable material for supercapacitors that exhibits excellent performance [42]. The method for producing N and S co-doping porous carbon nanosheets is depicted in Fig. 12.2.

Carbon-based networks in 3D configuration have abundant small cavities, extensive surface area, efficient electrical conductivity, and significant toughness resulting from their interconnected arrangement. For instance, Wu and his team created spongy and flexible 3D carbon materials using natural materials like watermelon and biomass. These materials contain small carbon fibers and nanospheres. The spongy carbon-based gels can react with chemicals and bend easily, making them a good support for making 3D composite materials. When the prepared 3D carbon structure was combined with Fe3O4 and calcinated, it showed excellent electrical efficiency. The prepared composite had the potential to store a high amount of energy, at 333 F/g, at a current density of 1 A/g within a potential window of -1.0 to 0 V in a 6 M KOH solution. Furthermore, the magnetite carbon aerogels (MCAs) can endure cycling

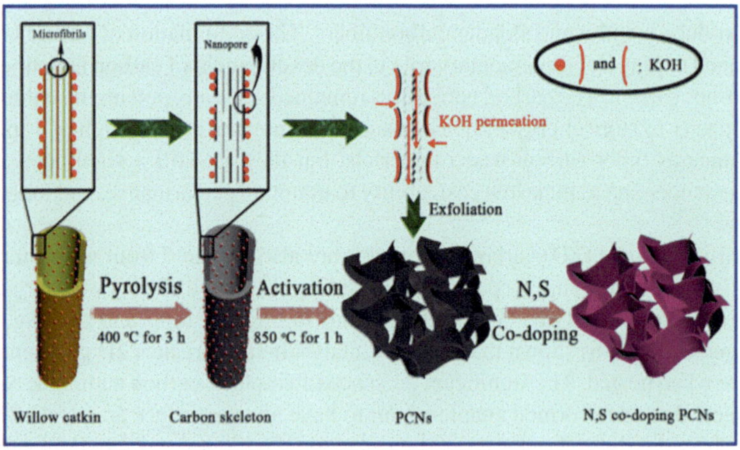

Fig. 12.2 Schematic diagram of the synthesis of N and S doping in porous carbon nanosheets. Reproduced with permission [42]

without losing strength, with 96% of their capacity remaining after 1000 charge and discharge cycles [43]. Pandey and his team used bentonite clay to degrade plastic waste and convert it to graphene nanosheet (GNS) for application in solar cells and supercapacitors (Fig. 12.3). In order to create graphene sheets, the plastic was subjected to two stages of high-temperature heating at 450 °C and 945 °C in the absence of oxygen. The dye-sensitized solar cells (DSSC) was constructed using graphene nanosheets in the photo-anode and a polymeric electrolyte. According to the acquired result, it had an 86.4% high fill factor and a high V_{oc} of 0.77 V, which was similar to results of computer calculations. The supercapacitor also showed that it can store a lot of energy (Ed) and release it quickly (Pd) with 38 Wh/kg and 1009.74 W/kg, respectively. Alternatively, using GNs as the main material for supercapacitor electrodes provided a high capacity of 398 F/g at a scan rate of 0.005 V/s[44].

Poochai and his colleagues proposed the idea of transforming waste coffee grounds into a nanoporous carbon material suitable for use in a unique supercapacitor battery [45]. Carbon nanotube (CNT) incorporated nanoporous carbon was created through a simple ball-milling process and then spread onto a carbon nanofiber paper using a doctor blade technique. Due to the placement of CNT structures between the carbon flakes, the developed material had a large surface area and small pore size. This combination had a better ability to charge and discharge quickly, more stable performance over multiple cycles, and held onto its charge for longer compared to nanoporous carbon and CNT electrodes. Abbas and his team developed the process of producing MWCNTs using plastic packaging and used them in supercapacitors. Plastic waste was used to make MWCNTs in three steps: heating, breaking down with a catalyst, and cooling. Treatment with nitric acid was used to optimize the functionality of the supercapacitor by modifying the MWCNTs

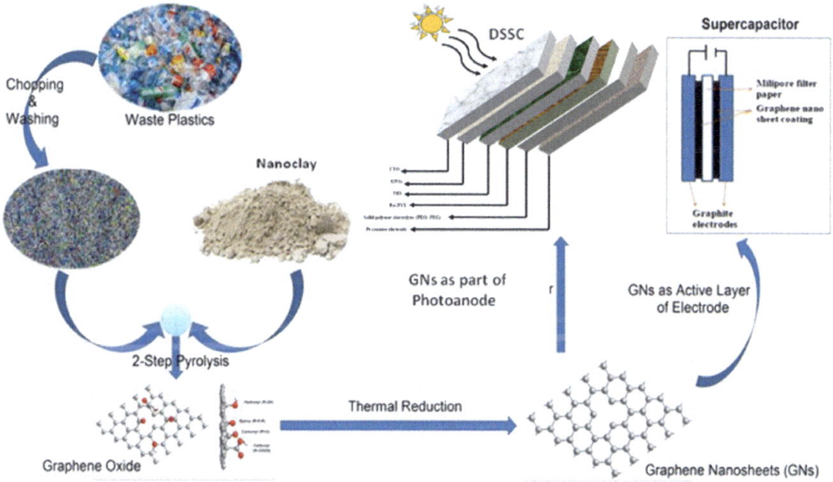

Fig. 12.3 A diagram showing how to make GNs from plastic waste and use them in solar cells and supercapacitors [44]

after their production. MWCNTs produced from plastic packaging waste outperform commercial MWCNTs in their capacity to store electricity, regardless of the amount of electricity being utilized [46]. Mishra and collagenous successfully created a new form of MWCNTs from plastic waste through a chemical process, and the electrode functioned effectively within a flexible supercapacitor. The effectiveness of carbon nanotube electrodes in storing electricity was evaluated through the use of two different approaches: CV curve and galvanostatic charge–discharge (GCD) methods. CNTs created from recycled polypropylene show suitable capability for use as part of the electrode in an electrochemical double-layer capacitor. They show that CNTs created at 700 °C using a faster scan rate showed a more robust electric current when compared to those made at 600 °C and 800 °C [47]. In previous years, scientists constructed a robust supercapacitor using bacterial cellulose. It holds a lot of energy 238 F g^{-1} at a current density of 5 A g^{-1}. They discovered that attaching Ni-Co layered double hydroxide (LDH) nanosheets to the carbon fibers, significantly enhanced the storage capacity, increasing it from 238 F/g at a current density of 0.5 A/g to 1949 F/g at a current density of 1 A/g [48]. The utilization of a combination of nickel oxide and carbon black in small quantities was shown by Wen and his group to be able to convert used plastic into a supercapacitor. They observed that polypropylene, polyethylene, and polystyrene were turned into CNTs using carbon black and nickel as catalysts. Compared to commercially available CNTs on the market, the new materials have a higher energy storage capacity [49].

Table 12.1 presents various forms of CNMs derived from waste material and their functions in energy storage devices. The information shows how specific capacitance, current density, and surface area are influenced by outside activating agents. The information indicates that both 3D materials possess a significant surface area featuring numerous small pores. It also includes 2D materials with different sizes of pores and varying amounts of material. The combination of these features, the device's adaptability, and its various shapes enhance the efficiency of the materials in storing electrical charge. In contrast, the presence of specific alkali metal ions contributes to the improved efficiency of supercapacitor electrodes. The performance of supercapacitors is significantly influenced by the various shapes and types of electrodes, such as hydrogels and flexible configurations [8].

12.3 Conclusion

Over the last decades, scientists and researchers have been working toward finding an equilibrium between repurposing waste materials for the creation of nanomaterials and using them for different applications in nanoscience and nanotechnology. By effectively managing this substantial problem, it can bring about positive changes in various industries and pave the way for new advancements in science and technology, which are essential for the continual progress of society. Waste-based CNMs have become highly significant for their role in energy storage and energy conservation. This article looks at how waste materials can be used to make new carbon-based

Table 12.1 Waste-derived Carbon nanomaterial and their electrochemical performance

CNT source	Preparation technique	Electrode Composition	Result	References
Waste plastic	Chemical vapor deposition (CVP)	Cluster network CNT/phase change material	An optimal thermal performance level was attained with the incorporation of 5% CNT in the composite material	[50]
Fish scale	Hydrothermal	Porous nanosheet structure/KOH	Low-cost synthesis method, The electrodes' ability to store energy is greatly enhanced by the high compacting density in both three-electrode and two-electrode systems	[51]
Waste plastic	Pyrolysis-CVD	CNT/$LiNiCoMnO_2$ active material and poly(vinylidene fluoride)	According to electrochemical tests, CNTs produced on the FeMo/MgO catalyst exhibited superior performance compared to commercial carbon black and CNT	[52]
Rose	Calcination	A 3D porous structure consisting of interconnected carbon nanosheets	High specific surface area (1911 m2 g^{-1}), cycling stability of 99% capacitance retention after 25,000 cycles even at a high specific capacitance of 208 F g^{-1}	[53]
Flax fiber	Calcination	Carbon cloth/MnO_2	Carbon cloth shows specific capacitance of 0.78 F g^{-1}, after combination with MnO_2 specific capacitance of MnO_2 reaches 683.73 F g-1 at 2 A g^{1}	[54]

(continued)

Table 12.1 (continued)

CNT source	Preparation technique	Electrode Composition	Result	References
Coconut shell	Hydrothermal	Mesoporous carbon/ $ZnCl_2$	High specific surface area (2440 m^2 g^{-1}), Stable energy density at 7.6 W h kg^{-1} was observed (90% of 8.5 W h kg^{-1}) at a high power density of ~ 4.5 kW kg^{-1}	[55]
Coconut Shells	Calcination	3D porous graphene-like sheets	High specific surface area (1506.19 m2 g^{-1}) and conductivity (32.14 S cm^{-}1), specific capacitance of 91.15 F g^{-1} at a current density of 0.2 A g^{-1}	[56]
Biomass reed flowers	Hydrothermal/ calcination	Hierarchically Porous Carbon/Cu foil/Ketjen black/ carboxymethyl cellulose	The material is excellent for lithium-ion battery production as it remains stable during cycling and retains a high energy capacity even after 1000 cycles	[57]
Rice straw	Hydrothermal	Porous carbon composite materials/ KOH	Large specific surface area (1122 m2 g^{-1}), high specific capacitance, excellent rate retention, cycling stability with 96% capacitance retention after 10,000 cycles	[58]
Rice husk	Calcination	Porous carbon nanosheets/rice husk-derived carbons	Specific capacitance of 315 F g^{-1} at 0.1 A g^{-1} and an excellent rate capability of 189 F g^{-1} at 50 A g^{-1}	[59]

(continued)

Table 12.1 (continued)

CNT source	Preparation technique	Electrode Composition	Result	References
Rice husk	Pyrolysis	Porous carbon/ graphite composites activated by KOH and NaOH	Specific capacitance larger than 200 F g^{-1}	[60]
Cotton stalk	KOH-chemical activation method	Flute type micropores activated carbon	Large specific surface area (1964.46 m^2 g^{-1}), Specific capacitance of 254 F g^{-1} at a current density of 0.2 A g^{-1}, energy density of 18.14 W h kg^{-1}, power density of 450.37 W kg^{-1}, long cyclic stability of 96% capacitance retention after 10,000 cycle	[61]

materials. It will focus on the newest advances in this area. The purpose of this chapter is to compile and outline the most recent progress in using waste materials for producing new carbon-based materials for energy storage applications. Different approaches and materials have been used to create carbon nanotubes (CNTs) with unique physicochemical properties, morphologies, and structures to improve their performance in energy storage applications. The waste-based CNMs are being used in various supercapacitors and have been proven effective in studies from initial to advanced levels. The discussion focused on examining the format and various techniques of learning to select the most suitable materials for specific applications.

References

1. Abedi M, Asadi M, Mehrzadeh M, Rahiminezhad Z, Ghasemi Y, Alimardani V (2024) Two-Dimensional (2D)-based hybrid composites for cancer diagnosis and therapy. In: Talreja N, Chauhan D, Ashfaq M (eds) Two-dimensional hybrid composites: synthesis, properties and applications. Springer Nature Singapore, Singapore, pp 295–327. https://doi.org/10.1007/978-981-99-8010-9_11
2. Abedi M, Rahiminezhad Z, Mehrzadeh M, Alimardani V, Ghasemi Y, Chauhan D, Ashfaq M, Viswanathan MR, Talreja N (2023) Two-dimensional (2D)-nanolayer based functional materials in the advanced oxidation process. In: Materials technology for the energy and environmental Nexus, vol 1. IOP Publishing, pp 10-1–10-38
3. Nardecchia S, Carriazo D, Ferrer ML, Gutiérrez MC, del Monte F (2013) Three dimensional macroporous architectures and aerogels built of carbon nanotubes and/or graphene: synthesis and applications. Chem Soc Rev 42(2):794–830. https://doi.org/10.1039/C2CS35353A

4. Cao X, Yin Z, Zhang H (2014) Three-dimensional graphene materials: preparation, structures and application in supercapacitors. Energy Environ Sci 7(6):1850–1865. https://doi.org/10.1039/C4EE00050A
5. Huang X, Qi X, Boey F, Zhang H (2012) Graphene-based composites. Chem Soc Rev 41(2):666–686. https://doi.org/10.1039/C1CS15078B
6. Ma Q, Yu Y, Sindoro M, Fane AG, Wang R, Zhang H (2017) Carbon-Based functional materials derived from waste for water remediation and energy storage. Adv Mater 29(13):1605361. https://doi.org/10.1002/adma.201605361
7. Falkowski P, Scholes RJ, Boyle E, Canadell J, Canfield D, Elser J, Gruber N, Hibbard K, Högberg P, Linder S, Mackenzie FT, Moore B 3rd, Pedersen T, Rosenthal Y, Seitzinger S, Smetacek V, Steffen W (2000) The global carbon cycle: a test of our knowledge of earth as a system. Science 290(5490):291–296. https://doi.org/10.1126/science.290.5490.291
8. Tatrari G, Karakoti M, Tewari C, Pandey S, Bohra BS, Dandapat A, Sahoo NG (2021) Solid waste-derived carbon nanomaterials for supercapacitor applications: a recent overview. Mater Adv 2(5):1454–1484. https://doi.org/10.1039/D0MA00871K
9. Kukovitskii EF, Chernozatonskii LA, L'Vov SG, Mel'nik NN (1997) Carbon nanotubes of polyethylene. Cheml Phys Lett 266(3):323–328. https://doi.org/10.1016/S0009-2614(97)00020-1
10. Zhang J, Du J, Qian Y, Xiong S (2010) Synthesis, characterization and properties of carbon nanotubes microspheres from pyrolysis of polypropylene and maleated polypropylene. Mater Res Bull 45(1):15–20. https://doi.org/10.1016/j.materresbull.2009.09.007
11. Liu J, Jiang Z, Yu H, Tang T (2011) Catalytic pyrolysis of polypropylene to synthesize carbon nanotubes and hydrogen through a two-stage process. Polym Degrad Stab 96(10):1711–1719. https://doi.org/10.1016/j.polymdegradstab.2011.08.008
12. Guo S, Dong S (2011) Graphene nanosheet: synthesis, molecular engineering, thin film, hybrids, and energy and analytical applications. Chem Soc Rev 40(5):2644–2672. https://doi.org/10.1039/C0CS00079E
13. Pumera M (2011) Graphene-based nanomaterials for energy storage. Energy Environ Sci 4(3):668–674. https://doi.org/10.1039/C0EE00295J
14. Clement-Nyns K, Haesen E, Driesen J (2010) The impact of charging plug-in hybrid electric vehicles on a residential distribution grid. IEEE Trans Power Syst 25(1):371–380. https://doi.org/10.1109/TPWRS.2009.2036481
15. Abedi M, Rahiminezhad Z, Mehrzadeh M, Alimardani V, Ghasemi Y, Chauhan D, Ashfaq M, Viswanathan MR, Talreja N (2023) Sulfide-based 2D nanomaterials for photocatalytic activity. In: Materials technology for the energy and environmental Nexus, vol 1. IOP Publishing, pp 1:7-1–7-28
16. Sun Y, Liu N, Cui Y (2016) Promises and challenges of nanomaterials for lithium-based rechargeable batteries. Nat Energy 1(7):16071. https://doi.org/10.1038/nenergy.2016.71
17. Galos J, Pattarakunnan K, Best AS, Kyratzis IL, Wang C-H, Mouritz AP (2021) Energy storage structural composites with integrated lithium-ion batteries: a review. Adv Mater Technol 6(8):2001059. https://doi.org/10.1002/admt.202001059
18. Iqbal S, Khatoon H, Pandit AH, Ahmad S (2019) Recent development of carbon based materials for energy storage devices. Mater Sci Energy Technol 2(3):417–428
19. Molaiyan P, Dos Reis GS, Karuppiah D, Subramaniyam CM, García-Alvarado F, Lassi U (2023) Recent progress in biomass-derived carbon materials for Li-ion and Na-ion batteries—a review. Batteries 9(2):116
20. Tian W, Wang L, Huo K, He X (2019) Red phosphorus filled biomass carbon as high-capacity and long-life anode for sodium-ion batteries. J Power Sources 430:60–66
21. Jiang J, Zhu J, Ai W, Fan Z, Shen X, Zou C, Liu J, Zhang H, Yu T (2014) Evolution of disposable bamboo chopsticks into uniform carbon fibers: a smart strategy to fabricate sustainable anodes for Li-ion batteries. Energy Environ Sci 7(8):2670–2679. https://doi.org/10.1039/C4EE00602J
22. Alvira D, Antorán D, Manyà JJ (2022) Plant-derived hard carbon as anode for sodium-ion batteries: a comprehensive review to guide interdisciplinary research. Chem Eng J 447:137468

23. Huang S, Li Z, Wang B, Zhang J, Peng Z, Qi R, Wang J, Zhao Y (2018) N-Doping and defective nanographitic domain coupled hard carbon nanoshells for high performance lithium/sodium storage. Adv Func Mater 28(10):1706294
24. Guizani C, Jeguirim M, Valin S, Limousy L, Salvador S (2017) Biomass chars: the effects of pyrolysis conditions on their morphology, structure, chemical properties and reactivity. Energies 10(6):796
25. Chen W, Wan M, Liu Q, Xiong X, Yu F, Huang Y (2019) Heteroatom-doped carbon materials: synthesis, mechanism, and application for sodium-ion batteries. Small Methods 3(4):1800323
26. Prompun P, Ratchahat S, Kaveevivitchai W, Kooamornpattana W (2022) Carbon Nanotube (CNTs) production from waste cooking oil as anode material for Li-ion batteries. J Phys: Conf Ser 2175(1):012041. https://doi.org/10.1088/1742-6596/2175/1/012041
27. Schauerman CM, Ganter MJ, Gaustad G, Babbitt CW, Raffaelle RP, Landi BJ (2012) Recycling single-wall carbon nanotube anodes from lithium ion batteries. J Mater Chem 22(24):12008–12015. https://doi.org/10.1039/C2JM31971C
28. Zhang L, Wang Y, Peng B, Yu W, Wang H, Wang T, Deng B, Chai L, Zhang K, Wang J (2014) Preparation of a macroscopic, robust carbon-fiber monolith from filamentous fungi and its application in Li–S batteries. Green Chem 16(8):3926–3934. https://doi.org/10.1039/C4GC00761A
29. Li M, Zhang H, Xiao T, Wang S, Zhang B, Chen D, Su M, Tang J (2018) Low-cost biochar derived from corncob as oxygen reduction catalyst in air cathode microbial fuel cells. Electrochimica Acta 283:780–788. https://doi.org/10.1016/j.electacta.2018.07.010
30. Xiong C, Li M, Zhao W, Duan C, Dai L, Shen M, Xu Y, Ni Y (2020) A smart paper@polyaniline nanofibers incorporated vitrimer bifunctional device with reshaping, shape-memory and self-healing properties applied in high-performance supercapacitors and sensors. Chem Eng J 396:125318. https://doi.org/10.1016/j.cej.2020.125318
31. Korenblit Y, Rose M, Kockrick E, Borchardt L, Kvit A, Kaskel S, Yushin G (2010) High-Rate electrochemical capacitors based on ordered mesoporous silicon carbide-derived carbon. ACS Nano 4(3):1337–1344. https://doi.org/10.1021/nn901825y
32. Xiong C, Dang W, Nie S, Qin C, Li D, Dai L, Shen M, Xu Y, Ni Y (2021) Fabrication of high value cellulose nanofibers@Ni foam by non carbonization: various application developed during the preparation. Cellulose 28(3):1455–1468. https://doi.org/10.1007/s10570-020-03597-6
33. Xiong C, Li M, Nie S, Dang W, Zhao W, Dai L, Ni Y (2020) Non-carbonized porous lignin-free wood as an effective scaffold to fabricate lignin-free Wood@Polyaniline supercapacitor material for renewable energy storage application. J Power Sources 471:228448. https://doi.org/10.1016/j.jpowsour.2020.228448
34. Burke A (2007) R&D considerations for the performance and application of electrochemical capacitors. Electrochimica Acta 53(3):1083–1091. https://doi.org/10.1016/j.electacta.2007.01.011
35. Jiang H, Lee PS, Li C (2013) 3D carbon based nanostructures for advanced supercapacitors. Energy Environ Sci 6(1):41–53. https://doi.org/10.1039/C2EE23284G
36. Huang J, Sumpter BG, Meunier V (2008) A universal model for nanoporous carbon supercapacitors applicable to diverse pore regimes, carbon materials, and electrolytes. Chemistry 14(22):6614–6626. https://doi.org/10.1002/chem.200800639
37. Qu D, Shi H (1998) Studies of activated carbons used in double-layer capacitors. J Power Sources 74(1):99–107
38. Qian W, Sun F, Xu Y, Qiu L, Liu C, Wang S, Yan F (2014) Human hair-derived carbon flakes for electrochemical supercapacitors. Energy Environ Sci 7(1):379–386. https://doi.org/10.1039/C3EE43111H
39. Feng H, Hu H, Dong H, Xiao Y, Cai Y, Lei B, Liu Y, Zheng M (2016) Hierarchical structured carbon derived from bagasse wastes: a simple and efficient synthesis route and its improved electrochemical properties for high-performance supercapacitors. J Power Sour 302:164–173. https://doi.org/10.1016/j.jpowsour.2015.10.063

40. Liu WJ, Tian K, He YR, Jiang H, Yu HQ (2014) High-yield harvest of nanofibers/mesoporous carbon composite by pyrolysis of waste biomass and its application for high durability electrochemical energy storage. Environ Sci Technol 48(23):13951–13959. https://doi.org/10.1021/es504184c
41. Sun L, Tian C, Li M, Meng X, Wang L, Wang R, Yin J, Fu H (2013) From coconut shell to porous graphene-like nanosheets for high-power supercapacitors. J Mater Chem A 1(21):6462–6470. https://doi.org/10.1039/C3TA10897J
42. Li Y, Wang G, Wei T, Fan Z, Yan P (2016) Nitrogen and sulfur co-doped porous carbon nanosheets derived from willow catkin for supercapacitors. Nano Energy 19:165–175. https://doi.org/10.1016/j.nanoen.2015.10.038
43. Wu X-L, Wen T, Guo H-L, Yang S, Wang X, Xu A-W (2013) Biomass-Derived sponge-like carbonaceous hydrogels and aerogels for supercapacitors. ACS Nano 7(4):3589–3597. https://doi.org/10.1021/nn400566d
44. Pandey S, Karakoti M, Surana K, Dhapola PS, SanthiBhushan B, Ganguly S, Singh PK, Abbas A, Srivastava A, Sahoo NG (2021) Graphene nanosheets derived from plastic waste for the application of DSSCs and supercapacitors. Sci Rep 11(1):3916. https://doi.org/10.1038/s41598-021-83483-8
45. Poochai C, Srikhaow A, Lohitkarn J, Kongthong T, Tuantranont S, Tuantranont S, Primpray V, Maeboonruan N, Wisitsoraat A, Sriprachuabwong C (2021) Waste coffee grounds derived nanoporous carbon incorporated with carbon nanotubes composites for electrochemical double-layer capacitors in organic electrolyte. J Energy Storage 43:103169. https://doi.org/10.1016/j.est.2021.103169.
46. Aumber A, Yap Mei Y, Faisal S, Zhao J, Andrei V, Qingyu Y, Grzegorz L, Tuti Mariana L (2021) Multiwall carbon nanotubes derived from plastic packaging waste as a high-performance electrode material for supercapacitors. Int J Energy Res 45(13):19611–19622. https://doi.org/10.1002/er.6967
47. Mishra N, Shinde S, Vishwakarma R, Kadam S, Sharon M, Sharon M (2013) MWCNTs synthesized from waste polypropylene plastics and its application in super-capacitors. In: Sathiyamoorthy D, Sharon M, Ramadurai K (eds) Carbon materials 2012 (CCM12): carbon materials for energy harvesting, environment, nanoscience and technology, pp 228–236
48. Lai F, Miao Y-E, Zuo L, Lu H, Huang Y, Liu T (2016) Biomass-Derived nitrogen-doped carbon nanofiber network: a facile template for decoration of ultrathin nickel-cobalt layered double hydroxide nanosheets as high-performance asymmetric supercapacitor electrode. Small 12(24):3235–3244. https://doi.org/10.1002/smll.201600412.
49. Wen X, Chen X, Tian N, Gong J, Liu J, Rümmeli MH, Chu PK, Mijiwska E, Tang T (2014) Nanosized carbon black combined with Ni_2O_3 as "universal" catalysts for synergistically catalyzing carbonization of polyolefin wastes to synthesize carbon nanotubes and application for supercapacitors. Environ Sci Technol 48(7):4048–4055. https://doi.org/10.1021/es404646e
50. Wang Y, Bailey J, Zhu Y, Zhang Y, Boetcher SKS, Li Y, Wu C (2022) Application of carbon nanotube prepared from waste plastic to phase change materials: the potential for battery thermal management. Waste Manage 154:96–104. https://doi.org/10.1016/j.wasman.2022.10.003
51. Liu M, Niu J, Zhang Z, Dou M, Wang F (2018) Potassium compound-assistant synthesis of multi-heteroatom doped ultrathin porous carbon nanosheets for high performance supercapacitors. Nano Energy 51:366–372. https://doi.org/10.1016/j.nanoen.2018.06.037
52. Nam E, Park G, Nam JY, Park S, Jo Y, Kim J, Park BG, Baek K, Kang SJ, Ra HW, Park Y, Seo MW, Lee KJ, An K (2023) Upcycling of plastic waste into carbon nanotubes as efficient battery additives. Green Chem 25(20):8007–8018. https://doi.org/10.1039/D3GC02100A
53. Zhao C, Huang Y, Zhao C, Shao X, Zhu Z (2018) Rose-derived 3D carbon nanosheets for high cyclability and extended voltage supercapacitors. Electrochimica Acta 291:287–296. https://doi.org/10.1016/j.electacta.2018.09.136
54. He S, Chen W (2015) Application of biomass-derived flexible carbon cloth coated with MnO_2 nanosheets in supercapacitors. J Power Sour 294:150–158. https://doi.org/10.1016/j.jpowsour.2015.06.051

55. Jain A, Xu C, Jayaraman S, Balasubramanian R, Lee JY, Srinivasan MP (2015) Mesoporous activated carbons with enhanced porosity by optimal hydrothermal pre-treatment of biomass for supercapacitor applications. Microporous Mesoporous Mater 218:55–61. https://doi.org/10.1016/j.micromeso.2015.06.041
56. Xia J, Zhang N, Chong S, Li D, Chen Y, Sun C (2018) Three-dimensional porous graphene-like sheets synthesized from biocarbon via low-temperature graphitization for a supercapacitor. Green Chem 20(3):694–700. https://doi.org/10.1039/C7GC03426A
57. Zhao W, Wen J, Zhao Y, Wang Z, Shi Y, Zhao Y (2020) Hierarchically porous carbon derived from biomass reed flowers as highly stable Li-ion battery anode. Nanomaterials 10(2):346
58. Liu S, Zhao Y, Zhang B, Xia H, Zhou J, Xie W, Li H (2018) Nano-micro carbon spheres anchored on porous carbon derived from dual-biomass as high rate performance supercapacitor electrodes. J Power Sour 381:116–126. https://doi.org/10.1016/j.jpowsour.2018.02.014
59. Dong S, He X, Zhang H, Xie X, Yu M, Yu C, Xiao N, Qiu J (2018) Surface modification of biomass-derived hard carbon by grafting porous carbon nanosheets for high-performance supercapacitors. J Mater Chem A 6(33):15954–15960. https://doi.org/10.1039/C8TA04080J
60. Guo Y, Qi J, Jiang Y, Yang S, Wang Z, Xu H (2003) Performance of electrical double layer capacitors with porous carbons derived from rice husk. Mater Chem Phys 80(3):704–709. https://doi.org/10.1016/S0254-0584(03)00105-6
61. Tian X, Ma H, Li Z, Yan S, Ma L, Yu F, Wang G, Guo X, Ma Y, Wong C (2017) Flute type micropores activated carbon from cotton stalk for high performance supercapacitors. J Power Sour 359:88–96. https://doi.org/10.1016/j.jpowsour.2017.05.054

Chapter 13
Waste-Derived Carbon Nanomaterials for Solar Cell Applications

Ramakrishna Madaka, Beauty Pandey, Dipak Kumar Sahoo, Mahesh Peddigari, Jagadeeswara Rao Dasari, Buddha Deka Boruah, and Jatindra Kumar Rath

Abstract This chapter explores the innovative use of waste-derived carbon nanomaterials in solar cells, addressing both environmental and energy challenges. It discusses the potential of these materials to improve solar cell efficiency and reduce costs while utilizing waste sources like biomass, plastics, and industrial by-products. The synthesis methods for these nanomaterials, including pyrolysis and hydrothermal carbonization, are examined. The chapter details how waste-derived carbon nanomaterials can enhance various solar cell components, such as electron and hole transport layers, in perovskite, organic, and dye-sensitized cells. Specific examples are presented, like graphene quantum dots from coal and carbon nanotubes from plastic waste, along with their performance improvements. The chapter also addresses challenges in using these materials, including purity and scalability issues. Environmental and economic impacts are evaluated, highlighting the potential for waste reduction and creating a circular economy in the solar energy sector. Future trends, such as integration with quantum dots and applications in flexible solar cells, are explored. The chapter concludes by emphasizing the transformative potential

R. Madaka (✉) · B. Pandey · D. K. Sahoo
School of Sciences, Woxsen University, Hyderabad, Telangana 502345, India
e-mail: ramakrishna.madaka@woxsen.edu.in

M. Peddigari
Department of Physics, IIT Hyderabad, Hyderabad, Telangana 502284, India

J. R. Dasari
Department of Physics, S.V. Government Degree College, Parvathipuram, A.P, India

B. D. Boruah
Institute for Materials Discovery (IMD), University College London (UCL), London WC1E 7JE, UK

J. K. Rath
Department of Physics, Indian Institute of Technology Madras, Chennai, Tamil Nadu 600036, India

Debye Institute, Utrecht University, P.O. Box 80000, 3508 TA Utrecht, The Netherlands

J. K. Rath
e-mail: jkr@iitm.ac.in

N. Talreja et al. (eds.), *Waste-Derived Carbon Nanostructures*, Nanostructure Science and Technology, https://doi.org/10.1007/978-3-031-75247-6_13

of waste-derived carbon nanomaterials in advancing sustainable solar technology, calling for continued research and development in this promising field.

Keywords Waste derived · Nanomaterials · Solar cells · Sustainable materials · Environmental impact · Carbon · Quantum dots

13.1 Introduction

The unorthodox industrial expansion and exponential population growth encountered over the last century have increased energy utilization and production of different kinds of waste. This explosive growth not only leads to high demands for resources but also requires their enhancements. Furthermore, the world is undergoing an energy crisis and is facing climate challenges together with environmental sustainability issues, stimulating massive research globally. Primarily, our energy economy has been fossil fuel dependent where the non-renewable energy sources were burnt to produce electricity. Consequently, it has resulted in the rapid depletion of fossil reserves, global warming, and deteriorating environmental conditions. Owing to these, scientists, researchers, and industrialists are constantly on the lookout for technologies that could resolve the rapidly rising energy and environmental concerns sustainably. Subsequently, attention has been directed toward clean and renewable energy sources like solar, water, wind, etc. to address these inherent problems arising from traditional energy sources. As the world seeks viable alternatives, the natural abundance and accessibility of sunlight, coupled with its capacity to catalyze environmentally friendly reactions that transform solar energy into chemical, electrical, or solar fuels, have made it look like a possible solution. This transition not only aligns with global strategies to combat environmental pollution and climate change but also secures a sustainable energy future. Interestingly, solar photovoltaic (PV) technologies have thus become one of the most promising alternate energy solutions, attracting significant attention and investment from scientists, industry, and society to achieve a decarbonized and affordable economy in the years to come [1–12].

Together with sustainable energy generation and environmental remediation, the fast pace of development, coupled with urbanization, industrialization, and population explosion, has triggered the importance of addressing various other hazards. One such global challenge that necessitates urgent intervention is the generation of large quantities of waste whose amounts are skyrocketing to staggering levels with each passing year. Investigations have revealed that around seven to eight billion tons of annual trash are produced all around the world. The escalating problem of such huge amounts of waste poses serious health and environmental scares causing a massive challenge that needs to be addressed by both the scientists as well as governments. Plastic waste is a significant contributor due to its non-biodegradable nature and potential to release toxic chemicals into the environment. In 2011 alone [13], global plastic production reached a staggering 280 million tons, with China contributing 23%, the European Union 21%, the United States 16%, and Japan 5%.

Improper disposal of plastic waste can lead to the release of harmful substances, such as bisphenol A (BPA) and phthalates, into the environment, potentially causing adverse health effects in humans and wildlife [14]. Over the past 50 years, plastic pollution has accelerated due to the widespread use of plastics across various industries, including packaging, construction, transportation, and consumer electronics. According to a study, approximately 8.3 billion metric tons of plastic had been produced globally as of 2015, with only 9% being recycled and 12% incinerated [15]. Recent research accounts for plastic waste for roughly 10% of the total global waste generated annually. The rapid growth of the electronics industry and the short lifespan of electronic devices have led to a surge in e-waste generation. A report by the United Nations University (2020) estimated that approximately 53.6 million metric tons of e-waste were generated globally in 2019 [16]. E-waste contains hazardous substances, including heavy metals (e.g., lead, cadmium, and mercury), which can contaminate soil and water resources, posing risks to human health and the environment [17]. Biomedical waste, generated by healthcare facilities, poses significant health risks due to its potential to transmit infectious diseases and expose individuals to hazardous materials. According to a study by the World Health Organization (WHO), around 85% of biomedical waste is considered non-hazardous, while the remaining 15% is regarded as hazardous and requires proper treatment and disposal [18]. Improper management of biomedical waste can lead to the spread of infectious diseases and the contamination of water sources [19]. The construction and demolition industry generates a substantial amount of waste, including concrete, wood, metals, and other materials. A study by the European Commission (2020) estimated that the construction sector accounts for approximately 35% of the total waste generated in the European Union [20]. Improper disposal of construction and demolition waste can lead to soil and water contamination, as well as air pollution from dust particles [21]. Additionally, the burning of biomass and agricultural waste releases harmful toxins and pollutants into the atmosphere. The nature of these emissions varies depending on the type of biomass feedstock but often includes nitrogen oxides, carbon monoxide, and sulfur dioxide. Moreover, waste biomass can trigger forest fires and contribute to a 25–30% annual increase in global greenhouse gas production. Reports estimate that crop yields worldwide generate a massive 146 billion tons of solid waste each year, further compounded by another 100 billion tons from household waste. Managing this colossal volume of waste in an environmentally responsible manner presents a formidable challenge. Effective waste management strategies, including source reduction, recycling, and proper disposal methods, are crucial to mitigate the environmental and health impacts of these waste streams on a global scale. Additionally, governments, industries, and communities must collaborate to implement comprehensive policies and initiatives that promote sustainable waste management practices, reduce plastic consumption, and encourage the development of innovative solutions for waste valorization and resource recovery. Concurrently, overcoming the present scenario desires approach with scientific innovation and development which can not only solve but also maximize the waste utilization through technological advancements. For scientific community, the field of nanoscience and nanotechnology has emerged as a promising solution, offering the potential to address both waste management and the global energy crisis simultaneously. Waste management technologies need to be revolutionized to comprehensively tackle the issue of different types of waste. Concurrently, the scientific community also faces the urgent requirement to develop eco-friendly and cost-effective energy solutions to mitigate the world's energy crisis.

Broadly, converting waste materials into valuable nanomaterials is an ethical and beneficial solution for waste management because it [1, 3–6, 11, 22]:

(a) Reduces Environmental Burden: Converting wastes into useful nanomaterials helps reduce the environmental burden posed by the huge amounts of solid waste generated globally each year. It prevents wastes from ending up in landfills or being burned, which causes pollution and greenhouse gas emissions.
(b) Enables Waste Valorization: This approach enables the valorization of waste materials that would otherwise have little or no value. It turns an environmental liability into a valuable resource by deriving high-performance carbon nanomaterials from the waste.
(c) Aligns with Circular Economy: Producing nanomaterials from waste fits the principles of a circular economy by closing the material loop and extracting maximum value from waste before final disposal. It promotes sustainable waste utilization.
(d) Facilitates Advanced Applications: The unique properties of carbon nanomaterials like nanotubes and graphene enable advanced applications in areas like energy storage, catalysis, electronics, and biomedical devices. This technological valorization benefits society.
(e) Reduces Reliance on Virgin Materials: Deriving nanomaterials from waste reduces the need to extract virgin raw materials from the environment through mining or drilling operations that can be environmentally damaging. In essence, this zero-waste valorization approach is an ethical solution as it addresses the escalating waste crisis through sustainable material recovery and enables useful applications of the derived nanomaterials for societal benefit.

Several research efforts have focused on efficient waste management strategies and the production of carbon nanomaterials (CNMs), such as carbon nanotubes (CNTs), graphene, and carbon quantum dots (CQDs), from solid waste sources for various industrial applications, particularly in the energy sector. CNMs, being carbon-based nanomaterials, possess exceptional properties and versatile potential for numerous applications.

Several reasons aid to the conversion of waste materials into valuable carbon-based nanomaterials. Some of them include [4, 21, 23–30]:

(a) Carbon-Rich Composition: Many waste materials, such as plastics, biomass, tires, and certain industrial wastes, contain high amounts of carbon in their composition. This carbon-rich nature makes them suitable precursors for the synthesis of carbon-based nanomaterials like carbon nanotubes, graphene, carbon nanofibers, and carbon dots.

(b) Presence of Graphitic Structures: Some waste materials, like used tires, waste plastics, and coal tar, already possess graphitic structures or aromatic carbon rings in their molecular framework. These graphitic structures can be exfoliated or rearranged through chemical or thermal processes to produce graphene or other carbon nanomaterials.
(c) Low Cost and Abundant Availability: Waste materials are readily available in large quantities and at low or even negative cost (if disposal costs are considered). This abundant and inexpensive supply of precursor materials makes the production of carbon nanomaterials from waste economically viable.
(d) Versatile Conversion Methods: Various conversion methods, such as pyrolysis, chemical vapor deposition, hydrothermal treatment, and electrochemical exfoliation, can be employed to transform waste materials into carbon nanomaterials. The choice of method depends on the nature of the waste and the desired nanocarbon product.
(e) Unique Properties: The carbon nanomaterials derived from waste often exhibit unique properties like high surface area, excellent electrical conductivity, thermal stability, and mechanical strength, which make them valuable for applications in energy storage, catalysis, composites, and electronics.
(f) Environmental Benefits: Converting waste into valuable nanomaterials not only reduces the environmental burden of waste disposal but also provides a sustainable route for material recovery and resource efficiency, aligning with the principles of a circular economy. By leveraging the carbon-rich nature and graphitic structures present in various waste materials, along with the availability of diverse conversion methods and the potential for unique properties, waste can be effectively transformed into valuable carbon-based nanomaterials, providing both economic and environmental benefits.

13.2 Properties of WDCNMs

Water-dispersible carbon nanomaterials (WDCNMs) exhibit a wide range of unique properties that make them attractive for various applications. Some of the key properties of WDCNMs are:

(a) **High Surface Area**

WDCNMs often possess an extremely high specific surface area due to their nanoscale dimensions and porous structures. For instance, Sun et al. synthesized hierarchical porous carbon nanosheets from waste coffee grounds, achieving a remarkably high surface area [31]. Such high surface areas are advantageous for applications like supercapacitors, catalysis, and adsorption, where a large reactive surface is desirable for efficient charge storage, catalytic activity, or adsorption of molecules.

(b) **Excellent Electrical Conductivity**

The unique structural features of WDCNMs, such as the conjugated π-electron system in graphene and the seamless cylindrical structure of carbon nanotubes, contribute to their excellent electrical conductivity. Chai et al. reported the use of waste-derived graphene nanosheets as efficient counter electrodes in dye-sensitized solar cells, achieving power conversion efficiencies comparable to those of conventional platinum-based electrodes [32]. The high electrical conductivity of these nanomaterials makes them promising for applications in energy storage, electronics, and conductive composites.

(c) **Thermal Stability**

Many WDCNMs exhibit remarkable thermal stability, owing to their strong carbon–carbon bonds and stable molecular structures. Researchers demonstrated the use of waste-derived carbon nanotubes in perovskite solar cells, where their thermal stability up to 600 °C contributed to improved device performance and stability [33]. This property enables the use of WDCNMs in high-temperature environments or as reinforcing fillers in composites for thermal management applications.

(d) **Mechanical Strength**

Certain WDCNMs, particularly carbon nanotubes and graphene, possess exceptional mechanical strength and stiffness due to their strong covalent bonds and unique molecular arrangements. Scientists reported the synthesis of waste-derived graphene nanocomposites with enhanced mechanical properties, such as increased tensile strength and Young's modulus, making them suitable for construction applications [34]. These mechanical properties make WDCNMs attractive for reinforcing composites, structural materials, and protective coatings.

(e) **Tunable Optical Properties**

Waste-derived carbon quantum dots (CQDs) and graphene quantum dots (GQDs) exhibit tunable optical properties, including fluorescence and photoluminescence, by adjusting their size, surface functionalization, and synthesis conditions. Perumal et al. reported the synthesis of fluorescent CQDs from biomass waste with tunable emission wavelengths, enabling their potential use in bioimaging, sensing, and optoelectronic devices [35]. These tunable optical properties open up possibilities for various applications in optics, photonics, and biomedical fields.

(f) **Chemical Stability**

Many WDCNMs, such as carbon nanotubes and graphene, exhibit remarkable chemical stability and resistance to harsh environments, making them suitable for applications in catalysis, energy storage, and environmental remediation. Poonia et al. (2022) highlighted the potential of WDCNMs for sustainable energy applications due to their chemical stability and resistance to corrosion and degradation [36]. This property ensures the long-term performance and durability of CNM-based devices and materials in various operating conditions.

The unique combination of these properties, often superior to those derived from conventional sources, makes WDCNMs attractive for a wide range of applications while promoting sustainable waste management and resource efficiency.

13.3 Advantages of Using WDCNMs in Solar Cells

WDCNMs offer several advantages when used in solar cells, making them a promising alternative to traditional materials. By utilizing solid waste as a precursor, the production of CNMs can be substantially increased, thereby providing a sustainable solution for waste valorization and addressing the growing demand for advanced nanomaterials. In the context of solar energy, WDCNMs have shown promising applications in improving the performance and reducing the cost of solar cells. For instance, research has demonstrated the use of waste-derived graphene and carbon nanotubes as efficient electrode materials in dye-sensitized solar cells (DSSCs) and perovskite solar cells (PSCs), enhancing their power conversion efficiency and stability [33, 37, 38]. Additionally, carbon quantum dots derived from biomass waste have been explored as effective light-harvesting and charge transport materials in quantum dot-sensitized solar cells (QDSSCs) [39–41]. The synthesis of CNMs from waste sources not only addresses the issue of waste management but also offers a cost-effective and sustainable approach to producing high-performance nanomaterials for solar energy applications. Some of the benefits of using WDCNMs in solar cells are:

(a) **Cost-Effectiveness**

One of the major advantages of using WDCNMs in solar cells is their cost-effectiveness. Traditional materials used in solar cell manufacturing, such as silicon, indium, and precious metal catalysts, can be expensive and contribute significantly to the overall cost of the devices. By utilizing waste materials as precursors for CNM synthesis, the production cost can be substantially reduced. Waste materials like plastics, biomass, and agricultural residues are abundantly available and often considered a disposal problem. By repurposing these waste streams as feedstocks for CNM production, the manufacturing process becomes more economically viable, as the raw materials are readily accessible and inexpensive, or even free in some cases. For example, the use of waste-derived carbon nanotubes in perovskite solar cells, highlighting their cost-effectiveness compared to conventional materials like platinum and other noble metals typically used as electrode materials, was demonstrated [33].

(b) **Improved Charge Transport and Collection**

CNMs, such as graphene and carbon nanotubes, possess exceptional electrical conductivity and charge transport properties due to their unique structural features and delocalized π-electron systems. When incorporated into solar cell architectures, these CNMs can facilitate efficient charge carrier transport and collection, leading to enhanced power conversion efficiencies. For instance, the use of waste-derived graphene nanosheets as efficient counter electrodes in dye-sensitized solar cells (DSSCs) was reported [32, 37]. The high electrical conductivity of graphene allowed for efficient charge transport, resulting in power conversion efficiencies comparable to those of conventional platinum-based electrodes.

(c) **Enhanced Light-Harvesting and Absorption**

Certain CNMs, particularly carbon quantum dots (CQDs), exhibit unique optical properties, including tunable light absorption and emission. When integrated into solar cell devices, CQDs can act as efficient light-harvesting and sensitizing agents, improving overall light absorption and energy conversion efficiency. The potential of waste-derived CQDs as light-harvesting materials in quantum dot-sensitized solar cells (QDSSCs) was demonstrated [32]. The tunable optical properties of CQDs allowed for efficient light absorption and charge transfer, contributing to improved device performance.

(d) **Improved Stability and Durability**

CNMs, such as carbon nanotubes and graphene, are known for their exceptional thermal stability and chemical resistance. These properties can contribute to enhanced stability and durability of solar cell devices when CNMs are incorporated into their architectures. The use of waste-derived carbon nanotubes in perovskite solar cells, where their thermal stability up to 600 °C led to improved device performance and stability, even under harsh environmental conditions was reported [32].

(e) **Environmentally Friendly and Sustainable**

Utilizing waste materials as precursors for CNM synthesis aligns with the principles of a circular economy and sustainable development. By valorizing waste streams, the production of CNMs for solar cell applications reduces the environmental burden associated with waste disposal and promotes resource efficiency. This approach not only addresses the issue of waste management but also contributes to the development of sustainable energy technologies, supporting the transition toward a more environmentally friendly and renewable energy future.

(f) **Versatile Applications**

CNMs can be integrated into various solar cell technologies, including dye-sensitized solar cells (DSSCs), perovskite solar cells (PSCs), quantum dot-sensitized solar cells (QDSSCs), and organic solar cells. Their unique properties make them suitable for applications in different components of solar cells, such as electrodes, charge transport layers, and light-harvesting materials. For example, waste-derived graphene has been explored as an efficient electrode material in DSSCs [33], while carbon nanotubes have shown promise as charge transport layers in PSCs [32], and CQDs have been utilized as light-harvesting agents in QDSSCs [35].

13.4 Challenges of Using WDCNMs in Solar Cells

Despite these advantages, challenges remain in terms of scalable production methods, purification techniques, and potential toxicity concerns associated with CNMs. Ongoing research efforts are focused on addressing these challenges and optimizing the integration of WDCNMs into solar cell architectures to fully leverage their potential and contribute to the development of cost-effective and sustainable solar energy technologies. To summarize, some of the challenges associated with using WDCNMs in solar cells that require to be addressed are:

(a) **Scalable and Consistent Production**

One of the major challenges is the scalable and consistent production of CNMs from waste sources. The composition and properties of waste materials can vary significantly, leading to inconsistencies in the quality and performance of the derived CNMs. Developing standardized and reproducible synthesis methods that can handle the heterogeneity of waste feedstocks is crucial for the large-scale implementation of CNMs in solar cell manufacturing. Additionally, scaling up the production processes from laboratory-scale to industrial-scale can be challenging, as it requires optimizing parameters such as reaction times, temperatures, and purification techniques to maintain consistent product quality and yield [42].

(b) **Purification and Quality Control**

WDCNMs often contain impurities and defects, which can adversely affect their performance in solar cell applications. Efficient purification techniques are necessary to remove these impurities and ensure high-quality CNMs with desired properties. However, purification processes can be complex, time-consuming, and costly, potentially offsetting the cost-effectiveness advantage of using WDCNMs. Establishing rigorous quality control measures and characterization techniques is essential to ensure that the CNMs meet the required specifications for solar cell integration, such as purity, size distribution, surface functionalization, and electrical properties [32].

(c) **Integration and Compatibility**

Integrating WDCNMs into existing solar cell architectures and manufacturing processes can be challenging. The compatibility of CNMs with other components, such as charge transport layers, electrodes, and encapsulation materials, needs to be carefully evaluated to ensure optimal device performance and stability. Additionally, the incorporation of CNMs may require modifications to the fabrication processes, which can involve adjustments in deposition techniques, annealing conditions, or device assembly methods. Ensuring seamless integration without compromising the overall efficiency and reliability of the solar cells is crucial [33].

(d) **Potential Toxicity and Environmental Concerns**

While CNMs offer environmental benefits in terms of waste valorization and sustainable material production, their potential toxicity and environmental impacts need to be thoroughly assessed. Some CNMs, particularly carbon nanotubes, have raised concerns regarding their potential for inducing toxicity or environmental persistence due to their nanoscale size and unique properties. Comprehensive studies are necessary to evaluate the potential risks associated with the production, handling, and disposal of WDCNMs throughout their lifecycle. Implementing appropriate safety measures and developing strategies for responsible and sustainable production and disposal of CNMs is essential [35].

(e) **Limited Knowledge and Expertise**

The field of WDCNMs for solar cell applications is relatively new and interdisciplinary, combining expertise from materials science, nanotechnology, waste management, and renewable energy. There is a need for more research and knowledge exchange among different disciplines to fully understand the synergies and challenges involved in integrating CNMs into solar cell technologies. Building a skilled workforce with expertise in CNM synthesis, characterization, and integration into solar cell architectures is crucial for driving innovation and commercialization in this field [38].

Regardless of these challenges, the potential benefits of using WDCNMs in solar cells, such as cost-effectiveness, improved performance, and environmental sustainability, continue to drive research efforts to overcome these hurdles. Collaborative efforts among researchers, industry, and policymakers are essential to address these challenges and unlock the full potential of WDCNMs for sustainable and efficient solar energy technologies.

13.5 Synthesis of WDCNMs

13.5.1 Different Methods for Synthesizing WDCNMs

Various carbon nanomaterials, e.g., graphene, carbon nanotubes (CNTs), carbon quantum dots, etc., have been synthesized using different forms of waste materials employing several methods. Each of these methods has its own advantages and drawbacks in terms of what types of nanomaterials are produced, the extent to which they can be commercially produced at scale and in a cost-effective way as well as level of control over both morphology and properties. There are four main ways to prepare WDCNMs:

Hydrothermal Carbonization: This process is to heat biomass waste sources in subcritical water conditions, usually 180–250 °C and autogenous pressures for some hours to days [43]. Under these conditions, the waste biomass undergoes a series of hydrolysis, dehydration, decarboxylation, polymerization and aromatization reactions leading to carbonization and conversion to hydrochar. And this hydrochar rich in carbon may act as a precursor for preparing other WDCNMs. Other operations, such as pyrolysis and activation of the hydrochar, result in porous carbon materials, while acid treatment exfoliates layers of graphene from the hydrochar. CNTs can also be grown on hydrochar derived from waste via chemical vapor deposition (Zhang et al.). This method is low energy, simple, tends to be cheaper, and eco-friendly, thus, it is used as a pre-treatment prior to drying the biomass product. The slow reaction times required, on the other hand, usually restrict its scalability.

Thermal microwave-assisted heating: It consists of heating waste biomass dispersed in solvents by means of microwave radiation, typically with power levels varying between 300 and 1000 W and temperatures up to a max of 220 °C, obtained through dielectric heating. Microwave heating provides fast and homogeneous heating which speeds up the underlying hydrolysis and carbonization reactions developed through conventional hydrothermal treatment [44]. The primary output is water-soluble and photoluminescence carbon quantum dots generated under 5–15 min. Microwave power and time tuning provide an on-demand ability to control carbon dot size and emission properties. However, special equipment is needed to do this, and it only works for creating nano-sized carbon dots. Figure 13.1 Schematic diagram of all steps involved in the synthesis of graphene from rice husk by simple microwave.

Pyrolysis: Direct pyrolysis of waste biomass in the temperature range of 400–800 °C under low O_2 or an inert atmosphere yields carbon-rich solid residue called biochar. This carbon-rich solid may be prepared into WDCNMs [45]. Processing the pyrolyzed biochar with KOH, $ZnCl_2$, H_3PO_4, or CO_2 opens up pores and surface areas of up to 1000 m^2/g. The porous carbon can be intercalated to generate graphene nanoplatelets. CNTs are made by high-temperature pyrolysis (above 1400 °C). Pyrolysis, despite being simple and scalable, provides less shape controllability due to the fact that higher temperatures result in more disordered structures. Figure 13.2 illustrates the basic pyrolysis procedure involved in the synthesis of graphene using biomass waste.

Solvothermal Synthesis: Waste biomass precursors with capping agents are heated in high boiling point organic solvents like ethylene glycol to 240–300 °C in enclosed vessels under autogenous pressures [46]. This gives fluorescent carbon dots with a

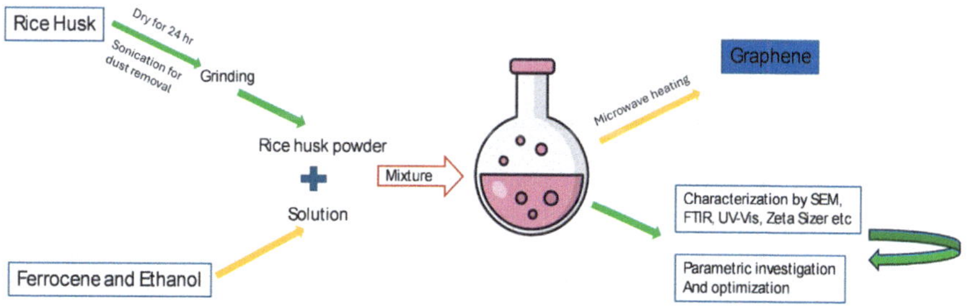

Fig. 13.1 Schematic diagram of all steps involved in the synthesis of graphene from rice husk by simple microwave

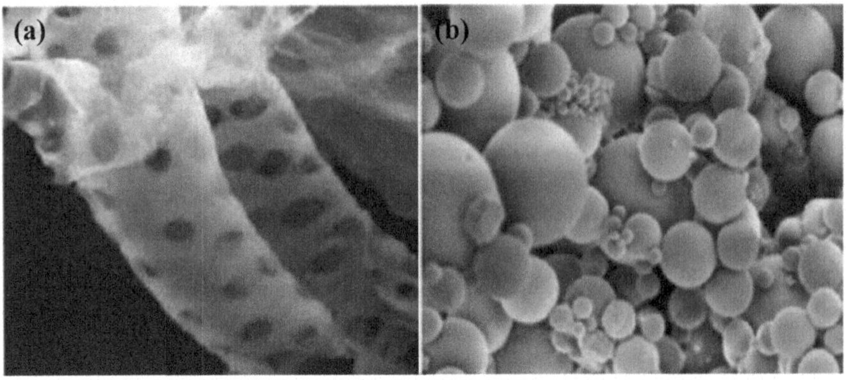

Fig. 13.2 SEM images of various nanomaterials generated from waste materials: **a** Porous carbon nanosheets synthesized from elm samaras plant waste. **b** Nano-polystyrene particles are generated from polystyrene waste. Reprinted with permission from reference [52]

dual core–shell structure within 2–5 nm of the diameter from one-pot route, which allows simultaneous carbonization and surface passivation process. The photoluminescence can be tuned across the entire visible range by a very basic method of reaction temperature and time. Such technology requires the expensive solvents and therefore the production of nanoscale carbon dots.

Chemical Vapor Deposition (CVD): The examples of utilizing waste-derived materials such as cellulose, hydrochar, and biocharm as available templates for CVD growth of graphene from hydrocarbon precursors can be proven by the examples [37, 43]. Common examples are Ni and Cu as metal catalysts that can reduce graphene growth temperature. Waste-derived carbon substrates are also capable to support the growth of vertically aligned CNT arrays by CVD. This provides an excellent control over morphology. But specific instrumentation is necessary to do this.

Every method has its own advantages and drawbacks. Simple, cost-effective hydrothermal carbonization, pyrolysis, and solvothermal methods are able to produce a wide range of WDCNMs from different waste feedstocks. The advantage of the microwave heating is in fact to achieve rapid synthesis for carbon quantum dots, while CVD represents the medium with best control over morphology.

13.5.2 Optimization of Synthesis Parameters

The properties of WDCNMs are extremely dependent on the synthesis conditions. The parameters that need to be optimized are temperature, pressure, reaction time, heating rate, reagents used in synthesis and post-synthesis processing. In hydrothermal carbonization, the extent of carbonization increases with temperature as hydrolysis and pyrolytic decomposition reactions accelerate [47], though, temperatures beyond 250 °C may ignite combustion of waste biomass. Optimum processing temperatures range from 180 to 220 °C. High autogenous pressures, e.g., 10–40 bar, are also conducive to carbonization. The carbon yield and surface area are increased by a factor of six through the decarboxylation and aromatization of waste biomass on prolongation of the reaction times from 2 to 12 h. Too long times are not (economically) viable though. Slow heating rates of 2–10 °C/min are more beneficial, as fast heating could contribute to the appearance of amorphous carbon.

The carbon dots with various photoluminescence properties and fine tunable size can be realized, by adjusting the microwave power and irradiation time, which is heavily depending on the chemical property of carbon sources used [44]. Smaller dots around 3 nm diameters with blue emission can be achieved in the low powers of 300W at short times of 5 min. When higher powers of 700–1000 W are utilized for the same duration of 15 min, larger dots emitting green colour are formed of about 5–8 nm in size. Higher radiation times typically lead to more carbonization and improved quantum yield. But, it also raises the danger of combustion.

The temperature and ramp rate during pyrolysis also play a significant role [45]. For carbonization of biomass, 500–700 °C are sufficient. To begin with cracking the wood into shards before passing to the gas phase for conversion, this requires extremely high temperatures. Such high temperatures, above 800 °C, will cause graphitization. During the heat treatment, carbonization only happens at slow ramp rates (~ 2–5 °C/min), while fast heating (> 10–20 °C/min) results in amorphous carbon. The second step, or post-activation process, tends to create porosity, especially for KOH activation, which invokes much greater micro- and mesopores than CO_2.

Parameters such as precursor source, substrate material, temperature, and pressure conditions, and the choice of catalyst affect graphene growth in CVD process. Lower process temperatures of 600–800 °C are preferred for waste-derived substrates to avoid damage [46]. The diffusion of the metal catalysts, such as Ni and Cu, is used as a catalyst to reduce the growth temperature.

All methods require extensive solvent, acid/alkali washes to ensure the removal of impurities like metal catalysts and ash. The volatiles are removed while drying followed by partial fragmentation of unstable functional groups during annealing. Finally, optimizing these synthesis parameters systematically facilitates the production of WDCNMs for a specific application.

13.5.3 Characterization of WDCNMs

A comprehensive characterization employing the whole range of analytical tools is needed to finally understand how the synthetic parameters determine the chemical, structural, and functional properties of WDCNMs, resulting from various synthesis modes. This then helps in establishing the proper application of the WDCNMs. Here are a few important characterization methods.

Electron Microscopy: Particle size is a crucial parameter for nanomaterials (NMs). Therefore, thoroughly investigating the size and shape of produced nanomaterials is essential for further applications. Scanning electron microscopy (SEM) characterizes the internal dispersion of nano-sized particles and can visualize the nanoporous surfaces of activated carbon and the shapes of carbon nanotubes. Berkmans et al. [48] confirmed 'Y'-shaped nanochannels ultrafine carbon tubes using SEM. Jain and Tripathi [49] also visualized nanostructured activated carbon with SEM, finding irregular and porous textures throughout. Chen et al. [50] synthesized porous carbon nanosheets from elm samaras plant waste, reporting a sandwich structure with two thin flake layers and porous tubes in between. Figure 13.2 presents SEM images of various nanostructured materials: Fig. 13.2a depicts carbon nanosheets from plant waste; [50] Fig. 13.2b has micro- and nano-polystyrene particles recycled from waste [51].

Energy dispersive X-ray spectroscopy (EDS) during SEM imaging can determine synthesized NM particle distribution patterns. Fang et al. [53] found uniform Ni distribution on Fe nanoparticle surfaces via EDS. They also examined other zerovalent metallic nanoparticles but detected only Fe, concluding that other metal loadings were too low for surface detection.

Nanostructured material particle/pore size information is essential for determining potential uses, obtained via transmission electron microscopy (TEM) imaging. Temperature significantly affects synthesized nanomaterial size and shape, allowing modification. Berkmans et al. [48] generated 100 nm solid carbon spheres via the 'arc discharge method' at ~ 1700 °C, confirmed by TEM. At ~ 2600 °C, these transformed into 95 nm non-branched and 364 nm Y-structured multi-walled carbon nanotubes. Xing and Zhang [54] recycled 55–120 nm (1000 °C) and 200 nm (1100 °C) lead nanoparticles from waste cathode ray tube glass, confirmed by TEM.

Spectroscopy: Fourier Transform Infrared Spectroscopy (FTIR) is a powerful analytical technique employed to study the vibrational characteristics of precursor materials and final products. Additionally, FTIR spectrograms can be utilized to detect the presence of impurities in the final product. In a study by Li et al. [26] nano-sized α-Fe_2O_3 red pigment powder was synthesized from waste ferrous sulfate. Upon analyzing the FTIR spectrogram of the final product, distinct peaks were observed at wavenumbers 1117.06 cm^{-1} and 1051.03 cm^{-1}, both of which were attributed to the ν_3 stretching vibration of SO_4^{2-} ions (Fig. 13.3a). Furthermore, an additional peak at 602.84 cm^{-1} corresponded to the ν_4 stretching vibration of SO_4^{2-} ions. The presence of these peaks indicated that the final α-Fe_2O_3 nanopowder contained residual sulfate impurities.

In another study, Mangalara and Varughese [51] produced polystyrene nanoparticles from polystyrene waste and characterized them using FTIR spectroscopy. The FTIR spectrogram revealed peaks at 3014 cm^{-1} and 2922 cm^{-1}, which were attributed to the stretching vibrations of aromatic C-H bonds. Additionally, a peak at 2839 cm^{-1} was observed, corresponding to the stretching vibration of aliphatic C-H bonds. Peaks in the regions of 1615–1580 cm^{-1} and 1510–1450 cm^{-1} were assigned to the stretching vibrations of aromatic C=C bonds (Fig. 13.3b). Based on the FTIR spectrogram, the authors concluded that no chemical changes occurred during the recycling process of polystyrene waste into polystyrene nanoparticles. UV–V is absorption spectroscopy is then used to determine the optical bandgap

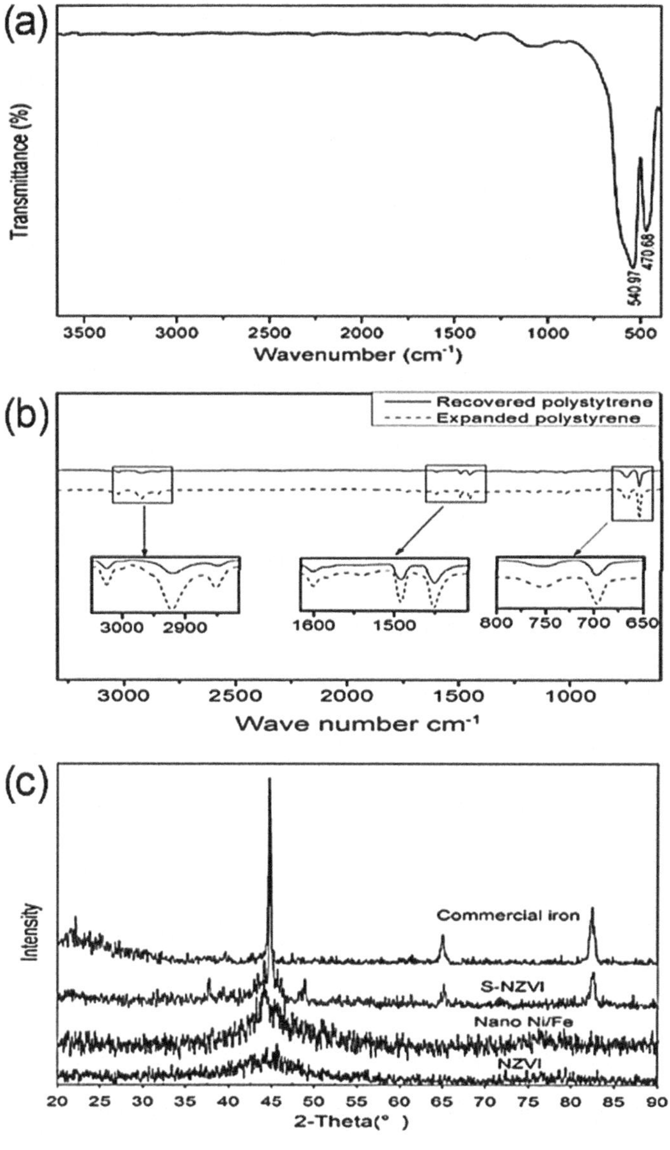

Fig. 13.3 Structural analysis of nanomaterials: **a** FTIR spectrogram of a-Fe_2O_3 NPs, [26] **b** FTIR spectrogram of expanded and recovered polystyrene, and **c** XRD data of synthesized FeNPs and commercial Fe powder. Reprinted with permission from reference [52]

and photoluminescence spectroscopy is carried out for obtaining the fluorescence efficiency and emission spectra of carbon dots.

X-Ray Techniques: X-ray diffraction (XRD) is used to study crystal structure and formation of graphitic domains based on peak position [55]. X-ray diffraction (XRD) measurements were employed to investigate the crystalline structure of nano-samples synthesized from various waste materials. For instance, Fang et al. [53] reported the XRD pattern of their synthesized material, which exhibited three prominent peaks corresponding to metallic iron (Fe) at 2θ values of 44.66°, 65.16°, and 82.36°. These peaks were assigned to the diffraction planes of Fe-110, Fe-200, and Fe-211, respectively, as illustrated in Fig. 13.3c. Notably, the characteristic diffraction peaks associated with iron oxides or hydroxides were not observed in the XRD pattern, indicating that the crystalline structure of the nanomaterial was attributed to the presence of zerovalent metallic iron (α-Fe^0) derived from the steel pickling waste liquor used as the precursor.

Several researchers have utilized X-ray fluorescence (XRF) analysis to determine the composition and quantify the various constituents present in the synthesized nanomaterials. Li et al. [26] reported the XRF results of their produced α-Fe_2O_3 powder in a tabulated form. According to their findings, the final nanoproduct contained several impurities, including 0.898% SiO_2, 0.486% TiO_2, and 0.112% MgO.

In order to characterize other structural properties of nanomaterials derived from waste sources, researchers commonly employ techniques such as Raman and UV–V is spectroscopy. For instance, graphene materials have been synthesized from various biowaste sources, including waste cookies, waste chocolates, waste grasses, waste plastics, dog feces, and roaches [56]. Raman spectroscopy has been utilized to determine the types of defects present in the graphene structures. Figure 13.4 presents the Raman spectra of graphene materials generated from the aforementioned biowaste sources. The patterns of the Raman spectrograms exhibit small or no D peaks, supporting the fact that there are very few defects in the graphene structures. However, the presence of a large 2D/G ratio, as observed in Fig. 13.4, indicates that the graphene derived from waste sources possesses a high-quality monolayer structure.

Dynamic light scattering (DLS) technique is also employed to determine the particle size distribution of waste-derived nanomaterials. Akter Jahan et al. [57] for example, reported the particle sizes of nano-hydroxyapatite derived from eggshell wastes using the DLS technique. The DLS data indicated different particle size distributions under various conditions, such as in the microemulsion phase, after oven drying (100 °C), and after calcination (600 °C).

Porosity analysis: While properties like surface area to pore volume or pore size are insignificant when particles are in bulk form, these properties start to change anomalously when particle sizes reach the nanometer scale. This behavior also applies to waste-derived nanomaterials (NMs). Therefore, evaluating the BET surface area of such NMs is crucial for their various applications. For instance, Hassan et al. [58] prepared eggshell waste-derived $CaCO_3$ nanoparticles (NPs) via a top-down approach. Combined mechano-chemical and ultrasonic irradiation techniques generated high surface area bio-$CaCO_3$ NPs, producing 10 nm particles with a BET surface

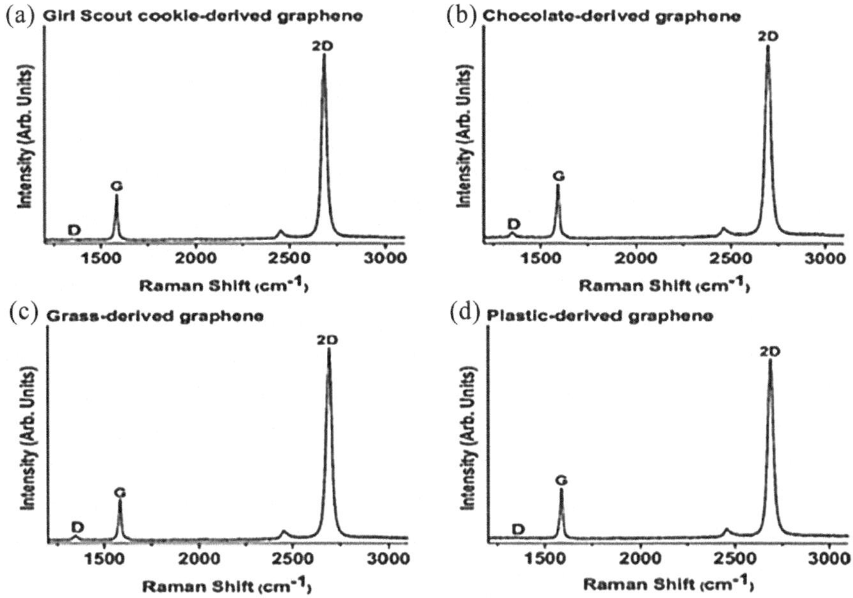

Fig. 13.4 Raman spectrograph of different wastes-derived graphene materials where the 2D to G peak intensity ratios indicate the monolayer structure of graphene. Reprinted with permission from reference [52]

area of about 44 m^2/g. In contrast, Amorim et al. [59] found that blast furnace flue dust-derived iron nanoparticles (FeNPs) exhibited a very low BET surface area of approximately 3.4 m^2/g with a total pore volume of 1.1×10^{-3} cm^3/g. These examples highlight the importance of evaluating the BET surface area when working with waste-derived NMs, as their properties can differ significantly from their bulk counterparts once at the nanometer scale.

Thermal and Electrical Analysis: Thermogravimetric Analysis (TGA) measures thermal stability & composition by recording wt% loss as a function of temperature [60]. Solution of Sheet Resistivity and Electrical Conductivity by Four-Point Probe Measurements.

In-situ environmental transmission electron microscopy (ETEM) enables for direct observation of the dynamic processes during waste biomass gasification and nanomaterial synthesis at different gas environments and temperatures [41]. This gives some understanding of the transformation mechanism.

Safety Evaluation: verify the cytocompatibility and biosafety via cytotoxicity tests. Model cell lines are tested for exposure to various concentrations of WDCNM, and the overall viability index, inflammatory response, changes in the morphology of cells, etc., are analyzed. Therefore, this detailed and systematic characterization provides basic knowledge about its property which can be utilized for solar cell applications.

13.6 Applications of WDCNMs in Solar Cells

The deposition of carbon-based paste as an electrode material on various substrates is primarily carried out using the doctor blade or screen-printing technique. The general configuration for the deposition of suitable electrodes for perovskite solar cells (PSCs) involves a transparent conducting substrate, on which an insulating layer is deposited. Subsequently, this insulating layer is covered with another layer of carbon paste. After depositing the carbon layer, a sintering process is generally conducted to create the complete electrode system, which is suitable for the infiltration of the perovskite ink. Finally, perovskite ink is drop-casted to fabricate the full PSCs.

The fabrication of electrodes for perovskite solar cells (PSCs) typically involves a temperature-dependent deposition process. Initially, a high-temperature step (400–500 °C) is required to deposit a mesoporous carbon paste layer onto an insulating substrate. This substrate can be composed of aluminum trioxide, zirconium dioxide, or titanium dioxide, with TiO_2 potentially serving as both a compact and mesoporous layer. The insulating layer plays a crucial role in preventing direct contact between the front and back electrodes, thereby minimizing photocurrent leakage. This method highlights the importance of temperature control and material selection in achieving optimal electrode performance for PSCs [61]. Among the insulating materials used in carbon-based electrodes for perovskite solar cells, zirconium dioxide (ZrO_2) is considered superior due to its larger pore size compared to aluminum trioxide (Al_2O_3) and titanium dioxide (TiO_2). This property allows for more efficient infiltration of the perovskite ink, leading to better formation of the perovskite phase. The thickness of the insulating layer is a critical factor affecting electrode performance, with optimal thicknesses varying by material—approximately 1 μm for ZrO_2 (as reported by Liu et al.) and 1.8 μm for Al_2O_3 (according to Barichello et al.). Importantly, researchers have established that the insulating layer thickness should not exceed the carrier diffusion length to maintain efficiency. These considerations have led to the widespread use of this method in fabricating mesoscopic perovskite solar cells [62–65].

The fabrication of carbon-based electrodes for perovskite solar cells presents a dichotomy between high-temperature and low-temperature deposition methods. While high-temperature sintering is effective, it's unsuitable for flexible substrates, which require lower processing temperatures. To address this limitation, researchers have developed a promising low-temperature approach using layer-by-layer deposition. This method involves screen-printing or doctor-blading carbon-based materials directly onto the hole transport material (HTM) or perovskite layer. It offers the additional flexibility of depositing the carbon layer on a separate substrate before transferring it to the cell. However, this seemingly simpler process often encounters contact issues between layers. Both high- and low-temperature methods have been extensively studied in recent years, with graphite and carbon black electrodes processed at various temperatures emerging as state-of-the-art options for electrode fabrication. The optimization of these methods involves careful consideration of carbon compositions, processing temperatures, and fabrication techniques, each of which plays a crucial role in determining the final electrode performance and compatibility with different cell architectures. The development of carbon-based electrodes for perovskite solar cells (PSCs) has been revolutionized by combining the high conductivity of graphite with the porous nature of carbon black. This innovative approach, first demonstrated in 2013 by Ku et al., marked a significant milestone in the fabrication of efficient and stable back electrodes for PSCs. Their groundbreaking work showcased a graphite/carbon black composite material that achieved impressive stability and power conversion efficiency (PCE) in hole transport material (HTM)-free PSCs, opening new avenues for electrode design.

Building upon this foundation, researchers sought to further enhance device performance through material refinement. A key advancement came with the introduction of spheroidal graphite as a top layer over the carbon composite. This modification proved transformative, addressing multiple aspects of electrode functionality:

1. Conductivity Enhancement: The spheroidal graphite significantly improved the overall conductivity of the electrode, facilitating more efficient charge transport.
2. Surface Morphology Optimization: Cross-sectional scanning electron microscopy (SEM) analysis revealed that spheroidal graphite effectively filled surface pores in the underlying carbon composite. This resulted in a markedly smoother electrode surface, which is crucial for optimal contact with adjacent layers and uniform current distribution.
3. Performance Boost: The integration of spheroidal graphite led to a substantial increase in power conversion efficiency, elevating it from 4.08 to 6.64%. This 63% improvement underscores the critical role of electrode composition and structure in overall device performance.
4. Stability Enhancement: Perhaps most importantly, the spheroidal graphite layer imparted hydrophobic properties to the electrode surface. This hydrophobicity acts as a protective barrier, preventing water molecules from penetrating the device structure. Water ingress is a major factor in the degradation of perovskite materials, so this added protection significantly enhances the long-term stability of the PSCs.

The synergistic combination of graphite and carbon black, further enhanced by spheroidal graphite, exemplifies the potential of carbon-based materials in PSC technology. This approach not only improves immediate performance metrics but also addresses the critical challenge of long-term stability, which is essential for the commercial viability of perovskite solar cells. As research in this area continues, we can expect further refinements in carbon electrode composition and structure, potentially leading to even higher efficiencies and more robust devices suitable for real-world applications. Long-term stability studies on carbon black-based monolithic perovskite solar cells (PSCs) have yielded impressive results, demonstrating the robustness of this technology. Researchers conducted an extensive 840-h test on methylammonium lead iodide/TiO_2-based solar cells featuring carbon black electrodes. These devices were stored in dry ambient conditions without encapsulation, simulating real-world exposure.

Throughout this extended period, the solar cells exhibited remarkable stability. Key performance parameters, including open-circuit voltage (V_{oc}), short-circuit current density (J_{sc}), and fill factor (FF), showed only minimal fluctuations. This consistency in performance is particularly noteworthy given the lack of protective encapsulation, which is often necessary for maintaining the integrity of perovskite devices. Perhaps most impressively, the PSCs maintained an efficiency of nearly 6.5% even after the full 840-h test period. This sustained performance underscores the inherent stability of carbon-based electrodes in perovskite solar cell architectures. These findings have significant implications for the future of PSC technology. They demonstrate that carbon-based PSCs not only offer excellent long-term stability but also present a user-friendly fabrication process. This combination of durability and ease of manufacture makes carbon-based PSCs particularly attractive for large-scale production. As the solar energy industry continues to seek reliable, efficient, and cost-effective technologies, these results suggest that carbon-based PSCs could play a crucial role in the widespread adoption of perovskite solar technology [66–71].

The carbon layer's thickness significantly impacts device performance. The carbon layer thickness can range from 5 to 15 μm, 9 μm proves optimal for device parameters. Layers thinner than 5 μm suffer from poor conductivity, whereas those exceeding 15 μm impede perovskite precursor penetration, both resulting in subpar performance. Graphite flake size influences device parameters. Flakes measuring 8 μm created larger pores, facilitating better perovskite precursor infiltration and fostering an efficient perovskite phase for high-performing PSCs. Corroborating these findings, carbon layer thickness effects on power conversion efficiencies (PCEs). Keeping TiO_2 and ZrO_2 thicknesses constant, they varied carbon layer thickness to 6.5, 15, 25, and 54 μm. The 54 μm layer yielded the lowest PCE at 4.3% due to restricted perovskite ink infiltration, while the 25 μm layer achieved the highest PCE of 10.7%.

Temperature significantly influences electrode performance in perovskite solar cells (PSCs). How annealing temperature affects commercially available carbon-based composite materials used as electrodes. Using screen-printed carbon paste, they found that high-temperature sintering improved inter-particle contact, enhancing conductivity and current flow. Electrodes sintered below 300 °C performed poorly due to inadequate perovskite precursor infiltration. In contrast, electrodes treated at 350 °C and 400 °C achieved optimal power conversion efficiencies (PCEs) of 8.4% and 12.4%, respectively. The study also noted that residual oxygen functionalities impact electrode performance in PSCs, though specific effects weren't detailed. The impact of annealing temperature on optimizing carbon-based electrodes for PSC applications, emphasizing the interplay between processing conditions, material characteristics, and device performance. The oxygen functionalities in carbon-based electrodes are one of the key factors. They prepared two carbon electrode samples: one with high oxygen content and a surface area of 186.4 $m^2\ g^{-1}$, and another with low oxygen content and a surface area of 112.5 $m^2\ g^{-1}$. Oxygen-rich carbon (ORC)

Table 13.1 WDCNs efficiencies [72]

Type	Jsc (mA/cm^2)	Voc (V)	FF	PCE (%)
Bare TiO_2	22.509	1.046	0.72	17.04
	21.9 ± 0.83	1.053 ± 0.007	0.72 ± 0.01	16.64 ± 0.60
TiO_2/s-SWCNTs	22.33	1.08	0.75	18.1
	21.99 ± 0.29	1.075 ± 0.005	0.74 ± 0.01	17.60 ± 0.43
TiO_2/m-SWCNTs	21.677	1.088	0.75	17.77
	21.445 ± 0.37	1.066 ± 0.019	0.74 ± 0.01	17.09 ± 0.70
TiO_2/2:1 s:m-SWCNTs	24.592	1.085	0.73	19.35
	24.185 ± 0.41	1.074 ± 0.015	0.71 ± 0.02	18.44 ± 0.89
TiO_2/1:2 s:m-SWCNTs	22.803	1.075	0.7	17.21
	22.52 ± 0.25	1.07 ± 0.004	0.70 ± 0.01	16.86 ± 0.48
TiO_2/1:1 s:m-SWCNTs	23.411	1.055	0.73	18.09
	22.80 ± 0.56	1.072 ± 0.017	0.73 ± 0.01	17.93 ± 0.14

increased the electrode's work function and enhanced interface contact between the carbon electrode and perovskite phase. ORC-based devices achieved a higher power conversion efficiency (PCE) of 15.7%, compared to 13.6% for oxygen-deficient carbon (ODC)-based devices. ORC devices also demonstrated faster photocurrent response and reduced hysteresis. Carbon nanoparticles are also synthesized from aloe vera at 1000 °C. Perovskite solar cells (PSCs) using these aloe vera carbon nanoparticles (AV-C) as electrodes achieved a PCE of 12.58% and exhibited high stability, retaining 85% of their initial efficiency. This work highlighted the potential of high-temperature-processed electrode systems for PSCs. These findings underscore the importance of carbon electrode composition and processing in optimizing PSC performance, particularly highlighting the benefits of oxygen-rich carbon and high-temperature synthesis methods (Table 13.1).

13.7 Performance of WDCNMs in Solar Cells

13.7.1 Power Conversion Efficiency

The global energy crisis and increasing demand for sustainable power sources have led to extensive research into renewable energy technologies. Solar energy has emerged as a promising solution due to its abundance, low carbon footprint, and potential to address the impending energy shortage. As traditional fossil fuel resources like coal and gas become depleted, solar cells (SCs) and photovoltaic technologies offer an attractive alternative for clean energy generation.

The development of solar cells has progressed through three main generations, each improving upon the efficiency, cost-effectiveness, and applicability of the technology. The first generation of solar cells was based primarily on silicon (Si), which remains an important material due to its abundance and suitable bandgap of 1.12 eV for absorbing photons in the visible spectrum.

Silicon-based first-generation solar cells can be categorized into three main types:

1. Single crystalline (monocrystalline) Si cells, with efficiencies around 28%;
2. Multicrystalline Si cells, with lower efficiencies of approximately 21%;
3. Amorphous Si cells, with substantially lower efficiencies around 16%.

Among these, multicrystalline Si solar cells have seen widespread application due to their balance of adequate efficiency and cost-effectiveness. The basic structure of Si-based solar cells typically involves a p–n junction formed between boron-doped (p-type) and phosphorus-doped (n-type) silicon layers. When exposed to light with energy exceeding the bandgap of Si, electron–hole pairs are generated, initiating the conversion of light into electricity.

Despite their widespread use, first-generation solar cells faced efficiency limitations. The highest laboratory efficiency achieved for monocrystalline Si among commercial photovoltaic technologies is 26.7%. This limitation, along with other factors like high production costs and material constraints, led to the development of second and third-generation solar cells (Fig. 13.5).

Recent advancements in materials science have opened up new avenues for improving solar cell performance. Carbon-based nanomaterials have garnered significant attention due to their unique electrical, material, and mechanical properties. Materials such as carbon nanotubes, graphene, fullerene, and carbon nanohybrids have been investigated for various components of solar cells, including electrodes, transport layers, active layers, and interfacial layers.

The incorporation of carbon nanomaterials into different types of solar cells has shown promise in enhancing both the efficiency and stability of photovoltaic technologies. Researchers have explored carbon-based solar cells in combination with various materials and architectures, including:

1. Silicon-based solar cells
2. III–V semiconductor solar cells
3. Dye-sensitized solar cells
4. Metal oxide solar cells
5. Perovskite solar cells
6. Organic solar cells
7. Quantum dot solar cells
8. Hybrid material solar cells

Each of these configurations leverages the unique properties of carbon nanomaterials to overcome the limitations of conventional materials and push the boundaries of solar cell performance.

The use of carbon nanomaterials in solar cells offers several potential advantages:

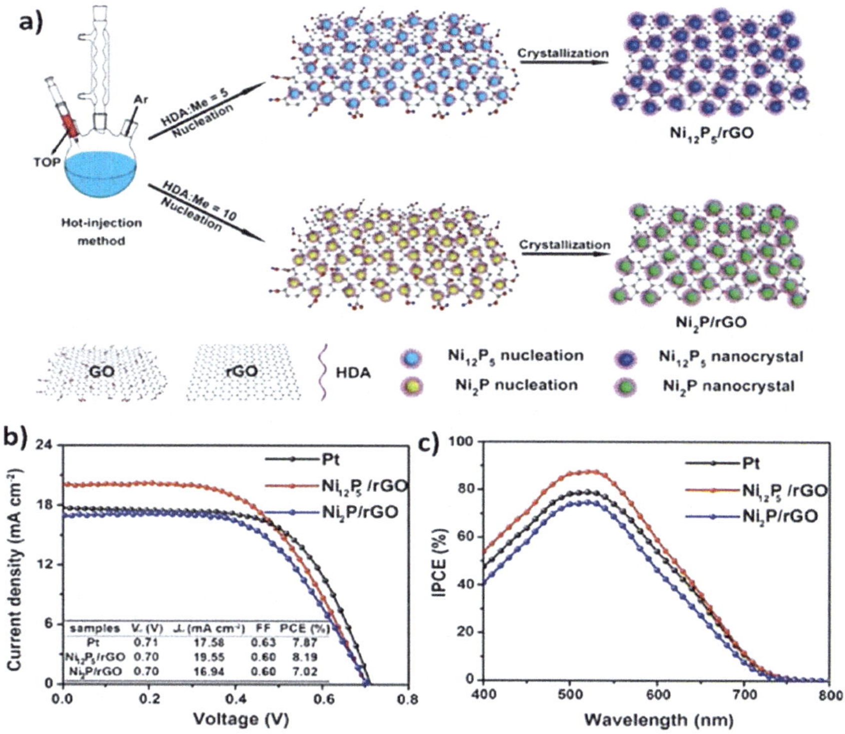

Fig. 13.5 **a** Schematic representation of $Ni_{12}P_5$/rGO and N_2P/rGO composites; **b** Current Density–Voltage curves of DSSCs based on Pt, $Ni_{12}P_5$/rGO, and Ni_2P/rGO CEs under AM 1.5 G. The inset indicates the detailed PV parameters of the DSSCs. **c** IPCE vs Wavelength of DSSCs based on Pt, Ni12P5/rGO, and Ni2P/rGO CEs. The image was reproduced with permission [72]

1. Enhanced electrical conductivity, potentially improving charge transport and collection;
2. Increased surface area, which can lead to better light absorption and charge generation;
3. Improved mechanical stability and flexibility, enabling new form factors and applications;
4. Potential for lower-cost production compared to some traditional materials;
5. Compatibility with solution-processing techniques, facilitating large-scale manufacturing.

However, the integration of carbon nanomaterials into solar cells also presents challenges that researchers must address. These include optimizing the interface between carbon materials and other components, ensuring long-term stability, and scaling up production processes while maintaining performance.

As research in this field progresses, scientists are working to overcome these challenges and further improve the efficiency and stability of carbon-based solar cells. The ultimate goal is to develop next-generation solar energy technologies that can meet the world's growing energy demands while minimizing environmental impact.

The review by Deshmukh et al. provides a comprehensive overview of recent progress in solar cells based on carbon nanomaterials. It covers developments in efficiency and stability across various types of carbon-based solar cells, offering insights into the potential of these materials for advanced photovoltaic technologies.

Looking ahead, the field of carbon-based solar cells faces several key challenges and opportunities:

1. Further improving efficiency to compete with and surpass conventional technologies;
2. Enhancing long-term stability and durability under real-world conditions;
3. Developing scalable and cost-effective manufacturing processes;
4. Optimizing the integration of carbon nanomaterials with other emerging photovoltaic materials;
5. Exploring new device architectures and configurations that leverage the unique properties of carbon nanomaterials.

As research in this area continues to advance, carbon-based solar cells have the potential to play a significant role in addressing global energy challenges. By combining the unique properties of carbon nanomaterials with innovative device designs and manufacturing techniques, researchers aim to develop highly efficient, stable, and cost-effective solar cells that can contribute to a sustainable energy future.

In conclusion, the integration of carbon nanomaterials into solar cell technologies represents a promising avenue for advancing photovoltaic performance and addressing global energy needs. Recent progress in this field has demonstrated the potential of carbon-based solar cells to overcome limitations of conventional materials and push the boundaries of efficiency and stability. Continued research and development in this area may lead to breakthrough technologies that can help meet the world's growing energy demands while minimizing environmental impact.

13.8 Future Directions

Development of new synthesis methods for WDCNMs: Water-dispersible carbon nanomaterials (WDCNMs) have garnered significant attention in recent years due to their unique properties and potential applications in various fields, including energy storage, electronics, and biomedical engineering. The development of new synthesis methods for WDCNMs is crucial for advancing their practical applications and improving their overall performance. One of the primary challenges in synthesizing WDCNMs is achieving uniform dispersion in aqueous media while maintaining their desirable properties. Traditional methods often involve harsh chemical treatments or the use of surfactants, which can negatively impact the material's performance. To address these issues, researchers have been exploring novel synthesis approaches. A promising avenue for WDCNM synthesis is the use of green chemistry principles. For instance, hydrothermal synthesis methods have shown great potential in producing WDCNMs with controlled morphology and surface chemistry. This approach involves the use of high-pressure and high-temperature conditions in aqueous solutions, allowing for the formation of carbon-based nanostructures with enhanced water dispersibility. Another innovative method is the use of biomass-derived precursors for WDCNM synthesis. This approach not only provides a sustainable source of carbon but also often results in materials with inherent hydrophilicity. For example, researchers have successfully synthesized WDCNMs from various biomass sources such as cellulose, lignin, and chitin. These bio-derived WDCNMs often exhibit excellent water dispersibility without the need for additional surface modifications. Plasma-assisted synthesis has also emerged as a promising technique for producing WDCNMs. This method allows for precise control over the surface chemistry and morphology of the nanomaterials. By tuning plasma parameters such as gas composition, power, and treatment time, researchers can create WDCNMs with tailored properties for specific applications. The development of in-situ functionalization methods during synthesis has gained traction as well. This approach allows for the simultaneous formation of carbon nanostructures and their surface modification, resulting in WDCNMs with enhanced dispersibility and functionality. For instance, the incorporation of nitrogen-containing precursors during synthesis can lead to the formation of N-doped WDCNMs with improved water dispersibility and electrochemical properties. Microfluidic-based synthesis methods have also shown promise in producing WDCNMs with controlled size and morphology. These techniques offer precise control over reaction conditions and allow for continuous production of nanomaterials. By manipulating flow rates, channel geometries, and reactant concentrations, researchers can fine-tune the properties of the resulting WDCNMs.

Improvement of the properties of WDCNMs: As research in the field of WDCNMs progresses, there is a constant drive to improve their properties to meet the demands of various applications. Several strategies have been employed to enhance the performance of WDCNMs across different domains. One key area of focus is improving the electrical conductivity of WDCNMs. This is particularly important for applications in energy storage and electronics. Researchers have explored various doping strategies to modify the electronic structure of WDCNMs. For instance, the incorporation of heteroatoms such as nitrogen, boron, or sulfur into the carbon framework has been shown to enhance conductivity and electrochemical activity. Surface functionalization is another crucial aspect of improving WDCNM properties. By attaching specific functional groups to the surface of WDCNMs, researchers can tailor their properties for targeted applications. For example, the addition of carboxyl or hydroxyl groups can enhance water dispersibility and provide reactive sites for further modifications. Polymer grafting has also been employed to improve the stability and processability of WDCNMs in aqueous environments. Enhancing the mechanical properties of WDCNMs is essential for their integration into composite materials and structural applications. Researchers have explored various strategies to improve the strength and toughness of WDCNMs, such as creating hierarchical structures or incorporating reinforcing agents. The development of WDCNM-based hydrogels and aerogels has opened up new possibilities for creating lightweight yet strong materials with excellent water retention properties.

For biomedical applications, improving the biocompatibility and biodegradability of WDCNMs is crucial. Researchers have investigated methods to reduce the potential toxicity of carbon nanomaterials while maintaining their desirable properties. This includes surface modifications with biocompatible polymers or biomolecules, as well as the development of carbon dots and other ultrasmall carbon nanostructures that can be easily cleared from biological systems. The optical properties of WDCNMs have also been a target for improvement, particularly for applications in sensing and imaging. Strategies such as controlling the size and surface chemistry of carbon dots have led to the development of highly fluorescent WDCNMs with tunable emission properties. These materials show great promise for bioimaging and environmental sensing applications. Enhancing the catalytic activity of WDCNMs is another area of active research. By incorporating metal nanoparticles or creating defect sites on the carbon surface, researchers have developed highly efficient WDCNM-based catalysts for various reactions, including oxygen reduction and hydrogen evolution. These materials show great potential for applications in fuel cells and electrocatalysis.

Scale-Up of WDCNMs Production: As the potential applications of WDCNMs continue to expand, there is an increasing need to scale up their production to meet industrial demands. Scaling up the synthesis of WDCNMs presents several challenges, including maintaining consistent quality, reducing production costs, and ensuring environmental sustainability. One approach to scaling up WDCNM production is the development of continuous flow synthesis methods. These techniques allow for the continuous production of nanomaterials, offering advantages in terms of process control and scalability. For instance, researchers have developed microreactor systems that can produce carbon dots and other WDCNMs with high consistency and throughput. Another strategy for scaling up production is the use of large-scale batch reactors. This approach has been particularly successful for hydrothermal synthesis methods. By optimizing reaction conditions and reactor designs, researchers have been able to produce WDCNMs on a kilogram scale while maintaining consistent quality. The development of more efficient purification and separation techniques is crucial for large-scale WDCNM production. Traditional methods such as centrifugation and dialysis can be time-consuming and challenging to scale up. Novel approaches, such as membrane-based separation techniques and continuous flow purification systems, are being explored to address these issues. Automation and process control play a vital role in scaling up WDCNM production. The integration of in-line monitoring systems and feedback control loops can help maintain consistent product quality during large-scale synthesis. Advanced characterization techniques, such as real-time particle size analysis and spectroscopic methods, are being developed to enable rapid quality control in industrial settings. The use of sustainable and cost-effective precursors is another important consideration in scaling up WDCNM production. Researchers are exploring the use of waste biomass and industrial by-products as carbon sources for WDCNM synthesis. This not only reduces production costs but also contributes to the circular economy by valorizing waste materials.

Energy efficiency is a crucial factor in large-scale WDCNM production. Researchers are developing more energy-efficient synthesis methods, such as microwave-assisted synthesis and low-temperature plasma processes. These approaches can significantly reduce energy consumption while maintaining or even improving product quality.

The development of modular and flexible production systems is gaining attention as a means to scale up WDCNM production. These systems allow for easy scale-up by simply adding more modules, providing flexibility in production capacity. This approach is particularly useful for producing a range of WDCNM types with different properties.

Commercialization of WDCNM-Based Solar Cells: The commercialization of WDCNM-based solar cells represents a promising avenue for harnessing the unique properties of these nanomaterials in the renewable energy sector. WDCNMs offer several advantages for solar cell applications, including high conductivity, tunable optical properties, and potential for low-cost production. However, several challenges need to be addressed to bring WDCNM-based solar cells to market. One of the primary focuses in commercializing WDCNM-based solar cells is improving their efficiency. While significant progress has been made in laboratory-scale devices, achieving commercially viable efficiencies remains a challenge. Researchers are exploring various strategies to enhance light absorption and charge transport in WDCNM-based solar cells. This includes optimizing the morphology of WDCNM active layers, developing novel device architectures, and creating hybrid systems that combine WDCNMs with other materials such as perovskites or quantum dots. Stability and lifetime are crucial factors for the commercial viability of solar cells. WDCNM-based devices must demonstrate long-term stability under real-world operating conditions. Researchers are investigating ways to improve the stability of these solar cells, such as developing more robust encapsulation methods and exploring the use of additives to enhance the structural integrity of the active layers. Scalable manufacturing processes are essential for the commercialization of WDCNM-based solar cells. Techniques such as roll-to-roll printing and spray coating are being adapted for the large-scale production of these devices. The development of ink formulations that allow for uniform deposition of WDCNMs on various substrates is a key area of research. Additionally, researchers are working on improving the compatibility of WDCNMs with existing solar cell manufacturing processes to facilitate their integration into production lines. Cost-effectiveness is a critical factor in the commercialization of any new solar cell technology. While WDCNMs offer the potential for low-cost production, efforts are ongoing to further reduce manufacturing costs. This includes optimizing material utilization, developing more efficient synthesis methods, and exploring the use of earth-abundant materials in WDCNM-based solar cells. Environmental sustainability is an important consideration in the commercialization of WDCNM-based solar cells. Researchers are focusing on developing environmentally friendly synthesis methods and ensuring that the materials used in these solar cells are non-toxic and recyclable. Life cycle assessments are being conducted to evaluate the overall environmental impact of WDCNM-based solar cells compared to conventional technologies. Standardization and certification are crucial steps in the commercialization process. Efforts are underway to develop standardized testing protocols and performance metrics specifically for WDCNM-based solar cells. This will facilitate fair comparisons between different technologies and build confidence among potential consumers and investors. Collaboration between academia and industry is playing a vital role in advancing the commercialization of WDCNM-based solar cells. Many research institutions are partnering with solar cell manufacturers to bridge the gap between laboratory discoveries and commercial products. These collaborations are helping to address practical challenges in scaling up production and integrating WDCNMs into existing solar cell technologies. Market positioning and identifying niche applications are important strategies for the initial commercialization of WDCNM-based solar cells. While these devices may not yet compete with traditional silicon solar cells in all applications, they may offer advantages in specific areas such as building-integrated photovoltaics, flexible and lightweight solar panels, or indoor light harvesting. As the field of WDCNM-based solar cells continues to advance, several startups and established companies are showing interest in commercializing this technology. Some are focusing on developing proprietary WDCNM materials optimized for solar cell applications, while others are working on novel device architectures that leverage the unique properties of WDCNMs.

In conclusion, the development of new synthesis methods, improvement of properties, scale-up of production, and commercialization of WDCNM-based solar cells represent interconnected challenges and opportunities in the field of carbon nanomaterials. As research progresses and industrial partnerships strengthen, we can expect to see significant advancements in WDCNM technology and its practical applications in the coming years. The successful commercialization of WDCNM-based solar cells could contribute to the diversification of renewable energy technologies and play a role in addressing global energy challenges.

13.9 Conclusion

The development, improvement, scale-up, and commercialization of water-dispersible carbon nanomaterials (WDCNMs) represent a rapidly evolving field with significant potential for various applications, particularly in solar cell technology. This comprehensive overview has highlighted several key points across these areas:

1. Development of New Synthesis Methods:
    - Green chemistry approaches, such as hydrothermal synthesis, have shown promise in producing WDCNMs with controlled properties.
    - Biomass-derived precursors offer a sustainable source for WDCNM synthesis, often resulting in inherently hydrophilic materials.
    - Plasma-assisted synthesis allows for precise control over surface chemistry and morphology.

- In-situ functionalization methods enable simultaneous formation and modification of WDCNMs.
- Microfluidic-based synthesis offers control over size and morphology with the potential for continuous production.

2. Improvement of WDCNM Properties:

- Electrical conductivity enhancement through heteroatom doping (e.g., nitrogen, boron, sulfur) has been a focus for energy storage and electronic applications.
- Surface functionalization strategies have been developed to tailor WDCNMs for specific applications and improve dispersibility.
- Mechanical property improvements have been achieved through hierarchical structures and composite formations.
- Biocompatibility and biodegradability enhancements have expanded the potential for biomedical applications.
- Optical property tuning, particularly in carbon dots, has opened up possibilities in sensing and imaging.
- Catalytic activity improvements have been realized through metal nanoparticle incorporation and defect site creation.

3. Scale-Up of WDCNM Production:

- Continuous flow synthesis methods have been developed to allow for consistent, high-throughput production.
- Large-scale batch reactors, particularly for hydrothermal synthesis, have achieved kilogram-scale production.
- Efficient purification and separation techniques, such as membrane-based systems, are being explored to address scalability challenges.
- Automation and process control integration are crucial for maintaining product quality in large-scale production.
- Sustainable and cost-effective precursors, including waste biomass, are being investigated to reduce production costs.
- Energy-efficient synthesis methods, like microwave-assisted and low-temperature plasma processes, are being developed.
- Modular and flexible production systems offer scalability and versatility in WDCNM production.

4. Commercialization of WDCNM-Based Solar Cells:

- Efficiency improvement remains a primary focus, with strategies including morphology optimization and novel device architectures.
- Stability and lifetime enhancement are crucial for commercial viability, involving improved encapsulation and structural integrity.
- Scalable manufacturing processes, such as roll-to-roll printing and spray coating, are being adapted for large-scale production.
- Cost-effectiveness is being addressed through optimized material utilization and more efficient synthesis methods.

- Environmental sustainability considerations include developing eco-friendly synthesis methods and ensuring recyclability.
- Standardization and certification efforts are underway to facilitate fair comparisons and build market confidence.
- Academia-industry collaborations are playing a vital role in bridging the gap between research and commercialization.
- Market positioning strategies focus on identifying niche applications where WDCNMs offer unique advantages.

Future Prospects of WDCNMs in Solar Cell Applications.

The future prospects of WDCNMs in solar cell applications appear promising, with several key trends and opportunities on the horizon:

1. Efficiency Breakthroughs: As research in WDCNM-based solar cells intensifies, we can expect significant improvements in device efficiency. The unique properties of WDCNMs, such as tunable bandgaps and high charge carrier mobility, provide a strong foundation for developing high-performance solar cells. Future breakthroughs may come from novel device architectures that fully leverage these properties or from hybrid systems that combine WDCNMs with other advanced materials.
2. Multifunctional Solar Cells: The versatility of WDCNMs opens up possibilities for developing multifunctional solar cells. Future devices may not only generate electricity but also serve additional purposes such as self-cleaning, structural reinforcement, or even data transmission. This multifunctionality could expand the application range of solar cells and create new market opportunities.
3. Flexible and Wearable Solar Technology: The mechanical properties of WDCNMs make them ideal candidates for flexible and wearable solar cells. As the demand for portable and integrated energy solutions grows, WDCNM-based solar cells could find applications in smart textiles, portable electronics, and building-integrated photovoltaics.
4. Improved Stability and Lifetime: Ongoing research into enhancing the stability of WDCNM-based solar cells is likely to yield significant improvements in device lifetime. This could involve developing new encapsulation techniques, creating more robust WDCNM structures, or designing self-healing mechanisms. As stability improves, the commercial viability of these solar cells will increase substantially.
5. Large-Scale Manufacturing Advancements: The coming years are likely to see significant progress in scaling up the production of WDCNM-based solar cells. Innovations in continuous manufacturing processes, quality control systems, and automation could dramatically reduce production costs and improve consistency. This could pave the way for widespread adoption of WDCNM-based solar technology.

6. Integration with Emerging Technologies: WDCNMs have the potential to synergize with other emerging technologies. For instance, their integration with artificial intelligence could lead to smart solar cells that adapt to changing environmental conditions. Combination with energy storage systems could result in highly efficient, self-sustaining energy solutions.
7. Sustainable Production and Circular Economy: Future developments are likely to focus on making the entire lifecycle of WDCNM-based solar cells more sustainable. This could involve using waste materials as precursors, developing water-based processing methods, and designing for easy recycling and material recovery at the end of the product life.
8. Niche Market Penetration: In the near term, WDCNM-based solar cells are likely to find success in niche markets where their unique properties offer distinct advantages. This could include applications in aerospace, where lightweight and flexible solar cells are highly valued, or in consumer electronics, where form factor and aesthetics are important considerations.
9. Standardization and Regulatory Framework: As the technology matures, we can expect the development of comprehensive standards and regulatory frameworks specific to WDCNM-based solar cells. This will be crucial for building consumer and investor confidence and facilitating wider market adoption.
10. Interdisciplinary Research and Development: The future of WDCNMs in solar cell applications will likely be shaped by interdisciplinary collaborations. Combining insights from materials science, physics, chemistry, and engineering could lead to unexpected breakthroughs and novel applications.

In conclusion, the field of WDCNM-based solar cells is at an exciting juncture. While challenges remain, particularly in terms of scaling up production and improving long-term stability, the unique properties of WDCNMs offer immense potential for advancing solar cell technology. As research progresses and industrial partnerships strengthen, we can anticipate significant developments in the coming years. The successful integration of WDCNMs into commercial solar cell products could play a crucial role in diversifying and enhancing renewable energy technologies, contributing to a more sustainable energy future.

References

1. Omar RA et al (2024) Waste-derived carbon nanostructures (WD-CNs): an innovative step toward waste to treasury. Environ Res 246:118096
2. Madaka R et al (2024) Two-dimensional transition metal oxides (TMOs) for solar cell applications. In: Talreja N, Chauhan D, Ashfaq M (eds) Two-dimensional hybrid composites: synthesis, properties and applications. Springer Nature Singapore, Singapore, pp 53–86
3. Mubarik S et al (2021) Synthetic approach to rice waste-derived carbon-based nanomaterials and their applications. Nanomanufacturing 1:109–159. https://doi.org/10.3390/nanomanufacturing1030010
4. Tatrari G et al (2021) Solid waste-derived carbon nanomaterials for supercapacitor applications: a recent overview. Mater Adv 2(5):1454–1484

5. Bai G et al (2023) Industrial waste-derived carbon materials as advanced electrodes for supercapacitors. Nanomaterials 13. https://doi.org/10.3390/nano13222924
6. Abdelbasir SM et al (2020) Waste-Derived nanoparticles: synthesis approaches, environmental applications, and sustainability considerations. Front Chem **8**
7. Ramakrishna M et al (2018) Hydrogenated amorphous silicon solar cells fabricated at low substrate temperature 110 °C on flexible PET substrate. AIP Conf Proc 1953(1):100040
8. Kanneboina V, Madaka R, Agarwal P (2021) Stepwise tuning of the doping and thickness of a-Si:H(p) emitter layer to improve the performance of c-Si(n)/a-Si:H(p) heterojunction solar cells. J Mater Sci: Mater Electron 32(4):4457–4465
9. Madaka R et al (2021) Tunnel recombination junction influence on the a-Si:H/SHJ tandem solar cell. Mater Today: Proc 39:1970–1973
10. Dsouza N et al (2023) Silicon surface passivation of industrial n-type CZ Si (111) by Al_2O_3 layers deposited by thermal ALD process for application in carrier selective contact solar cells. J Mater Sci: Mater Electron 34(14):1179
11. Yuan X et al (2021) Review on upgrading organic waste to value-added carbon materials for energy and environmental applications. J Environ Manage 296:113128
12. International Technology Roadmap for Photovoltaic-2018 Results; https://itrpv.vdma.org/en/ 201907-03-2020
13. Zhou H et al (2014) An overview of characteristics of municipal solid waste fuel in China: physical, chemical composition and heating value. Renew Sustain Energy Rev 36:107–122
14. Geyer R, Jambeck JR, Law KL (2017) Production, use, and fate of all plastics ever made. Sci Adv 3(7):e1700782
15. Lithner D, Larsson Å, Dave G (2011) Environmental and health hazard ranking and assessment of plastic polymers based on chemical composition. Sci Total Environ 409(18):3309–3324
16. Baldé CP et al (2020) The global e-waste monitor 2020: quantities, flows and the circular economy potential. United Nations University (UNU)/United Nations Institute for Training and Research (UNITAR)—co-hosted SCYCLE Programme, International Telecommunication Union (ITU) & International Solid Waste Association (ISWA), Bonn/Geneva/Rotterdam. https://ewastemonitor.info/wp-content/uploads/2020/11/GEM_2020_def_july1_low.pdf
17. Balde CP et al (2017) The global e-waste monitor—2017, United Nations University (UNU). International Telecommunication Union (ITU) & International Solid Waste Association (ISWA), Bonn/Geneva/Vienna. https://collections.unu.edu/eserv/UNU:6341/Global-E-waste_Monitor_2017__electronic_single_pages_pdf
18. World Health Organization (2018) Health-care waste. https://www.who.int/news-room/fact-sheets/detail/health-care-waste
19. Windfeld ES, Brooks MS-L (2015) Medical waste management—a review. J Environ Manage 163:98–108
20. European Commission (2020) Construction and Demolition Waste (CDW). https://ec.europa.eu/environment/topics/waste-and-recycling/construction-and-demolition-waste_en
21. Maha CM, Fujiwaraa T, Hob CS (2018) Environmental impacts of construction and demolition waste management alternatives. Chem Eng Trans 63:343–348
22. González-González RB et al (2023) Pharmaceutical pollution fingerprinting and waterbodies remediation using waste-derived carbon dots as sustainable advanced nanomaterials. Environ Res 238:117180
23. Spitalsky Z et al (2010) Carbon nanotube–polymer composites: chemistry, processing, mechanical and electrical properties. Prog Polym Sci 35(3):357–401
24. Carlton JM et al (2008) Comparative genomics of the neglected human malaria parasite *Plasmodium vivax*. Nature 455(7214):757–763
25. Sahoo NG et al (2010) Polymer nanocomposites based on functionalized carbon nanotubes. Prog Polym Sci 35(7):837–867
26. Li X et al (2016) Bacteria-assisted preparation of nano α-Fe_2O_3 red pigment powders from waste ferrous sulfate. J Hazard Mater 317:563–569
27. Xie X-L, Mai Y-W, Zhou X-P (2005) Dispersion and alignment of carbon nanotubes in polymer matrix: a review. Mater Sci Eng R Rep 49(4):89–112

28. Pandey B, Roy SC (2021) Carbon-based Nanomaterials as Photocatalysts and their performance attributes. In: Carbon nanostructures: fundamentals to applications. AIP Publishing LLC, p 0
29. Pandey B, Shankar D (2023) Recent trends in metallic nanocomposites for sensing and electrochemical devices. In: Tiwari SK, Kumar V, Thomas S (eds) Nanoparticles reinforced metal nanocomposites: mechanical performance and durability. Springer Nature Singapore, Singapore, pp 237–271
30. Pandey B, Rani S, Roy SC (2020) A scalable approach for functionalization of TiO_2 nanotube arrays with g-C_3N_4 for enhanced photo-electrochemical performance. J Alloy Compd 846:155881
31. Yun YS et al (2015) Hierarchically porous carbon nanosheets from waste coffee grounds for supercapacitors. ACS Appl Mater Interfaces 7(6):3684–3690
32. Chai Y et al (2023) Biomass-derived carbon for dye-sensitized solar cells: a review. J Mater Sci 58(14):6057–6075
33. Habisreutinger SN, Nicholas RJ, Snaith HJ (2017) Carbon nanotubes in perovskite solar cells. Adv Energy Mater 7(10):1601839
34. Ikram R et al (2023) Recycling waste sources into nanocomposites of graphene materials: overview from an energy-focused perspective. **12**(1)
35. Perumal S et al (2021) Sustainable synthesis of multifunctional carbon dots using biomass and their applications: a mini-review. J Environ Chem Eng 9(4):105802
36. Poonia K et al (2024) Sustainability, performance, and production perspectives of waste-derived functional carbon nanomaterials towards a sustainable environment: a review. Chemosphere 352:141419
37. Jantaping N et al (2016) Challenges and strategies of surface modification of electrogalvanized coatings for electron microscopy analysis. Micron 86:48–53
38. Saidaminov MI et al (2018) Suppression of atomic vacancies via incorporation of isovalent small ions to increase the stability of halide perovskite solar cells in ambient air. Nat Energy 3(8):648–654
39. Zhu S et al (2013) Highly photoluminescent carbon dots for multicolor patterning, sensors, and bioimaging. Angew Chem Int Ed 52(14):3953–3957
40. Liu J, Li R, Yang B (2020) Carbon dots: a new type of carbon-based nanomaterial with wide applications. ACS Cent Sci 6(12):2179–2195
41. Liu X et al (2017) Environmental transmission electron microscopy (ETEM) studies of single iron nanoparticle carburization in synthesis gas. ACS Catal 7(7):4867–4875
42. Azwar E et al (2018) Transformation of biomass into carbon nanofiber for supercapacitor application—a review. Int J Hydrogen Energy 43(45):20811–20821
43. Yuan Y, Wan C (2022) Dual application of waste grape skin for photosensitizers and counter electrodes of dye-sensitized solar cells. Nanomaterials 12. https://doi.org/10.3390/nano12030563
44. Coşkun Y et al (2022) Development of highly luminescent water-insoluble carbon dots by using calix[4]pyrrole as the carbon precursor and their potential application in organic solar cells. ACS Omega 7(22):18840–18851
45. Kim CK et al (2017) Anchovy-derived nitrogen and sulfur co-doped porous carbon materials for high-performance supercapacitors and dye-sensitized solar cells. RSC Adv 7(57):35565–35574
46. Yuan Y, Wan C (2022) Dual application of waste grape skin for photosensitizers and counter electrodes of dye-sensitized solar cells. Nanomaterials (Basel) 12(3)
47. Yu J et al (2022) Soybean root-derived heteroatoms co-doped porous carbon with ultra-high specific surface area for high performance supercapacitors. Diam Relat Mater 126:109044
48. Berkmans AJ et al (2014) Synthesis of branched, nano channeled, ultrafine and nano carbon tubes from PET wastes using the arc discharge method. 34(11):2139–2145
49. Jain A, Tripathi SK (2015) Nano-porous activated carbon from sugarcane waste for supercapacitor application. J Energy Storage 4:121–127
50. Chen C et al (2016) Three-dimensional scaffolding framework of porous carbon nanosheets derived from plant wastes for high-performance supercapacitors. Nano Energy 27:377–389

51. Mangalara SCH, Varughese S (2016) Green recycling approach to obtain nano- and microparticles from expanded polystyrene waste. ACS Sustain Chem Eng 4(11):6095–6100
52. Samaddar P et al (2018) Synthesis of nanomaterials from various wastes and their new age applications. J Clean Prod 197:1190–1209
53. Fang Z et al (2011) Degradation of the polybrominated diphenyl ethers by nanoscale zero-valent metallic particles prepared from steel pickling waste liquor. Desalination 267(1):34–41
54. Xing M, Zhang F-S (2011) Nano-lead particle synthesis from waste cathode ray-tube funnel glass. J Hazard Mater 194:407–413
55. Ali A, Chiang YW, RM (2022) Santos X-ray diffraction techniques for mineral characterization: a review for engineers of the fundamentals, applications, and research directions. Minerals 12. https://doi.org/10.3390/min12020205
56. Ruan G et al (2011) Growth of graphene from food, insects, and waste. ACS Nano 5(9):7601–7607
57. Akter Jahan S et al (2017) Nano-hydroxyapatite prepared from eggshell-derived calcium-precursor using reverse microemulsions as nanoreactor. Mater Today Proc 4(4, Part E):5497–5506
58. Hassan TA, Rangari VK, Jeelani S (2014) Value-Added biopolymer nanocomposites from waste eggshell-based $CaCO_3$ nanoparticles as fillers. ACS Sustain Chem Eng 2(4):706–717
59. Amorim CC et al (2013) Performance of blast furnace waste for azo dye degradation through photo-Fenton-like processes. Chem Eng J 224:59–66
60. Nasrollahzadeh M et al (2019) Plant-Mediated green synthesis of nanostructures: mechanisms, characterization, and applications. In: Nasrollahzadeh M et al (eds) Interface science and technology, chap 6. Elsevier, pp 199–322
61. Meng Z et al (2018) Investigation of Al_2O_3 and ZrO_2 spacer layers for fully printable and hole-conductor-free mesoscopic perovskite solar cells. Appl Surf Sci 430:632–638
62. Rajendran Lakshmi P, Babu SG (2024) Waste biomass-derived activated carbon-supported 0D, 1D, 2D, and 3D nanostructures of copper oxide for hydrogenation reaction: a study on the role of structural properties. Langmuir
63. Shamroza M et al (2021) Synthetic approach to rice waste-derived carbon-based nanomaterials and their applications. Nanomanufacturing
64. Shi Ying L et al (2015) Carbon quantum dots and their applications. Chem Soc Rev
65. Umesh SM et al (2022) Effect of various aqueous extracting agents on fluorescence properties of waste tea residue derived carbon dots (WTR-CDs): comparative spectroscopic analysis. Luminescence
66. Akshata P et al (2024) Polymer waste valorization into advanced carbon nanomaterials for potential energy and environment applications. Macromol Rapid Commun
67. Farai D et al (22020) Recovery of waste gold for the synthesis of gold nanoparticles supported on radially aligned nanorutile: the growth of carbon nanomaterials. RSC Adv
68. Jinjie Z et al (2024) Structural and compositional analysis of MOF-derived carbon nanomaterials for the oxygen reduction reaction. Chem Commun
69. Lindokuhle Precious M et al (2022) Lignocellulosic biomass waste-derived cellulose nanocrystals and carbon nanomaterials: a review. Int J Mol Sci
70. De Michaël V et al (2013) Carbon nanotubes: present and future commercial applications. Science
71. Muhammad Z et al (2023) Rambutan seed waste-derived nitrogen-doped carbon dots with l-aspartic acid for the sensing of Congo red dye. RSC Adv
72. Deshmukh MA et al (2021) Recent progress in solar cells based on carbon nanomaterials. Sol Energy 220:953–990

Chapter 14
Waste-Derived Carbon Nanomaterials for Microbial Fuel Cells

Mehdi Abedi, Vahid Alimardani, Younes Ghasemi, Radmehr Kazemi, and Mina Shafiee

Abstract Microbial fuel cells (MFCs) offer an environmentally friendly way to generate energy from renewable energy sources. This technology utilizes bacteria to convert organic waste into electricity and is beneficial for treating waste materials. Despite being in existence for a decade, microbial fuel cells are not yet capable of generating significant power for practical purposes due to the ineffective interaction between bacteria and cell materials. Developing anodes and cathodes from biomass presents an effective approach to resolving issues and optimizing the function of MFCs. These materials are advantageous due to their suitable structure, accessibility, cheap, functional groups, and porous nature. These features enhanced electrode performance in MFCs.

Keywords Microbial fuel cells · Waste material · Carbon nanotube · Carbon cathode · Carbon anode

14.1 Introduction

An increase in the global population, the development of more cities, new devices, new technologies, and the establishment of additional factories have led to a dramatic rise in energy usage around the world. A considerable amount of the world's energy supply is sourced from fossil fuels. The major issue with fossil fuels is their significant

M. Abedi (✉) · Y. Ghasemi · R. Kazemi
Pharmaceutical Sciences Research Center, Shiraz University of Medical Sciences, Shiraz, Iran
e-mail: mehdi.abedi.a@gmail.com

M. Abedi · M. Shafiee
Research and Development Department, Danesh Salamat Kowsar Co, Shiraz, Iran

V. Alimardani
Student Research Committee, Shiraz University of Medical Sciences, Shiraz, Iran

Y. Ghasemi
Department of Pharmaceutical Biotechnology, School of Pharmacy, Shiraz University of Medical Sciences, Shiraz, Iran

N. Talreja et al. (eds.), *Waste-Derived Carbon Nanostructures*, Nanostructure Science and Technology, https://doi.org/10.1007/978-3-031-75247-6_14

emission of greenhouse gases which are not renewable in a short time. As a result, it contributes to pollution, alterations in the climate, and the warming of the Earth. As concerns about fossil fuels rise, more individuals are exploring renewable energy alternatives, including solar, geothermal, wind, ocean wave energy, and biomass [1]. Wind and solar energy sources are advancing rapidly as popular alternatives to renewable energy. Nonetheless, the weather conditions in the region have a considerable effect on the performance of these renewable energy solutions. Moreover, large-scale utilization of solar and wind energy poses challenges due to high costs and potential negative impacts on the environment. These challenges emphasize the necessity for cost-effective and straightforward renewable energy options.

Among the different energy technologies designed to decrease environmental pollutants, fuel cells and batteries are showing significant market progress. In industrialized countries, fuel cells are increasingly favored. These technologies find applications in multiple domains, including powering automobiles, investigating outer space, and delivering electrical energy. Fuel cells are a novel technology that converts chemical energy into electrical energy by using fuel, such as hydrogen, and oxygen from the air. During this process, an electrochemical reaction occurs within the electrodes of the fuel cell. Fuel cells offer the benefits of producing high energy output and minimal pollution. As a result, they are an effective technology for producing clean energy. Despite the pursuit of clean and sustainable energy sources, microbial fuel cells (MFCs) have not garnered much interest from either government or industry. The challenges associated with the development and marketing of microbial fuel cells have led to the perception that they are neither practical nor economical [2].

Microbial fuel cells (MFCs) represent a novel technology that can convert the natural activities of microorganisms into electrical energy from chemicals. Technologies like MFC, falling under biological energy systems, can provide solutions to the economic and environmental challenges posed by wind and solar energy. This technology can help unlock energy stored in large organic molecules using a process that involves microorganisms [2]. Biomass materials provide important advantages for fuel cell technology. These materials come from renewable sources, such as leftover plants and wood fibers, and provide a green and sustainable option compared to traditional carbon sources. They can be changed into special carbon materials that work better as catalysts and conduct electricity well, which helps fuel cells work better and more efficiently. Using biomass materials follows the idea of a circular economy, helps reduce waste, and might cut down production costs. Integrating biomass resources into fuel cells can promote greater environmental sustainability and fuel cell functionality.

The performance of MFC can be influenced by various factors, such as the type of organic material utilized, the composition of bacteria present, internal resistance, electrode material characteristics, membrane properties, and ion concentration [3]. The functionality of the anode is a key factor in determining the potential power output of MFC systems. Generally, the surface of the anode material plays a significant role in determining its effectiveness. Therefore, it is crucial to develop innovative materials and designs for anode electrodes to improve the functionality of this technology.

An effective biofilm on the anode is beneficial for the performance of the MFCs, impacting the speed of the reaction. The formation of a suitable biofilm on the anode is crucial for the optimal functioning of MFCs, influencing the performance and rate of the reaction. Using nanomaterials for creating anodes can aid in the efficient transfer of electrons by small organisms in the anode chamber, resulting in increased power output [4]. Therefore, many researchers focused on increasing the surface area of the anode to promote bacterial growth and on scaling up MFC systems. To achieve optimal electrochemical efficiency and ensure successful electron conversion, the anode must possess specific properties that promote bacterial adhesion including excellent electrical conductivity, high surface area, biocompatibility, stability, durability, ease of access to materials, and cost [3]. This section focuses on the historical application of waste-derived carbon nanomaterials in assisting MFCs in generating energy.

14.2 Microbial Fuel Cell Mechanism

The crucial components in MFCs are the anode, cathode, and one or two chambers separated by a membrane [5]. By maintaining an anaerobic environment in the anode portion of the MFC, microorganisms facilitate the production of electrons and protons. Carbon dioxide is produced through the process of oxidation. The cathode receives oxygen or air and undergoes reduction, capturing released electrons. The anode is where hydrogen loses its electrons in a reaction. When fuel is turned into energy, it is changed at the anode and the oxidation agent is changed at the cathode. This procedure generates an electric current. Electrons flow through a wire, generating electricity that can be utilized for various purposes. When ions pass through the liquid, the circuit is completed [6]. Electrons move from the anode section to the cathode section through an outside circuit. In this process, they combine with protons and oxygen to form water, represented by the following reaction [7]:

$$O_2 + 4H^+ + 4e^- \rightarrow 2H_2O(E^0 = 1.23\ V). \quad (14.1)$$

Equation (14.1) illustrates that oxygen is consistently consumed to maintain the capacity for electricity generation. Oxygen can be supplied to the cathode chamber by injecting air into the water or by using an air–cathode system. Due to its superior ability to accept electrons relative to various other substances, oxygen is frequently recognized as a prominent electron receiver. The inadequate connection between oxygen and the electrode, along with the slow oxygen reduction occurring on typical carbon electrodes, is the factor that limits the effectiveness of oxygen in MFCs [8]. Although using electrodes with special coatings can optimize the cathodic reaction, catalysts are generally expensive and derived from rare materials [9]. A MFC utilizes small microorganisms, such as algae and bacteria, to decompose waste materials without the need for oxygen. This process also produces electricity in the form of electrons [10]. One significant advantage of MFCs compared to conventional low-temperature fuel cells is their ability to utilize complex organic waste as fuel in the anode chamber. MFCs operate effectively at standard temperatures and neutral pH conditions. By employing low-voltage electricity, exo-electrogenic bacteria can facilitate the decomposition of contaminants [11, 12].

Previously, various methods have been explored to enhance electricity generation and the efficiency of MFCs. This process includes the investigation of different microorganisms, alterations to membranes and electrode surfaces, and even genetic changes in bacteria [11, 13].

Organic waste is placed in the anode chamber which plays a significant role in the chemical reaction [5]. The illustration in Fig. 14.1 displays the positive electrode in the anode chamber, the negative electrode in the cathode chamber, and the use of a proton exchange membrane to maintain separation while allowing proton movement from the anode to the cathode.

The performance of MFCs is influenced by several elements, such as the material composition, temperature, microbial species, substrate materials, cell assembly, design, and type of materials utilized for the anode, cathode, and separator. The three main parts of a MFC are the anode, cathode, and proton exchange membrane. Over time, various materials for anodes and cathodes have been utilized to enhance the capacity and efficiency of MFCs [12].

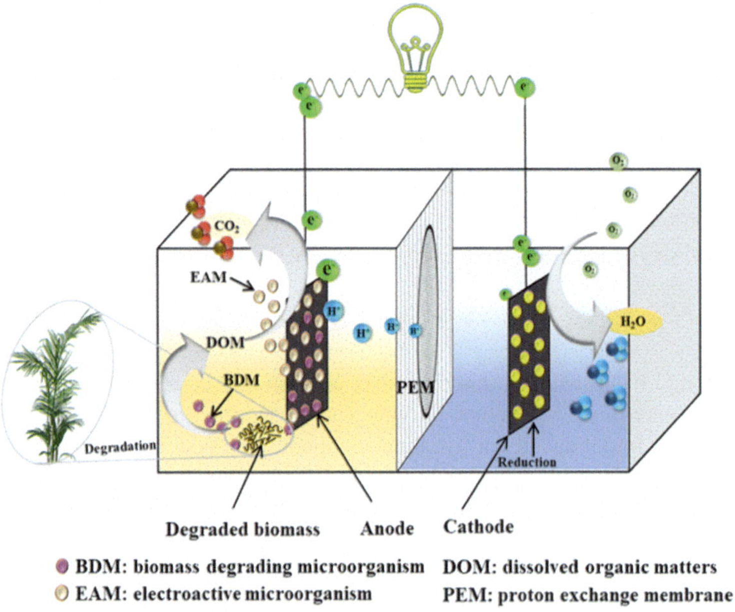

Fig. 14.1 Schematic representation of a microbial fuel cell. The figure was reproduced with permission under a common creative license [14]

14.3 Electrode Fabrication Techniques for Waste-Derived Electrode

Enhancing the surface area, pores, and conductivity of carbon anodes created from a variety of waste materials can be accomplished using different techniques, such as physical or chemical activation, hydrothermal carbonization, and molten salt carbonization [15]. Figure 14.2 represents different methods used for the preparation of biomass-derived carbon anode for microbial fuel cells.

Physical activation (pyrolysis): The process of physical activation involves heating up carbon to create carbonized anodes, and using gases, such as CO2, N2, or steam, to assist in removing carbon molecules and forming small pores in the material. However, it is difficult to combine steam and a gas carrier for physical activation because the process needs a very high temperature. However, using only direct pyrolysis for physical activation is not efficient because it takes a lot of time, requires a lot of energy, and may not be environmentally friendly [16]. In the pyrolysis process, some elements in natural materials react and act as natural additives, which helps improve the storage and movement of ions. Also, when the temperature gets really high, the water evaporates and creates tiny holes in the carbon nanomaterial [17].

Chemical activation: The use of powerful acids like phosphoric acid, sulfuric acid, and nitric acid or bases like potassium hydroxide and sodium hydroxide can reduce the temperature needed for pyrolysis, shorten the time it takes, and increase the amount of carbon produced. This can also create more surface area and cause

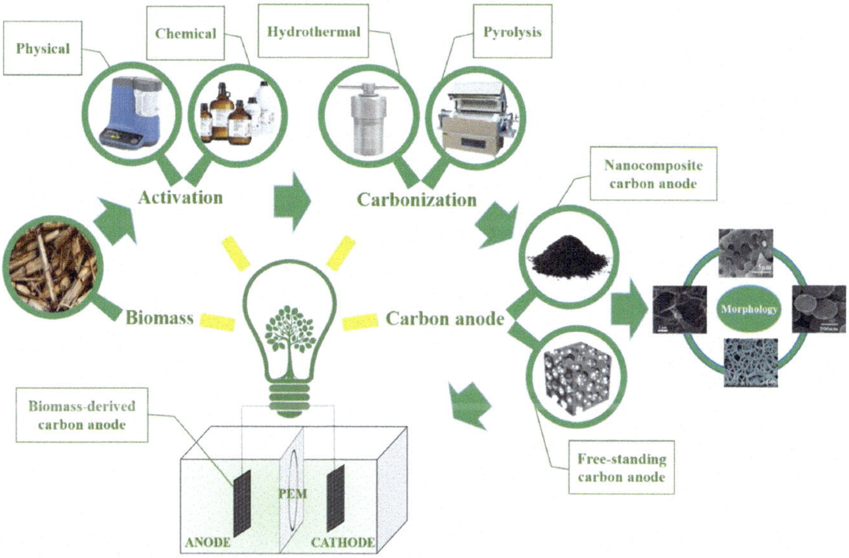

Fig. 14.2 Schematic representation of Biomass-derived Carbon Anode for Microbial Fuel Cells. The figure was reproduced with permission under a common creative license [15]

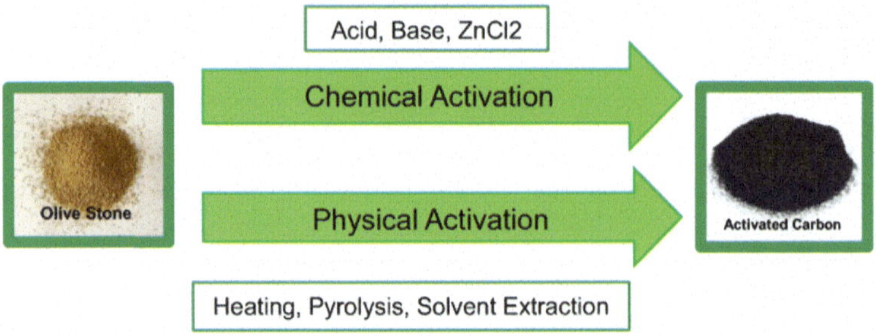

Fig. 14.3 Schematic representation of physical and chemical biomass activation. The figure was reproduced with permission under a common creative license [19]

the structure to become disordered and form pores during the pyrolysis process [18]. However, chemically activated carbon anodes do not effectively conduct electricity and have low density, resulting in limited energy storage in MFCs. Figure 14.3 displays a schematic representation of the physical and chemical biomass activation.

Hydrothermal carbonization: Biomass is transformed into carbon materials using heat and a special solution in a process known as thermochemical treatment. Under high temperature and pressure and in the absence of air, it occurs within an enclosed system [20]. The carbon anodes made through this process are comparable to coal and are perfect for transforming plant material into carbon anodes [21]. However, hydrothermal carbonization has some disadvantages, such as less pores and a low specific surface area, which may not be a good method for making high-quality anode carbon [16]. Figure 14.4 represents a schematic diagram for biomass carbonization in the hydrothermal process.

Molten salt: Carbonization is a way to heat things up so they turn into carbon. The process entails using low-temperature melting substances, such as salt, along with heat to create a carbon nanocomposite [23]. This method of carbonization enhances the durability of the carbon material, enabling it to endure high temperatures without shrinking. It also aids in preserving its initial form, leading to the creation of additional carbon material [21]. The use of molten salt for material carbonization is expensive, largely due to the cost of grinding and the potential for contamination from grinding materials [23].

14.4 Waste-Derived Carbon Materials for Anode Preparation

Currently, scientists are modifying the anode electrode by testing out different materials to improve bacteria adhesion and promote electron transfer to the anode surface. Modifying the surface of traditional anode materials has been discovered to enhance the performance of MFCs. This is due to the alteration of the physical and chemical

Fig. 14.4 Schematic representation of Hydrothermal carbonization process. The figure was reproduced under a common creative license [22]

properties of the anode electrodes, creating a larger surface area for bacteria adhesion and facilitating electron mobility [24]. Carbon obtained from natural biomass is characterized by a structure with numerous tiny cavities and a specific arrangement. This can assist in the movement of charged molecules across the anode and also aid in the growth and retention of bacteria [17]. A variety of renewable materials, such as plants, marine waste, and household and industrial waste, can be utilized to create a natural, conductive carbon for MFCs. The efficiency and effectiveness of MFCs can be influenced by the physical characteristics and surface area of conductive nanocomposite carbon derived from waste materials [25]. It was suggested that the distinct physical and chemical properties of crop and forest residue, comprised mostly of natural elements and few artificial components, could be used to manufacture carbon anodes with perforations [20].

Sato and his colleagues created an innovative hydroponic fuel cell system (MFC-Hyp), which is compatible with a soil-less agricultural system. They employed a ceramic membrane to isolate the separate components of the system. The MFC-Hyp moves nutrients from wastewater in the MFC to water in the hydroponic container (Hyp) using a ceramic barrier, without needing outside energy. To evaluate the effectiveness of this system, they utilized wastewater generated from potato processing as a source of energy and nutrients while cultivating garlic chives without soil. After 49 days, the *Allium tuberosum* leaves in the MFC-Hyp treatment were found to be 142% larger than those grown in the Hyp treatment without the MFC. It was evident that the nutrients found in wastewater from the MFC significantly enhanced the growth of the plants in the Hyp. When using *Allium tuberosum*, the MFC-Hyp achieved a peak power density of 130 ± 45 mW/m^2. According to the outcomes of this research, the MFC-Hyp system could serve as a highly promising technology for achieving "carbon neutrality." It can play an important role in a varied system for energy, water, and food around the world [26]. Marassi conducted a study where dairy wastewater was inoculated with specific microbes, including *S. oneidensis* and *C. butyricum*, to enhance the performance of an air–cathode microbial fuel cell. The air–cathode MFC underwent testing in two separate phases, which lasted 75 days and 30 days, respectively. It demonstrated significant potential for energy generation and excelled in pollutant removal, achieving a 96% decrease in biochemical oxygen demand (BOD) and a 94% decrease in chemical oxygen demand (COD) for 75 days. Concurrently, the air–cathode MFC demonstrated excellent performance in eliminating various substances, achieving complete removal of nitrate (NO_3^-), 95% of phosphorus (P), 75% of sulfate, and 47% of organic nitrogen (N). The success of the system can be attributed to the team's collaboration and the efficient utilization of the air cathode [27].

Zhang and his team examined the feasibility of employing carbon produced from lignin as electrode material in microbial fuel cells. Their research revealed that employing carbon electrodes produced from lignin enhanced the efficiency of the microbial fuel cell, leading to improved power generation [28].

Modified biochar catalysts or composites are more effective for electricity generation compared to standard biochar. The additional atoms lead to an imbalanced surface charge on the catalyst, enhancing the mobility of electrons. Composites or dopants can be incorporated by either initially preparing the materials or by applying physical and chemical treatments to the biochar post-production. The goal is to spread the charge more evenly over the catalyst surface. This facilitates a decrease in resistance during charge transfer, enhancing the oxygen reduction reaction [29]. In a study, Sun explored the potential of repurposing waste materials to serve as a renewable carbon source for generating nitrogen-doped carbon nanotubes. The nanotubes provided crucial support for catalysts in proton exchange membrane fuel cells, leading to improvements in the oxygen reduction efficiency. The maximum power density achieved was 1600 mW/m^2, surpassing the conventional platinum catalyst which exhibited a power density of 1393 mW/m^2. The decrease in power levels with NCNTs was less than with platinum catalysts after 25 cycles. This means that MFCs using prepared carbon nanotube as the catalyst can produce electricity longer than those using platinum catalyst [30]. Hung and his team created a type of material using coffee waste-derived activated carbons (CWAC) that can be used as an anode in MFCs to help Escherichia coli bacteria generate electricity. By changing the size of the pores in CWAC, they were able to make it produce more power density (3927 mW m^{-2}) than regular activated carbon (975 mW m^{-2}). The CWAC's increased power strength comes from its excellent conductivity and appropriately sized holes, resulting in fast electron movement and effective bacteria adhesion. Additionally, the scientists examined the long-term performance of the MFC using the CWAC anode. It continued operating for over 100 h at a 2000 mW m^{-2} power level without requiring additional nutrients. The results indicate that CWAC has the potential to

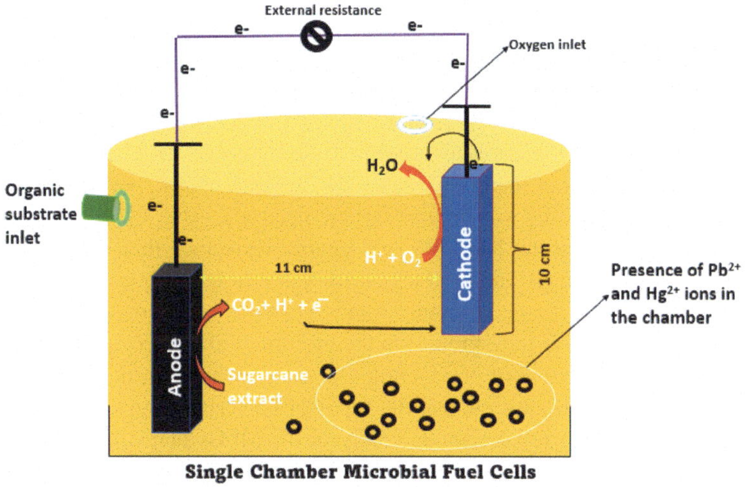

Fig. 14.5 Systematic presentation of single chamber MFC, used in present work. The figure was reproduced with permission under a common creative license [32]

be a suitable material for eco-friendly energy systems and could lower the cost of devices [31]. Aleid and his team study waste sugarcane extract as an organic substrate for the development of MFCs. The MFC operated for 79 days using a concentration of 250 mg/L for both lead and mercury ions. The data on electrochemical-generated electricity suggests that the slow development of power generation and the application of biofilm on surfaces occur over a period of time. In contrast, Pb^{2+} was removed at 82% and Hg^{2+} was removed at 74.85%. Recent research using biofilm analysis and a literature review suggests that certain kinds of bacteria can produce energy and remove metals [32]. A single-chambered MFC was developed, as systematically drawn in Fig. 14.5.

Utilizing biomass waste, such as biochar, appears to be a promising choice for creating a cost-effective and eco-friendly anode. Several research studies have indicated that repurposing leftover plant materials, such as kenaf stems, cardboard, palm kernel shells, and sewage sludge combined with fly ash, can be beneficial in creating anode electrodes for MFCs. Lin and his team found that carbon made from the husk of the Trapa natans plant could be an effective and eco-friendly material for electrodes in MFCs that use Canna indica. Results from the tests demonstrated that MFCs employing Trapa natans husk-derived carbon-coated graphite felt for electrodes can attain a maximum power density of 55 mW/m^2, which was significantly higher than the MFC employing only graphite felt electrodes (22 mW/m^2) [33]. Kim and his team examined the development and utilization of carbon catalysts made from bamboo, which incorporate both sulfur and nitrogen (SNBCs), intended for fuel cells and metal-air batteries. The main material used in the study was bamboo, which was heated to high temperatures between 700 and 900 °C to create charcoal. The SNBCs showed better performance than earlier carbon-based catalysts and were similar to Pt/

C catalysts. Various testing methods validated the effective incorporation of sulfur into the bamboo-derived carbon catalyst (S, N co-doped SNBCs). The investigation demonstrated that catalysts made from biomass can be successfully utilized in operational devices like fuel cells [34]. Li and his team developed a sludge carbon electrode with a high surface area, excellent electrical conductivity, and non-toxicity to living organisms. This made the electrode produce higher power density and lower power loss compared to a regular carbon electrode. After the biofilm formed, the sludge carbon electrode had 72% less charge resistance than the non-carbon electrode. Following the formation of the biofilm, the sludge carbon electrode exhibited a 72% decrease in charge resistance relative to the non-carbon electrode. Many different types of bacteria were present in the biofilm, with proteobacteria being the most dominant and capable of conducting electricity. Changing the sludge carbon conditions slightly could also change the types of bacteria in the community [35]. Using waste materials to make anode catalysts for MFCs helps make them work better and produce more energy while cleaning up wastewater. Chen and his team discovered a way to use the natural 3D structure of microbes in fuel cells. Kenaf (Hibiscus cannabinus) plant was converted into carbon and utilized as a material in MFCs. The current generated by the carbon was comparable to that of other 3D anodes created in diverse manners. Due to it being both renewable and cost-effective, this material is well suited for use in large microbial bioelectrochemical systems [36]. Enhancing the performance and reducing the material costs are essential for the advancement of microbial bioelectrochemical systems, such as microbial fuel cells. Chen and his team suggest using layered corrugated carbon (LCC) as a cheap, but effective electrode material. They make LCC by carbonization of corrugated cardboard. The cardboard was sliced into the proper size pieces and then bonded with a specialized adhesive made from corn starch. The cardboard materials were subjected to high-temperature burning in a furnace at 100 °C with nitrogen gas for an hour. Subsequently, the LCC samples were adhered to graphite plates of identical size for the purpose of the current collection. By adding more corrugated layers, the current flow will be amplified. 200 A m^{-2} is generated at three layers and 390 A m^{-2} is generated at six layers of corrugated material [37].

Table 14.1 displays the potential use of carbon derived from waste material in modifying the surface of an electrode in a microbial fuel cell.

14.5 Waste-Derived Carbon Materials for Cathode Preparation

Utilizing organic materials, such as plants and trees, can effectively produce carbon with varying atom compositions. Nitrogen and phosphorus found in plants can be utilized to create catalysts that facilitate oxygen addition in reactions. Utilizing plant material can result in the production of carbon and other elements needed for the catalyst. Similar to the production of anodes, carbon cathode catalysts can also be

Table 14.1 Summary of natural materials used to build electrodes in microbial fuel cells (MFCs)

Waste material	Electrode	Power density	References
Palm kernel shell	Anode and cathode	1730 mW/m^3	[38]
Sugarcane	Anode	59,940 mW/m^3	[39]
Wood	Anode	483 mW/m^2	[40]
Fly ash/ Sewage sludge	Anode	3200 mW/m^2	[41]
Wood-based biochars	Anode	457 mW/m^2	[42]
Chestnut shells	Anode	759 mW/m^2	[43]
Cocklebur fruits	Anode	0.572 mW/m^3	[44]
Sludge carbon	Anode	2165 mW/m^2	[45]
Sewage sludge	Anode	1069 W/m^2	[46]
kapok fibers	Anode	1738.1 mW/m^2	[47]
Candle soot	Anode	1650 mW/m^2	[20]
Mongo wood	Anode	589.8 mW/m^2	[48]
Rice plant	Anode	41.41 mW/m^2	[49]
Loofah sponge	Anode	4083 mW/m^3	[50]
crumb rubber	Anode	421 mW/m^2	[51]
Spirogyra	Anode	408 mW/m^2	[52]
Loofah sponge	Anode	61,700 mW/m^3	[53]
Pinecone	Anode	10,880 mW/m^3	[40]
Bacterial cellulose	Anode	1747 mW/m^2	[54]
Kapok fibers	Anode	1738 mW/m^2	[47]
Almond shell	Anode	4346 mW/m^2	[55]
Waste water sludge	Anode	2165 mW/m^2	[51]
Onion	Cathode	742 mW/m^2	[56]
Lotus leaves	Cathode	511 mW/m^2	[57]
Alfalfa leaf	Cathode	1328 mW/m^2	[58]
Chlorella pyrenoidosa	Cathode	2068 mW/m^2	[59]
Kraft paper	Cathode	830 mW/m^2	[60]
Watermelon rind	Cathode	262 mW/m^3	[61]
Plant moss	Cathode	703 mW/m^2	[62]
Chitin	Cathode	705 mW/m^2	[63]
Bamboo	Cathode	1056 mW/m^2	[64]
Bamboo tube	Cathode	40,400 W/m^3	[65]
Dandelion seed	Cathode	975 mW/m^2	[66]
Sugarcane	Cathode	120 mW/m^2	[67]
Cornstalk	Cathode	1122 mW/m^2	[68]
Livestock sewage sludge	Cathode	1273 mW/m^2	[69]
Corncob	Cathode	458 mW/m^2	[70]

(continued)

Table 14.1 (continued)

Waste material	Electrode	Power density	References
Pectin	Cathode	1161 mW/m^2	[71]
Cocoon silk	Cathode	800 mW/m^2	[72]
Bananas	Cathode	528 mW/m^2	[73]

derived from animal waste and sewage sludge due to their high nitrogen content [17]. Through the pyrolysis process, nitrogen is incorporated into the porous carbon structure. During pyrolysis, organic matter breaks down without water or oxygen at high temperatures with an inert gas. Deng converted animal waste into biocarbon, which, when processed with KOH, displayed a strong potential to assist in the oxygen reduction reaction, comparable to Pt/C, in a specific solution. The carbon catalyst they created was found to generate a significant amount of power density (1273 mW/m^2), comparable to the Pt/C catalyst C (1294 mW/m^2), and maintained its efficacy for 3 months. After 3 months, the Pt/C catalyst yielded a power output of 462 mW/m^2 when prepared [69]. Yuan and collagenous-prepared nitrogen-doped carbon sheets were made from chitin using two steps: hydrothermal treatment and pyrolysis. The new catalyst produced the same amount of power density (705 ± 5 mW/m^2) as the Pt/C C (727 ± 5 mW/m^2) catalyst in MFCs [63]. Zhou et al. utilized spider silk in the creation of carbon nanofibers, which have potential applications as a sustainable electrocatalyst. This technique was straightforward and included the carbonization of spider silk. By employing this technique, the catalyst achieved a power density of 1800 mW/m^2, surpassing the 704 mW/m^2 power density of Pt/C in MFCs. Moreover, the prepared catalyst exhibited continuous strength for a prolonged duration, up to 90 days [74]. Li and his team made biocarbon from fresh corncobs by drying them in the sun for 5 days and then grinding them into powder and paralyzing them in a nitrogen atmosphere. Prepared MFCs by this strategy achieved 458 mW/m^2 of power density [70]. In addition to pyrolysis, a hydrothermal method is also used to prepare the catalyst. Zhu and his team found an easy way to make carbon nanodots for use in electronics. They used soy milk and heated it at 180 °C for 3 h. During a hydrothermal process, tiny organic molecules are transformed into carbon particles, resulting in the formation of carbon nanoparticles [75]. A distinctive form of carbon material was produced by Zhou and his team through hydrothermal treatment and then followed by pyrolysis [62]. They found that the hydrothermal process triggers the formation of carbon nanodots that accumulate on the porous carbon. This made the surface area higher and made it easier to reach the active sites. The catalysts made had a power density of 703 mW/m^2 in MFCs. Yuan and his team explored the potential of using sewage sludge biochar in an MFC to produce electricity. Yuan and his team explored the potential of using sewage sludge biochar in an MFC to produce electricity. They discovered an abundance of nitrogen and iron on the surface of the cathode electrode when they carbonized biochar at 900° and observed numerous microorganisms. The carbonized cathode at 990° in MFCs yielded an equivalent power output to the Pt cathode, reaching a peak power density of 500. The new cathode stayed strong

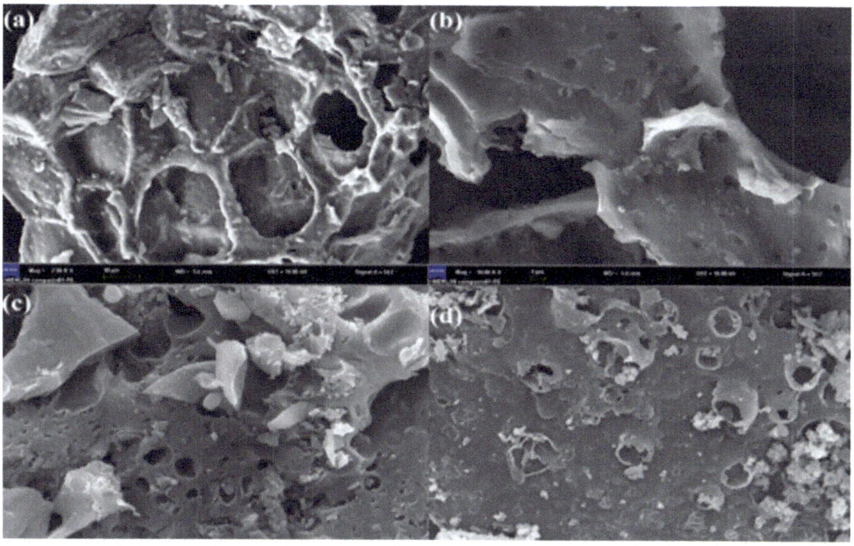

Fig. 14.6 SEM images of carbonized watermelon rind carbon. Reproduced with permission [61]

and could handle methanol well [76]. A cost-effective and sustainable technique for generating nitrogen-doped carbon with numerous pores has been uncovered by Zhong and his team. Watermelon rind is the primary component they rely on due to its high nitrogen content and stability. Watermelon rind biochar serves as an auxiliary material in air-powered microbial fuel cells. The pyrolysis product in this research exhibits a redox peak current density of f 0.19 mA/cm^2, comparable to that of the Pt/C catalyst. The Watermelon rind carbon catalyst features an increased presence of C-N bonds and a greater concentration of pyridinic and graphitic nitrogen within its carbon framework. As a result, it boasts an impressive active surface area of 658 m^2/g, making it highly effective in the oxygen reduction reaction via a four-electron mechanism [61]. As illustrated in Fig. 14.6, the activated carbon samples exhibit a minuscule structure characterized by a smooth surface that is populated with well-spaced large pores. This significantly improves the ion mass transport, electrical conductivity of biochar carbon, and its effectiveness in facilitating chemical reactions.

14.6 Conclusion

Utilizing microbial-powered fuel cells is an effective method of generating energy by converting various types of waste into usable energy. The material utilized in the anode is a crucial factor in MFCs. The performance of the MFC can be influenced by the surface structure, composition, and shape of the surface property. Even though expensive and not very powerful anode materials are usually used in MFCs, their high cost and low power output make them not very practical for use in MFCs. Thus far, scientists have investigated techniques for producing anodes with high surface areas, excellent electrical conductivity, and the capability to function effectively with microorganisms, to improve the power production of MFCs. This indicates that utilizing carbon-based waste materials for the production of carbon anodes could offer a cheaper and more sustainable option for energy production. A better comprehension of making carbon anodes from biomass is necessary to enhance the efficiency of MFCs. Furthermore, the process of producing carbon anodes from biomass materials often involves high temperatures and poses difficulties in ensuring their performance. The high heating costs make it challenging to use carbon anodes made from biomass. In the future, researchers must develop innovative ways to lower the cost of biomass carbon anodes and manage their structure and surface. This is important because heat treatment is a key factor. Nanocomposite carbon anodes typically require binding materials for production, but this can reduce their stability. Future research should focus on enhancing plant-derived materials, improving MFC efficiency, and simplifying the fabrication of anodes for practical use. Furthermore, employing enzymes to activate carbon materials innovatively may be a viable method for creating cost-effective and eco-friendly materials [15].

References

1. Ahanchi M, Jafary T, Yeneneh AM, Rupani PF, Shafizadeh A, Shahbeik H, Pan J, Tabatabaei M, Aghbashlo M (2022) Review on waste biomass valorization and power management systems for microbial fuel cell application. J Cleaner Prod 380:134994. https://doi.org/10.1016/j.jclepro.2022.134994
2. Unuofin JO, Iwarere SA, Daramola MO (2023) Embracing the future of circular bio-enabled economy: unveiling the prospects of microbial fuel cells in achieving true sustainable energy. Environ Sci Pollut Res 30(39):90547–90573. https://doi.org/10.1007/s11356-023-28717-0
3. Abd-Elrahman NK, Al-Harbi N, Basfer NM, Al-Hadeethi Y, Umar A, Akbar S (2022) Applications of nanomaterials in microbial fuel cells: a review. Molecules 27(21). https://doi.org/10.3390/molecules27217483
4. Salar-García M, Ortiz-Martínez V (2019) Advanced research in nanosciences for water technology. Springer
5. Poureshghi F, Calay RK, Das S (2023) Enhanced performance of microbial fuel cells using electrochemically treated carbon felt anode. Results Chem 6:101203. https://doi.org/10.1016/j.rechem.2023.101203
6. Molahalli V, Sharma A, Bijapur K, Soman G, Chattham N, Hegde G (2024) Low-cost bio-waste carbon nanocomposites for sustainable electrochemical devices: a systematic review. Mater Today Commun 38:108034. https://doi.org/10.1016/j.mtcomm.2024.108034
7. Ucar D, Zhang Y, Angelidaki I (2017) An overview of electron acceptors in microbial fuel cells. Front Microbiol 8:643
8. Rhoads A, Beyenal H, Lewandowski Z (2005) Microbial fuel cell using anaerobic respiration as an anodic reaction and biomineralized manganese as a cathodic reactant. Environ Sci Technol 39(12):4666–4671
9. Zhou M, Chi M, Luo J, He H, Jin T (2011) An overview of electrode materials in microbial fuel cells. J Power Sources 196(10):4427–4435

10. Al Moinee A, Sanzida N (2020) Performance analysis of different anode materials of a double chambered microbial fuel cell. Chem Eng Res Bull 26–31
11. Logan BE, Call D, Cheng S, Hamelers HV, Sleutels TH, Jeremiasse AW, Rozendal RA (2008) Microbial electrolysis cells for high yield hydrogen gas production from organic matter. Environ Sci Technol 42(23):8630–8640
12. Roy H, Rahman TU, Tasnim N, Arju J, Rafid MM, Islam MR, Pervez MN, Cai Y, Naddeo V, Islam MS (2023) Microbial fuel cell construction features and application for sustainable wastewater treatment. Membranes (Basel) 13(5). https://doi.org/10.3390/membranes13050490
13. Santoro C, Arbizzani C, Erable B, Ieropoulos I (2017) Microbial fuel cells: from fundamentals to applications. A review. J Power Sources 356:225–244
14. Moradian JM, Fang Z, Yong Y-C (2021) Recent advances on biomass-fueled microbial fuel cell. Bioresources Bioprocess 8(1):14. https://doi.org/10.1186/s40643-021-00365-7
15. Moradian JM, Wang S, Ali A, Liu J, Mi J, Wang H (2022) Biomass-Derived carbon anode for high-performance microbial fuel cells. Catalysts 12(8):894
16. Gao Z, Zhang Y, Song N, Li X (2017) Biomass-derived renewable carbon materials for electrochemical energy storage. Mater Res Lett 5(2):69–88
17. Yang W, Chen S (2020) Biomass-Derived carbon for electrode fabrication in microbial fuel cells: a review. Ind Eng Chem Res 59(14):6391–6404. https://doi.org/10.1021/acs.iecr.0c00041
18. Kim KS, Zhao Y, Jang H, Lee SY, Kim JM, Kim KS, Ahn J-H, Kim P, Choi J-Y, Hong BH (2009) Large-scale pattern growth of graphene films for stretchable transparent electrodes. Nature 457(7230):706–710
19. Saleem J, Shahid UB, Hijab M, Mackey H, McKay G (2019) Production and applications of activated carbons as adsorbents from olive stones. Biomass Convers Biorefinery 9(4):775–802. https://doi.org/10.1007/s13399-019-00473-7
20. Long W, Fang B, Ignaszak A, Wu Z, Wang Y-J, Wilkinson D (2017) Biomass-derived nanostructured carbons and their composites as anode materials for lithium ion batteries. Chem Soc Rev 46(23):7176–7190
21. Wang J, Nie P, Ding B, Dong S, Hao X, Dou H, Zhang X (2017) Biomass derived carbon for energy storage devices. J Mater Chem A 5(6):2411–2428
22. Petrović J, Ercegović M, Simić M, Koprivica M, Dimitrijević J, Jovanović A, Janković Pantić J (2024) Hydrothermal carbonization of waste biomass: a review of hydrochar preparation and environmental application. Processes 12(1):207
23. Arendt R, Rosolowski J, Szymaszek J (1979) Lead zirconate titanate ceramics from molten salt solvent synthesized powders. Mater Res Bull 14(5):703–709
24. Fan X, Zhou Y, Jin X, Song RB, Li Z, Zhang Q (2021) Carbon material-based anodes in the microbial fuel cells. Carbon Energy 3(3):449–472
25. Luo X, Chen S, Hu T, Chen Y, Li F (2021) Renewable biomass-derived carbons for electrochemical capacitor applications. SusMat 1(2):211–240
26. Sato C, Apollon W, Luna-Maldonado AI, Paucar NE, Hibbert M, Dudgeon J (2023) Integrating microbial fuel cell and hydroponic technologies using a ceramic membrane separator to develop an energy-water-food supply system. Membranes (Basel) 13(9). https://doi.org/10.3390/membranes13090803
27. Marassi RJ, Queiroz LG, Silva DCVR, Silva FTD, Silva GC, Paiva TCBD (2020) Performance and toxicity assessment of an up-flow tubular microbial fuel cell during long-term operation with high-strength dairy wastewater. J Cleaner Prod 259:120882. https://doi.org/10.1016/j.jclepro.2020.120882
28. Zhang W, Qiu X, Wang C, Zhong L, Fu F, Zhu J, Zhang Z, Qin Y, Yang D, Xu CC (2022) Lignin derived carbon materials: current status and future trends. Carbon Res 1(1):14. https://doi.org/10.1007/s44246-022-00009-1
29. Chakraborty I, Sathe SM, Dubey BK, Ghangrekar MM (2020) Waste-derived biochar: applications and future perspective in microbial fuel cells. Bioresour Technol 312:123587. https://doi.org/10.1016/j.biortech.2020.123587
30. Feng L, Yan Y, Chen Y, Wang L (2011) Nitrogen-doped carbon nanotubes as efficient and durable metal-free cathodic catalysts for oxygen reduction in microbial fuel cells. Energy Environ Sci 4(5):1892–1899. https://doi.org/10.1039/C1EE01153G

31. Hung Y-H, Liu T-Y, Chen H-Y (2019) Renewable coffee waste-derived porous carbons as anode materials for high-performance sustainable microbial fuel cells. ACS Sustain Chem Eng 7(20):16991–16999. https://doi.org/10.1021/acssuschemeng.9b02405
32. Aleid GM, Alshammari AS, Ahmad ARD, Hussain F, Oh S-E, Ahmad A, Ibrahim MNM, Umar K (2023) Advancement in microbial fuel cells technology by using waste extract as an organic substrate to produce energy with metal removal. Processes 11(8):2434
33. Lin F-Y, Lin Y-Y, Li H-T, Ni C-S, Liu C-I, Guan C-Y, Chang C-C, Yu C-P, Chen W-S, Liu T-Y, Chen H-Y (2022) Trapa natans husk-derived carbon as a sustainable electrode material for plant microbial fuel cells. Appl Energy 325:119807. https://doi.org/10.1016/j.apenergy.2022.119807
34. Lv X, Chen M, Kimura H, Du W, Yang X (2023) Biomass-Derived carbon materials for the electrode of metal-air batteries. Int J Mol Sci 24(4). https://doi.org/10.3390/ijms24043713
35. Li M, Li Y-W, Yu X-L, Guo J-J, Xiang L, Liu B-L, Zhao H-M, Xu M-Y, Feng N-X, Yu P-F, Cai Q-Y, Mo C-H (2020) Improved bio-electricity production in bio-electrochemical reactor for wastewater treatment using biomass carbon derived from sludge supported carbon felt anode. Sci Total Environ 726:138573. https://doi.org/10.1016/j.scitotenv.2020.138573
36. Chen S, He G, Hu X, Xie M, Wang S, Zeng D, Hou H, Schröder U (2012) A three-dimensionally ordered macroporous carbon derived from a natural resource as anode for microbial bioelectrochemical systems. ChemSusChem 5(6):1059–1063. https://doi.org/10.1002/cssc.201100783
37. Chen S, He G, Liu Q, Harnisch F, Zhou Y, Chen Y, Hanif M, Wang S, Peng X, Hou H, Schröder U (2012) Layered corrugated electrode macrostructures boost microbial bioelectrocatalysis. Energy Environ Sci 5(12):9769–9772. https://doi.org/10.1039/C2EE23344D
38. Offei F, Thygesen A, Mensah M, Tabbicca K, Fernando D, Petrushina I, Daniel G (2016) A viable electrode material for use in microbial fuel cells for tropical regions. Energies 9(1):35
39. Zhou Y, Zhou G, Yin L, Guo J, Wan X, Shi H (2017) High-Performance carbon anode derived from sugarcane for packed microbial fuel cells. ChemElectroChem 4(1):168–174. https://doi.org/10.1002/celc.201600510
40. Xu Z-A, Ma X-M, Shi Y-T, Moradian JM, Wang Y-Z, Sun G-F, Du J, Ye X-M, Yong Y-C (2022) Surface activated natural wood biomass electrode for efficient microbial electrocatalysis: Performance and mechanism. Int J Energy Res 46(6):8480–8490. https://doi.org/10.1002/er.7705
41. Jia Y, Feng H, Shen D, Zhou Y, Chen T, Wang M, Chen W, Ge Z, Huang L, Zheng S (2018) High-performance microbial fuel cell anodes obtained from sewage sludge mixed with fly ash. J Hazard Mater 354:27–32. https://doi.org/10.1016/j.jhazmat.2018.04.008
42. Huggins T, Wang H, Kearns J, Jenkins P, Ren ZJ (2014) Biochar as a sustainable electrode material for electricity production in microbial fuel cells. Bioresour Technol 157:114–119. https://doi.org/10.1016/j.biortech.2014.01.058
43. Chen S, Tang J, Jing X, Liu Y, Yuan Y, Zhou S (2016) A hierarchically structured urchin-like anode derived from chestnut shells for microbial energy harvesting. Electrochim Acta 212:883–889
44. Yai C, Chen M, Qian Y, Zhang L, Lu M, Xie X, Huang L, Huang W (2019) Packed anode derived from cocklebur fruit for improving long-term performance of microbial fuel cells. Sci China-Mater 62(5):645–652
45. Zhu K, Xu Y, Yang X, Fu W, Dang W, Yuan J, Wang Z (2022) Sludge derived carbon modified anode in microbial fuel cell for performance improvement and microbial community dynamics. Membranes 12(2):120
46. Yuan Y, Liu T, Fu P, Tang J, Zhou S (2015) Conversion of sewage sludge into high-performance bifunctional electrode materials for microbial energy harvesting. J Mater Chem A 3(16):8475–8482. https://doi.org/10.1039/C5TA00458F
47. Zhu H, Wang H, Li Y, Bao W, Fang Z, Preston C, Vaaland O, Ren Z, Hu L (2014) Lightweight, conductive hollow fibers from nature as sustainable electrode materials for microbial energy harvesting. Nano Energy 10:268–276

48. Li M, Li Y-W, Cai Q-Y, Zhou S-Q, Mo C-H (2020) Spraying carbon powder derived from mango wood biomass as high-performance anode in bio-electrochemical system. Biores Technol 300:122623
49. Khudzari JM, Gariépy Y, Kurian J, Tartakovsky B, Raghavan GSV (2019) Effects of biochar anodes in rice plant microbial fuel cells on the production of bioelectricity, biomass, and methane. Biochem Eng J 141:190–199. https://doi.org/10.1016/j.bej.2018.10.012
50. Zhang S, Bao X, Wan F, Xie Y, Liu Q, Zhou B, Xia S (2018) Application of loofah sponge three-dimensional cage-shape activated carbon fiber with high strength in an air-cathode microbial fuel cell. J Renew Sustain Energy 10(4)
51. Wang H, Davidson M, Zuo Y, Ren Z (2011) Recycled tire crumb rubber anodes for sustainable power production in microbial fuel cells. J Power Sources 196(14):5863–5866
52. Shi Y-T, Yu Y-Y, Xu Z-A, Lian J, Yong Y-C (2019) Superior carbon belts from Spirogyra for efficient extracellular electron transfer and sustainable microbial energy harvesting. J Mater Chem A 7(12):6930–6938. https://doi.org/10.1039/C9TA00332K
53. Zheng J, Cheng C, Zhang J, Wu X (2016) Appropriate mechanical strength of carbon black-decorated loofah sponge as anode material in microbial fuel cells. Int J Hydrogen Energy 41(48):23156–23163. https://doi.org/10.1016/j.ijhydene.2016.11.003
54. Zou L, Qiao Y, Wu Z-Y, Wu X-S, Xie J-L, Yu S-H, Guo J, Li CM (2016) Tailoring unique mesopores of hierarchically porous structures for fast direct electrochemistry in microbial fuel cells. Adv Energy Mater 6(4):1501535. https://doi.org/10.1002/aenm.201501535
55. Li M, Ci S, Ding Y, Wen Z (2019) Almond shell derived porous carbon for a high-performance anode of microbial fuel cells. Sustain Energy Fuels 3(12):3415–3421
56. Li D, Deng L, Yuan H, Dong G, Chen J, Zhang X, Chen Y, Yuan Y (2018) N, P-doped mesoporous carbon from onion as trifunctional metal-free electrode modifier for enhanced power performance and capacitive manner of microbial fuel cells. Electrochimica Acta 262:297–305. https://doi.org/10.1016/j.electacta.2017.12.164
57. Ye W, Tang J, Wang Y, Cai X, Liu H, Lin J, Van der Bruggen B, Zhou S (2019) Hierarchically structured carbon materials derived from lotus leaves as efficient electrocatalyst for microbial energy harvesting. Sci Total Environ 666:865–874. https://doi.org/10.1016/j.scitotenv.2019.02.300
58. Deng L, Yuan Y, Zhang Y, Wang Y, Chen Y, Yuan H, Chen Y (2017) Alfalfa leaf-derived porous heteroatom-doped carbon materials as efficient cathodic catalysts in microbial fuel cells. ACS Sustain Chem Eng 5(11):9766–9773. https://doi.org/10.1021/acssuschemeng.7b01585
59. Fan Z, Li J, Zhou Y, Fu Q, Yang W, Zhu X, Liao Q (2017) A green, cheap, high-performance carbonaceous catalyst derived from Chlorella pyrenoidosa for oxygen reduction reaction in microbial fuel cells. Int J Hydrogen Energy 42(45):27657–27665. https://doi.org/10.1016/j.ijhydene.2017.07.177
60. Yang W, Li J, Fu Q, Zhang L, Zhu X, Liao Q (2017) A simple method for preparing a binder-free paper-based air cathode for microbial fuel cells. Bioresour Technol 241:325–331. https://doi.org/10.1016/j.biortech.2017.05.063
61. Zhong K, Li M, Yang Y, Zhang H, Zhang B, Tang J, Yan J, Su M, Yang Z (2019) Nitrogen-doped biochar derived from watermelon rind as oxygen reduction catalyst in air cathode microbial fuel cells. Appl Energy 242:516–525. https://doi.org/10.1016/j.apenergy.2019.03.050
62. Zhou L, Fu P, Wen D, Yuan Y, Zhou S (2016) Self-constructed carbon nanoparticles-coated porous biocarbon from plant moss as advanced oxygen reduction catalysts. Appl Catal B: Environ 181:635–643. https://doi.org/10.1016/j.apcatb.2015.08.035
63. Yuan H, Deng L, Cai X, Zhou S, Chen Y, Yuan Y (2015) Nitrogen-doped carbon sheets derived from chitin as non-metal bifunctional electrocatalysts for oxygen reduction and evolution. RSC Adv 5(69):56121–56129. https://doi.org/10.1039/C5RA05461C
64. Yang W, Li J, Ye D, Zhu X, Liao Q (2017) Bamboo charcoal as a cost-effective catalyst for an air-cathode of microbial fuel cells. Electrochimica Acta 224:585–592. https://doi.org/10.1016/j.electacta.2016.12.046
65. Yang W, Li J, Zhang L, Zhu X, Liao Q (2017) A monolithic air cathode derived from bamboo for microbial fuel cells. RSC Adv 7(45):28469–28475. https://doi.org/10.1039/C7RA04571A

66. Tang J, Wang Y, Zhao W, Zeng RJ, Liu T, Zhou S (2019) Biomass-derived hierarchical honeycomb-like porous carbon tube catalyst for the metal-free oxygen reduction reaction. J Electroanal Chem 847:113230. https://doi.org/10.1016/j.jelechem.2019.113230
67. Bose D, Sridharan S, Dhawan H, Vijay P, Gopinath M (2019) Biomass derived activated carbon cathode performance for sustainable power generation from microbial fuel cells. Fuel 236:325–337. https://doi.org/10.1016/j.fuel.2018.09.002
68. Sun Y, Duan Y, Hao L, Xing Z, Dai Y, Li R, Zou J (2016) Cornstalk-Derived nitrogen-doped partly graphitized carbon as efficient metal-free catalyst for oxygen reduction reaction in microbial fuel cells. ACS Appl Mater Interfaces 8(39):25923–25932. https://doi.org/10.1021/acsami.6b06895
69. Deng L, Yuan H, Cai X, Ruan Y, Zhou S, Chen Y, Yuan Y (2016) Honeycomb-like hierarchical carbon derived from livestock sewage sludge as oxygen reduction reaction catalysts in microbial fuel cells. Int J Hydrogen Energy 41(47):22328–22336. https://doi.org/10.1016/j.ijhydene.2016.08.132
70. Li M, Zhang H, Xiao T, Wang S, Zhang B, Chen D, Su M, Tang J (2018) Low-cost biochar derived from corncob as oxygen reduction catalyst in air cathode microbial fuel cells. Electrochim Acta 283:780–788. https://doi.org/10.1016/j.electacta.2018.07.010
71. Ma Y, You S, Jing B, Xing Z, Chen H, Dai Y, Zhang C, Ren N, Zou J (2019) Biomass pectin-derived N, S-enriched carbon with hierarchical porous structure as a metal-free catalyst for enhancing bio-electricity generation. Int J Hydrogen Energy 44(31):16624–16638. https://doi.org/10.1016/j.ijhydene.2019.04.158
72. Fu P, Zhou L, Sun L, Huang B, Yuan Y (2017) Nitrogen-doped porous activated carbon derived from cocoon silk as a highly efficient metal-free electrocatalyst for the oxygen reduction reaction. RSC Adv 7(22):13383–13389. https://doi.org/10.1039/C7RA00433H
73. Yuan H, Deng L, Qi Y, Kobayashi N, Tang J (2014) Nonactivated and activated biochar derived from bananas as alternative cathode catalyst in microbial fuel cells. Sci World J 2014:832850. https://doi.org/10.1155/2014/832850
74. Zhou L, Fu P, Cai X, Zhou S, Yuan Y (2016) Naturally derived carbon nanofibers as sustainable electrocatalysts for microbial energy harvesting: a new application of spider silk. Appl Catal B: Environ 188:31–38. https://doi.org/10.1016/j.apcatb.2016.01.063
75. Zhu C, Zhai J, Dong S (2012) Bifunctional fluorescent carbon nanodots: green synthesis via soy milk and application as metal-free electrocatalysts for oxygen reduction. Chem Commun 48(75):9367–9369. https://doi.org/10.1039/C2CC33844K
76. Yuan Y, Yuan T, Wang D, Tang J, Zhou S (2013) Sewage sludge biochar as an efficient catalyst for oxygen reduction reaction in an microbial fuel cell. Bioresour Technol 144:115–120. https://doi.org/10.1016/j.biortech.2013.06.075

Chapter 15
Waste-Derived Graphene: A New Avenue for Supercapacitors

Shailendra Rajput, Sabyasachi Parida, Ankit K. Srivatsva, and Swasti Saxena

Abstract The primary obstacle facing supercapacitor technology at the moment is their failure to demonstrate that they can function on par with batteries while also lowering their cost. Increasing the cell voltage or capacitance will increase the energy density. The capacitance is defined by the contact between the material of the electrode and the electrolyte, whereas the latter is determined by the stability of the electrolyte. Therefore, by carefully selecting an electrolyte that enhances the electrode material and designing the electrode material rationally, it is possible to greatly improve the energy density of the supercapacitor capacitance. This will maximize the adsorption of ions from the electrolyte resulting in an electrode/electrolyte connection with a high-potential window. A novel way to improve the efficiency of storage of energy for supercapacitors and batteries is provided by graphene. Graphene-based materials, particularly supercapacitors (SCs), are being investigated extensively as potential active electrode materials for energy storage and conversion devices. The key features that make them reliable and helpful conduits for charge transport are their large electrochemically active surface area, hierarchical porous structure, superior compressibility, high mechanical stability, and great conductivity. Waste materials can be widely employed as precursors for the synthesis of graphene, including sugarcane bagasse, rice husk, tea waste, coconut shell, peanut shell, lotus stem, and hemp fiber. Numerous techniques have been devised for the synthesis of graphene, with chemical vapor deposition, hydrothermal reaction, chemical activation, and chemical exfoliation being heavily employed in the handling of agricultural waste. This

S. Rajput (✉)
College of Engineering, Xi'an International University, Xi'an 710077, China
e-mail: shailendra@xaiu.edu.cn

S. Parida
Department of Physics, C.V. Raman Global University, Bhubaneswar 752054, India
e-mail: sabyasachi@cvrce.edu.in

A. K. Srivatsva
Department of Physics, Indrashil University, Mehsana 382740, India

S. Saxena
Department of Physics, Pandit Deendayal Energy University, Gandhinagar 382007, India

N. Talreja et al. (eds.), *Waste-Derived Carbon Nanostructures*, Nanostructure Science and Technology, https://doi.org/10.1007/978-3-031-75247-6_15

chapter describes in detail how waste-derived graphene can be used as an electrode material for supercapacitors.

Keywords Graphene · Supercapacitor · Electrode · Energy storage · Waste

15.1 Introduction

Lately, there has been a lot of interest in the use of waste-derived graphene as a viable material for supercapacitors. Because of their high power density, quick charging times, and extended cycle life, supercapacitors—also referred to as electrochemical capacitors—have become an essential energy storage technology. Graphene, a two-dimensional carbon allotrope known for its remarkable electrical conductivity and surface area is one of the materials used in the production of supercapacitors, but its high cost and restricted availability prevent them from being widely used. In order to overcome these obstacles, scientists are focusing on waste-derived graphene as an affordable and environmentally friendly substitute for supercapacitor electrodes. Researchers have concentrated on two primary approaches to overcome this problem: raising the cell voltage and capacitance [1, 2].

1. **Increasing Capacitance**: The capacitance of a supercapacitor is determined by the surface area of the electrode material and the efficiency of the electrolyte–electrode interface in storing charge. Rational design of electrode materials is crucial for maximizing surface area and optimizing the interface properties.

    - **Electrode Material Design**: By engineering nanostructures or introducing defects into the electrode material, researchers can increase its surface area, providing more active sites for charge storage. For example, graphene-based materials with high surface-to-volume ratios have shown promising results in enhancing capacitance.
    - **Electrolyte-Electrode Interface**: The interface between the electrode material and the electrolyte plays a critical role in charge storage. Selecting an electrolyte that complements the electrode material is essential for optimizing ion adsorption and minimizing charge transfer resistance at the interface. Ionic liquids or organic electrolytes with appropriate redox potentials can enhance the capacitance of supercapacitors by widening the electrochemical window and improving charge storage efficiency [3].

2. **Increasing Cell Voltage**: The cell voltage of a supercapacitor is limited by the stability of the electrolyte. Higher cell voltages can significantly enhance the energy density of supercapacitors, enabling them to store more energy per unit volume or mass.

    - **Electrolyte Stability**: Traditional aqueous electrolytes have limited operating voltages due to water electrolysis at high voltages. Non-aqueous electrolytes, such as organic electrolytes or ionic liquids, offer higher stability windows and can enable higher cell voltages. However, careful selection and design of electrolytes are required to ensure compatibility with the electrode materials and to prevent side reactions that could degrade performance or reduce cycle life [4].

By carefully designing electrode materials and selecting appropriate electrolytes, researchers aim to optimize the energy density of supercapacitors while maintaining their advantageous characteristics such as high power density and long cycle life. This approach requires a multidisciplinary effort, combining materials science, electrochemistry, and engineering principles to overcome the current limitations and unlock the full potential of supercapacitor technology for various applications, including energy storage systems, electric vehicles, and portable electronics.

Because of their high power density, quick charge/discharge rates, and extended cycle life, supercapacitors—also referred to as ultracapacitors or electrochemical capacitors—have attracted a lot of interest as energy storage technologies. However, the materials chosen for the electrodes have a big impact on how well supercapacitors work. Activated carbon and other conventional electrode materials have low energy densities and poor economic viability. Waste-derived graphene is a new alternative that uses waste materials to create electrodes based on graphene that have improved characteristics for use in supercapacitors. Indeed, graphene has created new opportunities to improve the efficiency of energy storage devices, such as batteries and supercapacitors. Here's how:

1. **Supercapacitors**: Because of its exceptional mechanical strength, high surface area, and superior electrical conductivity, graphene is a perfect material for supercapacitor electrodes. Researchers can greatly boost the specific capacitance and energy density of supercapacitors by introducing graphene into their designs. Electric cars and grid-level energy storage systems are two examples of applications that benefit from graphene-based supercapacitors' quick charging and discharging rates, extended cycle life, and increased power density.
2. **Batteries**: By overcoming significant drawbacks like poor mechanical degradation, restricted ion transport, and low electrical conductivity, graphene can also enhance battery performance. Materials based on graphene can be used in battery designs as freestanding electrodes, coatings, or electrode additions. Battery manufacturers can produce longer-lasting and faster-charging energy storage solutions for a variety of applications, such as consumer electronics, electric vehicles, and renewable energy storage systems, by incorporating graphene into their batteries. This will improve the charge/discharge rates, energy density, and cycle life of batteries. Figure 15.1 shows the schematic of a typical symmetry paper-based supercapacitor.

In summary, graphene presents a new opportunity to revolutionize energy storage technology by improving the performance and efficiency of both supercapacitors and batteries. As research and development efforts continue to advance, graphene-based energy storage devices are expected to play a significant role in the transition toward a cleaner and more sustainable energy future.

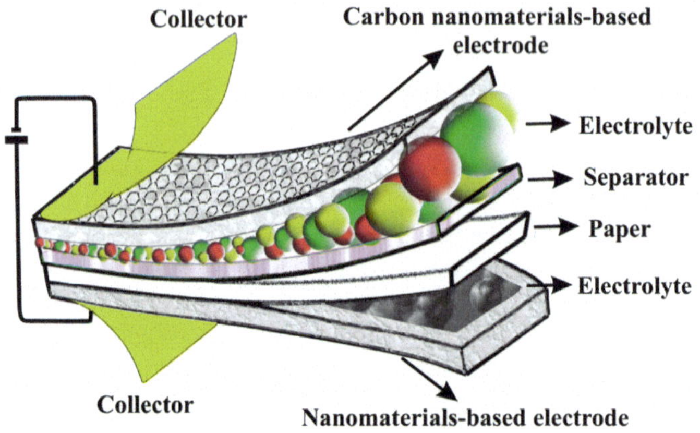

Fig. 15.1 Schematic of typical symmetry paper-based supercapacitor. The image was reproduced under a common creative license [30]

Graphene offers a new opportunity to boost the performance of energy storage for supercapacitors and batteries. The single layer of atoms of carbon organized in a two-dimensional honeycomb lattice is called graphene, and it has amazing features that make it a very desirable material for improving the efficiency of batteries and supercapacitors, among other energy storage devices.

1. **High Surface Area**: Graphene has an exceptionally high surface area-to-volume ratio due to its two-dimensional structure. This large surface area provides ample active sites for ions to adsorb onto the electrode surface, leading to increased energy storage capacity.
2. **Excellent Electrical Conductivity**: Graphene exhibits extraordinary electrical conductivity, allowing for efficient charge and discharge processes within energy storage devices. This property enables rapid electron transport across the electrode–electrolyte interface, resulting in fast charging and discharging rates.
3. **Mechanical Strength**: Graphene is one of the strongest materials known, with exceptional mechanical properties. Its high mechanical strength ensures the structural integrity and durability of electrodes in energy storage devices, even under harsh operating conditions.
4. **Flexibility and Tunability**: Graphene's properties can be tailored and optimized through various fabrication techniques, including doping, functionalization, and nanostructuring. This flexibility allows researchers to design graphene-based materials with specific characteristics optimized for energy storage applications.

In supercapacitors, graphene-based electrodes offer higher specific capacitance, longer cycle life, and lower equivalent series resistance compared to traditional electrode materials. These attributes result in supercapacitors with improved energy density, power density, and overall performance.

Similarly, in batteries, incorporating graphene into electrode designs can enhance ion transport kinetics, increase the reversible capacity, and improve the cycling stability of batteries. Graphene-based electrodes enable faster charging and discharging rates, higher energy density, and longer cycle life, addressing key limitations of conventional battery technologies.

15.2 Synthesis Methods of Waste-Derived Graphene

Waste-derived graphene can be synthesized through various methods, including chemical exfoliation, thermal reduction, and hydrothermal processes. Common sources of waste materials for graphene production include biomass, industrial by-products, and waste plastics. These methods offer a sustainable approach to graphene synthesis while mitigating environmental pollution and reducing production costs.

- **Chemical Exfoliation**: In this method, graphene is produced by exfoliating graphite or graphite oxide using chemical agents [5].
- **Thermal Reduction**: Thermal reduction involves heating carbon-containing precursors, such as biomass or waste plastics, at high temperatures in the absence of oxygen [6].
- **Hydrothermal Processes**: Hydrothermal methods involve the synthesis of graphene under high-pressure, high-temperature conditions in an aqueous solution containing carbon sources [7].

15.2.1 Common Sources of Waste Materials

- **Biomass**: Biomass refers to organic materials derived from plants, agricultural residues, or forestry waste [8].
- **Industrial By-products**: Industrial processes generate various by-products that often contain carbon-rich compounds suitable for graphene production [9].
- **Waste Plastics**: Plastics are widely used in various industries and contribute significantly to environmental pollution [10].

15.2.2 Sustainable Approach to Graphene Synthesis

- Utilizing waste materials for graphene synthesis offers a sustainable alternative to conventional methods that rely on pristine graphite or carbon sources [11].
- By repurposing waste materials, this approach reduces the environmental impact associated with waste disposal and minimizes the extraction of raw materials from natural resources.
- Additionally, waste-derived graphene production can contribute to circular economy principles by converting waste into value-added products, thereby

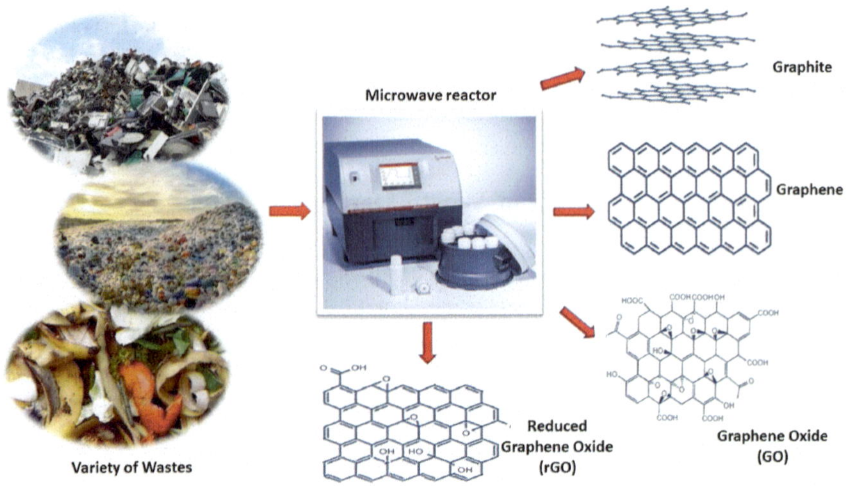

Fig. 15.2 Synthesis methods of waste-derived graphene

reducing production costs and promoting resource efficiency. Figure 15.2 shows the synthesis methods of waste-derived graphene.

15.3 Environmental Pollution Mitigation and Cost Reduction

- Waste-derived graphene synthesis mitigates environmental pollution by repurposing waste materials that would otherwise end up in landfills or incinerators.
- By utilizing waste streams as feedstocks, this approach reduces the need for energy-intensive processes and minimizes greenhouse gas emissions associated with traditional graphene production methods.
- Moreover, waste-derived graphene synthesis offers cost advantages by utilizing low-cost or free raw materials, thereby reducing production expenses and making graphene-based materials more economically viable for various applications [12].

In fact, there are many of sustainable resources available for the synthesis of graphene, including waste products like sugarcane bagasse, husk of rice, tea waste, coconut shell, peanut shell, lotus stem, and hemp fiber. Graphene has been produced from these agricultural wastes using a variety of procedures, the most well-known ones being chemical vapor deposition (CVD), hydrothermal process reaction, chemical activation, and chemical exfoliation.

1. **Hydrothermal Reaction**: In this method, the waste biomass is subjected to hydrothermal treatment under high temperature and pressure conditions. This

process promotes the breakdown of organic matter into carbon-rich precursors, which can then be converted into graphene through subsequent thermal or chemical processes [13].

2. **Chemical Activation**: Chemical activation involves the treatment of biomass with activating agents, such as potassium hydroxide (KOH) or phosphoric acid (H3PO4), followed by carbonization at elevated temperatures. This process creates porous carbon structures with high surface areas, which can be further processed to obtain graphene materials [14].
3. **Chemical Exfoliation**: Chemical exfoliation techniques involve the intercalation of chemical agents, such as acids or oxidizing agents, between the layers of graphite in the biomass precursor. Subsequent mechanical or thermal treatment then leads to the exfoliation of graphene sheets from the precursor material.
4. **Chemical Vapor Deposition (CVD)**: CVD is a widely used technique for synthesizing high-quality graphene films on substrates. In this method, hydrocarbons or other carbon-containing gases are decomposed on a catalytic surface at high temperatures, resulting in the deposition of graphene layers. Waste biomass-derived precursors can be used as carbon sources in CVD processes to produce graphene films [15].

These methods offer sustainable and cost-effective approaches to graphene synthesis, utilizing waste biomass as feedstock while mitigating environmental pollution. By repurposing agricultural waste materials for graphene production, researchers can contribute to the development of green and environmentally friendly processes for advanced materials synthesis. Additionally, graphene derived from agricultural waste holds promise for various applications, including energy storage, electronics, composites, and environmental remediation. As research in this field continues to progress, further optimization of synthesis methods and exploration of novel precursor materials are expected to enhance the efficiency and scalability of waste-derived graphene production [16, 17]. Figure 15.3 shows a graphical illustration of the waste-driven nanocomposites.

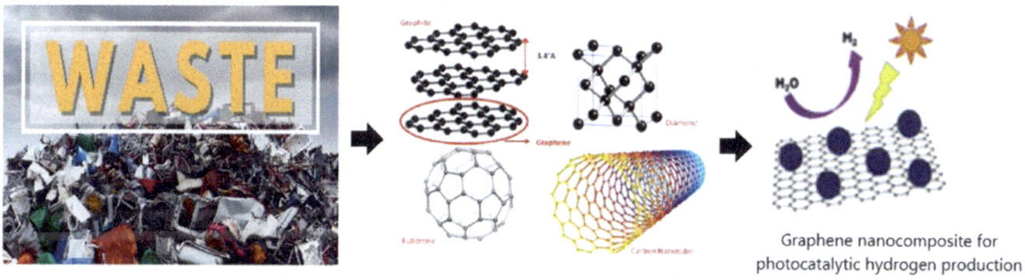

Fig. 15.3 Waste-driven nanocomposites

15.4 Properties of Waste-Derived Graphene

Waste-derived graphene exhibits properties comparable to pristine graphene, including high electrical conductivity, large surface area, and excellent mechanical strength. Moreover, the unique chemical composition of waste-derived graphene can introduce additional functionalities, such as heteroatom doping and surface functionalization, further enhancing its suitability for supercapacitor applications [3].

15.4.1 High Electrical Conductivity

Graphene, whether derived from waste or produced through conventional methods, is renowned for its exceptional electrical conductivity. Waste-derived graphene inherits this property, allowing it to efficiently transport electrons within a supercapacitor electrode.

The high electrical conductivity of graphene facilitates rapid charge/discharge cycles, making supercapacitors based on waste-derived graphene suitable for high-power applications [3].

15.4.2 Large Surface Area

Waste-derived graphene typically possesses a large surface area due to its two-dimensional structure composed of interconnected carbon atoms.

This large surface area provides ample active sites for ion adsorption from the electrolyte, thereby increasing the capacitance of the supercapacitor.

The enhanced surface area of waste-derived graphene contributes to higher energy storage capacity and improved performance of supercapacitors compared to conventional electrode materials.

15.4.3 Excellent Mechanical Strength

Graphene is one of the strongest materials known, with exceptional mechanical properties such as high tensile strength and flexibility. Waste-derived graphene inherits these mechanical characteristics, which are crucial for maintaining structural integrity and durability in supercapacitor electrodes. The excellent mechanical strength of waste-derived graphene ensures the long-term stability and reliability of supercapacitors, even under harsh operating conditions [3].

15.4.4 Unique Chemical Composition

Waste-derived graphene can possess a unique chemical composition compared to pristine graphene due to the nature of the waste materials used as precursors. This unique chemical composition may include heteroatom doping, where foreign atoms (e.g., nitrogen, oxygen) are incorporated into the graphene lattice, altering its electronic properties. Surface functionalization is another aspect where chemical groups are attached to the graphene surface, imparting specific functionalities such as improved wettability or enhanced electrochemical performance.

These additional functionalities introduced by the unique chemical composition of waste-derived graphene further enhance its suitability for supercapacitor applications by optimizing charge storage and transport processes [3]. Figure 15.4 shows the graphical representation of the properties of graphene.

15.5 Application in Supercapacitors

The integration of waste-derived graphene into supercapacitor electrodes has demonstrated significant improvements in electrochemical performance, including enhanced specific capacitance, improved rate capability, and prolonged cycling stability. These advancements are attributed to the interconnected network of graphene sheets, which facilitate rapid ion diffusion and electron transport within the electrode structure.

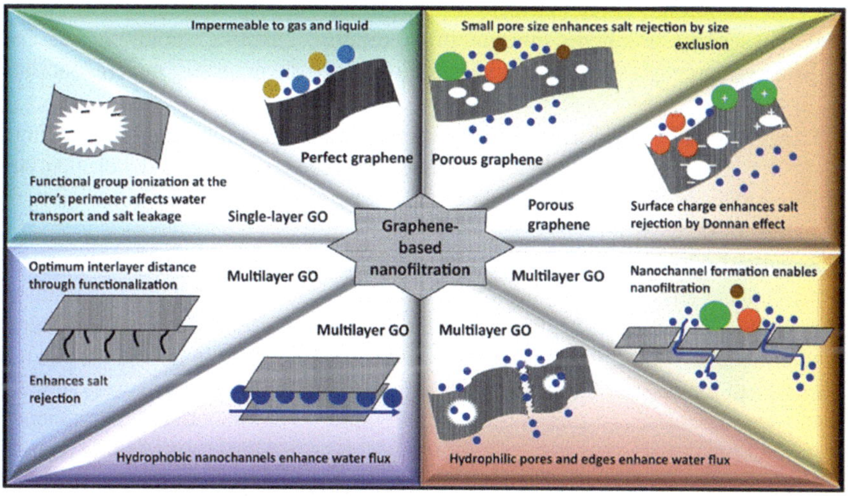

Fig. 15.4 Properties of graphene. The figure was reproduced with the permission [31]

15.5.1 Applicability of Waste-Derived Graphene as Supercapacitor Electrode Material

The growing demand for energy storage solutions has fueled extensive research into advanced electrode materials for supercapacitors. Waste-derived graphene, synthesized from various waste materials such as biomass, industrial by-products, and waste plastics, has emerged as a promising candidate due to its unique properties and sustainable production route. This chapter explores the applicability of waste-derived graphene as a supercapacitor electrode material, highlighting its advantages, challenges, and potential applications.

15.5.2 Advantages of Waste-Derived Graphene

1. **Sustainability**: Waste-derived graphene offers a sustainable alternative to conventional graphene production methods, as it repurposes waste materials that would otherwise contribute to environmental pollution. By utilizing waste streams as feedstocks, waste-derived graphene helps mitigate the environmental footprint associated with energy storage device manufacturing [9, 10].
2. **Cost-Effectiveness**: Waste materials are often available at low or no cost, making waste-derived graphene a cost-effective option for supercapacitor electrodes. This affordability enhances the economic viability of energy storage technologies, particularly in applications where cost is a significant factor [8, 11].
3. **High Electrical Conductivity**: Graphene exhibits exceptional electrical conductivity, facilitating efficient electron transport within supercapacitor electrodes. Waste-derived graphene inherits this property, enabling rapid charge and discharge processes and enhancing the overall performance of supercapacitors [3, 18].
4. **Large Surface Area**: The two-dimensional structure of graphene provides a large surface area, offering ample active sites for ion adsorption from the electrolyte. This enhances the specific capacitance of supercapacitors and improves their energy storage capacity [19, 20].
5. **Excellent Mechanical Strength**: Waste-derived graphene possesses excellent mechanical strength, ensuring the structural integrity and durability of supercapacitor electrodes. This mechanical robustness contributes to the long-term stability and reliability of energy storage devices [19, 20]. Figure 15.5 shows the graphical representation of the applications of supercapacitors.

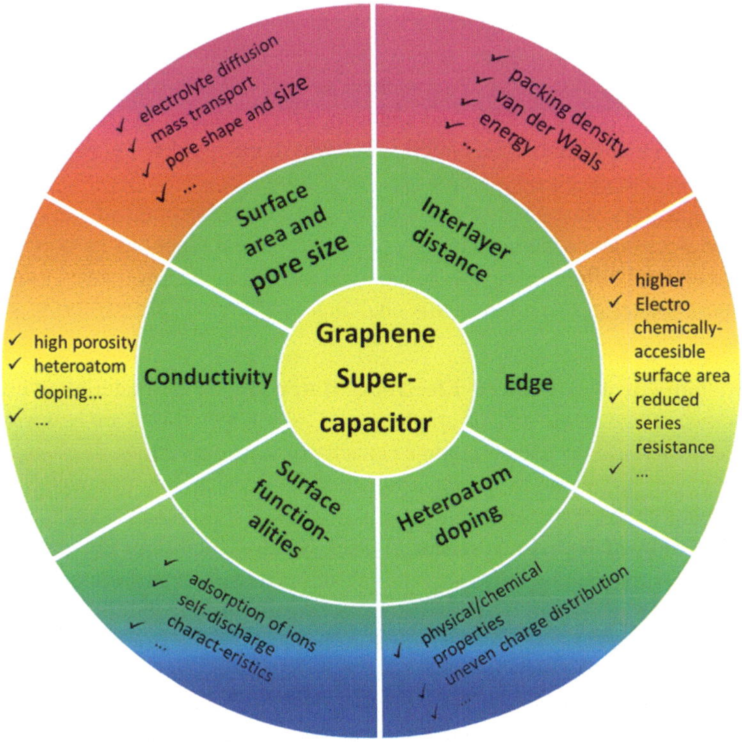

Fig. 15.5 Applications of supercapacitors. The image was reproduced with the permission [32]

15.6 Challenges and Considerations

1. **Synthesis Optimization**: Optimizing synthesis processes is crucial to ensure consistent quality, scalability, and cost-effectiveness of waste-derived graphene production. Fine-tuning reaction parameters, exploring novel precursors, and improving purification techniques are essential for achieving reproducible and high-quality graphene materials [1, 2].
2. **Structural Characterization**: Advanced characterization techniques are needed to understand the structural, morphological, and electrochemical properties of waste-derived graphene-based materials. Spectroscopy, microscopy, and electrochemical analysis provide insights into the structure–property relationships, enabling the tailoring of graphene properties for specific supercapacitor applications [3, 21].
3. **Long-Term Stability**: Enhancing the long-term stability of waste-derived graphene-based electrodes is critical for practical supercapacitor applications. Surface modification, heteroatom doping, and hybridization with other materials can mitigate degradation mechanisms and improve the durability of energy storage devices [20, 22].

The main challenge of supercapacitor technology lies in achieving a balance between energy density, power density, and cost-effectiveness to compete with conventional energy storage systems such as batteries. While supercapacitors excel in delivering high power quickly and enduring a large number of charge–discharge cycles, they typically lag behind batteries in terms of energy density, which refers to the amount of energy stored per unit mass or volume.

Supercapacitors store energy through the physical adsorption of ions onto the surface of electrodes, whereas batteries rely on chemical reactions to store energy. This fundamental difference results in supercapacitors having lower energy densities compared to batteries. To address this challenge, researchers are exploring various avenues, including:

1. **Development of Advanced Electrode Materials**: Novel materials with high specific surface areas, such as graphene and carbon nanotubes, are being investigated to increase the energy storage capacity of supercapacitors. These materials offer larger surface areas for ion adsorption, thus enhancing the overall energy density of supercapacitors [23].
2. **Exploration of Hybrid Systems**: Hybrid systems combining the strengths of supercapacitors and batteries are being developed to overcome the limitations of each technology individually. For example, hybrid supercapacitors, also known as asymmetric supercapacitors, integrate a high-energy–density battery-like electrode with a high power density capacitor-like electrode to achieve a balance between energy and power density [24].
3. **Advancements in Electrolytes**: Research is focused on developing electrolytes with higher conductivity and wider potential windows to improve the energy storage capacity and operating voltage of supercapacitors. Ionic liquids and pseudocapacitive electrolytes are being explored as alternatives to conventional aqueous and organic electrolytes [25].
4. **Innovations in Device Design**: Optimizing the design and configuration of supercapacitor devices can improve their energy storage performance. For instance, the use of three-dimensional electrode architectures and nanoscale engineering techniques can enhance ion transport kinetics and increase the effective surface area of electrodes [26].
5. **Cost Reduction Strategies**: Cost-effective manufacturing processes and scalable production techniques are essential for the widespread adoption of supercapacitors. Research efforts are focused on developing low-cost electrode materials, improving synthesis methods, and optimizing device fabrication processes to reduce manufacturing costs without compromising performance [27].

Addressing these challenges requires interdisciplinary collaboration and continued innovation across materials science, electrochemistry, engineering, and manufacturing. By overcoming the energy density limitation while maintaining the inherent advantages of supercapacitors, such as rapid charging/discharging rates and long cycle life, supercapacitor technology can realize its full potential and find broader applications in areas such as transportation, renewable energy storage, and electronics.

15.7 Applications and Future Outlook

1. **High-Performance Supercapacitors**: Waste-derived graphene-based electrodes offer high specific capacitance, rapid charge/discharge rates, and long cycle life, making them ideal for high-performance supercapacitors. These energy storage devices find applications in portable electronics, electric vehicles, renewable energy systems, and grid-level energy storage [22, 28].
2. **Emerging Technologies**: The integration of waste-derived graphene into supercapacitors paves the way for the development of next-generation energy storage technologies. Ongoing research aims to further enhance the electrochemical performance, scalability, and sustainability of waste-derived graphene-based electrodes, driving innovations in the field of energy storage [9, 18].

Waste-derived graphene holds immense potential as a sustainable, cost-effective, and high-performance material for supercapacitor electrodes. By leveraging waste materials as valuable resources, researchers can address environmental challenges, reduce production costs, and accelerate the transition toward a more sustainable energy future. Continued efforts in synthesis optimization, structural characterization, and long-term stability enhancement will further enhance the applicability of waste-derived graphene in energy storage devices, driving advancements in energy storage technology.

Application in Supercapacitors:

1. **Enhanced Specific Capacitance**:
    - Waste-derived graphene, with its large surface area and excellent electrical conductivity, significantly enhances the specific capacitance of supercapacitors.
    - The increased surface area provides more active sites for ion adsorption, leading to higher charge storage capacity per unit mass or volume.
    - Several studies have demonstrated the superior electrochemical performance of supercapacitors utilizing waste-derived graphene electrodes [22, 28].
2. **Improved Rate Capability**:
    - Supercapacitors based on waste-derived graphene exhibit improved rate capability, allowing for rapid charge and discharge processes.
    - The high electrical conductivity of waste-derived graphene enables efficient electron transport, reducing internal resistance and enabling fast charge/discharge kinetics.
    - This attribute is crucial for applications requiring high power output or regenerative braking systems in electric vehicles [18, 20].
3. **Prolonged Cycling Stability**:

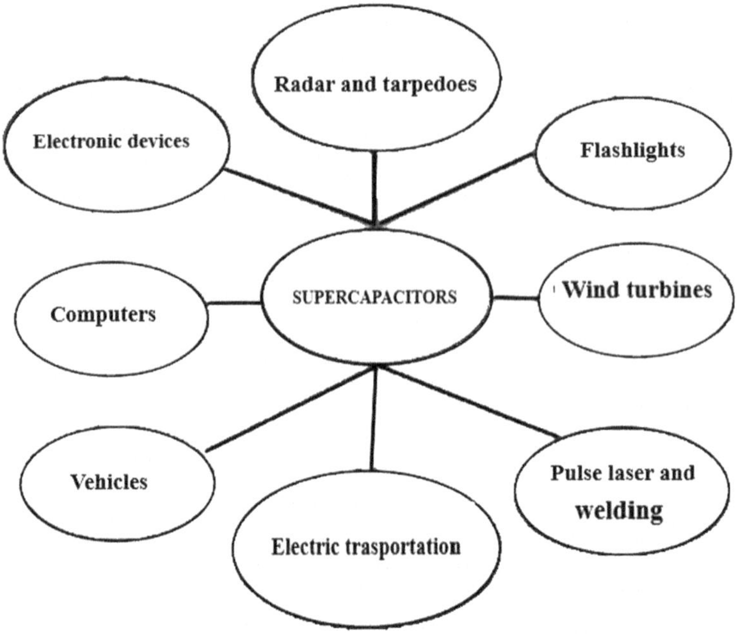

Fig. 15.6 Various applications of supercapacitors

- Waste-derived graphene-based supercapacitors demonstrate excellent cycling stability, maintaining high capacitance over thousands of charge/discharge cycles.
- The mechanical robustness and chemical stability of waste-derived graphene contribute to the long-term performance and durability of supercapacitors.
- Studies have reported negligible capacitance decay and excellent retention of electrochemical properties over extended cycling tests [21, 29] (Fig. 15.6).

15.7.1 Environmental and Economic Benefits

Utilizing waste materials for graphene production offers several environmental and economic benefits. By repurposing waste streams, this approach reduces reliance on finite natural resources and minimizes the environmental footprint associated with traditional graphene synthesis methods. Furthermore, the cost-effectiveness of waste-derived graphene makes supercapacitors more affordable and accessible, particularly in emerging markets and sustainable energy applications.

1. **Reduction of Environmental Footprint**:

- Utilizing waste materials for graphene production reduces the environmental impact associated with waste disposal and traditional graphene synthesis methods.
- Repurposing waste streams minimizes the need for landfill space and reduces pollution from incineration or improper disposal practices.
- By mitigating the environmental footprint, waste-derived graphene contributes to sustainable resource management and promotes a circular economy [10, 12].

2. **Conservation of Natural Resources**:
   - Waste-derived graphene reduces reliance on finite natural resources, such as graphite, which is commonly used as a precursor for graphene production.
   - By repurposing waste materials as feedstocks, this approach conserves natural resources and helps alleviate pressure on ecosystems and mineral reserves [8, 11].
3. **Cost-effectiveness and Accessibility**:
   - Waste-derived graphene offers a cost-effective alternative to conventional graphene production methods, as waste materials are often available at low or no cost.
   - The affordability of waste-derived graphene makes supercapacitors more economically viable, particularly in emerging markets and sustainable energy applications where cost is a significant factor [9, 19].

Challenges and Future Outlook:

Despite the promising prospects of waste-derived graphene in supercapacitor technology, several challenges remain to be addressed. These include optimizing synthesis processes to ensure consistent quality and scalability, improving the understanding of structure–property relationships, and enhancing the long-term stability of waste-derived graphene-based electrodes. Additionally, the development of standardized testing protocols and regulatory frameworks is essential to facilitate the commercialization and widespread adoption of waste-derived graphene-based supercapacitors.

1. **Optimization of Synthesis Processes**:
   - One of the primary challenges is to optimize synthesis processes for waste-derived graphene to ensure consistent quality, scalability, and cost-effectiveness.
   - Strategies such as fine-tuning reaction parameters, exploring novel precursors, and improving purification techniques are essential to achieve reproducible and high-quality graphene materials [1, 2].
2. **Understanding Structure–Property Relationships**:
   - Further research is needed to deepen our understanding of the complex structure–property relationships in waste-derived graphene-based materials.

- Investigating the influence of synthesis parameters, precursor characteristics, and post-treatment methods on the electrochemical performance of supercapacitor electrodes is crucial for optimizing material design and enhancing device performance [3, 21].

3. **Enhancement of Long-Term Stability**:

   - Improving the long-term stability of waste-derived graphene-based electrodes is critical for their practical application in supercapacitors.
   - Strategies such as surface modification, heteroatom doping, and hybridization with other materials can mitigate degradation mechanisms and enhance the durability of supercapacitor devices [20, 22].

4. **Development of Standardized Testing Protocols and Regulatory Frameworks**:

   - Establishing standardized testing protocols and regulatory frameworks is essential to facilitate the commercialization and widespread adoption of waste-derived graphene-based supercapacitors.
   - Consistent evaluation methods and quality control measures are necessary to ensure the reliability, safety, and performance consistency of supercapacitor devices for various applications [4, 18].

15.8 Conclusion

Waste-derived graphene holds immense potential as a sustainable and cost-effective material for supercapacitor electrodes. Through innovative synthesis methods and advanced characterization techniques, researchers can unlock the full capabilities of waste-derived graphene and accelerate its integration into next-generation energy storage devices. By leveraging waste materials as a valuable resource, we can mitigate environmental pollution, reduce production costs, and pave the way for a more sustainable energy future.

1. **Waste-derived graphene as a sustainable and cost-effective material for supercapacitor electrodes**:

   - Waste-derived graphene refers to graphene produced from various waste materials, such as biomass, industrial by-products, or waste plastics. This approach offers a sustainable alternative to conventional graphene production methods, as it repurposes waste streams that would otherwise contribute to environmental pollution.
   - By utilizing waste materials as precursors for graphene synthesis, researchers can reduce the reliance on finite natural resources and minimize the environmental footprint associated with traditional manufacturing processes.

- Moreover, waste-derived graphene offers cost advantages over pristine graphene, as waste materials are often available at low or no cost. This makes supercapacitor electrodes based on waste-derived graphene more economically viable, particularly in applications where cost is a significant factor [9, 10].

2. **Unlocking the full capabilities of waste-derived graphene through innovative synthesis methods and advanced characterization techniques**:

   - To fully exploit the potential of waste-derived graphene, researchers are developing innovative synthesis methods and employing advanced characterization techniques.
   - Synthesis methods such as chemical exfoliation, thermal reduction, and hydrothermal processes are being optimized to ensure consistent quality, scalability, and cost-effectiveness of waste-derived graphene production.
   - Advanced characterization techniques, including spectroscopy, microscopy, and electrochemical analysis, are employed to understand the structural, morphological, and electrochemical properties of waste-derived graphene-based materials. This enables researchers to tailor the properties of waste-derived graphene for specific supercapacitor applications [8, 11].

3. **Accelerating the integration of waste-derived graphene into next-generation energy storage devices**:

   - Waste-derived graphene holds significant promise for advancing energy storage technologies, particularly in the development of next-generation supercapacitors.
   - By harnessing the high electrical conductivity, large surface area, and excellent mechanical strength of waste-derived graphene, researchers can enhance the performance metrics of supercapacitor electrodes, including specific capacitance, rate capability, and cycling stability.
   - The integration of waste-derived graphene into supercapacitors can enable the development of high-performance energy storage devices with fast charging/discharging rates, long cycle life, and improved energy density [3, 18].

4. **Leveraging waste materials as a valuable resource to mitigate environmental pollution, reduce production costs, and pave the way for a more sustainable energy future**:

   - By repurposing waste materials for graphene production, researchers can mitigate environmental pollution by diverting waste streams from landfills or incinerators.
   - This approach contributes to resource conservation and promotes a circular economy by converting waste into value-added products.
   - Additionally, the cost-effectiveness of waste-derived graphene production reduces production costs and makes energy storage technologies more affordable and accessible, thereby accelerating the transition toward a more sustainable energy future [19, 20].

In summary, waste-derived graphene offers a sustainable and cost-effective solution for supercapacitor electrodes, with the potential to drive advancements in energy storage technology and contribute to a more sustainable energy landscape. Through innovative synthesis methods, advanced characterization techniques, and the utilization of waste materials as valuable resources, researchers can unlock the full capabilities of waste-derived graphene and pave the way for a greener and more sustainable future.

References

1. Zhang L, Zhao X (2016) Carbon-based materials as supercapacitor electrodes. Chem Soc Rev 45(15):4181–4196
2. Niu Z et al (2016) Review of carbon and graphene quantum dots for sensing and energy applications. Nano Res 9(5):1236–1268
3. Wang Y et al (2017) Supercapacitor devices based on graphene materials. J Mater Chem A 5(19):9199–9216
4. Zhu Y et al (2010) Graphene and graphene oxide: synthesis, properties, and applications. Adv Mater 22(35):3906–3924
5. Novoselov KS et al (2004) Electric field effect in atomically thin carbon films. Science 306(5696):666–669
6. Stankovich S et al (2007) Synthesis of graphene-based nanosheets via chemical reduction of exfoliated graphite oxide. Carbon 45(7):1558–1565
7. Zhang Y et al (2010) Graphene film–reduced graphene oxide composites as ultrahigh-performance supercapacitor electrodes. Nano Lett 10(9):3249–3255
8. Chen W et al (2014) Sustainable nitrogen-doped carbonaceous materials from biomass derivatives. Carbon 72:114–123
9. Wang J et al (2015) Synthesis of graphene sheets from biomass pyrolysis with high yield and their application in oil adsorption. Carbon 82:417–428
10. Lu P et al (2018) Sustainable preparation of graphene from waste biomass and its application in energy storage. ACS Sustain Chem Eng 6(4):5159–5166
11. Zhuo S et al (2018) Sustainable preparation of graphene from solid carbon sources. Nat Commun 9(1):1–9
12. Yu D et al (2014) Scalable synthesis of hierarchically structured carbon nanotube–graphene fibres for capacitive energy storage. Nat Nanotechnol 9(7):555–562
13. Ghosh K, Jayaraman V (2020) Graphene synthesis from agricultural waste biomass: a review. In: Green synthesis, characterization, and applications of nanomaterials. Elsevier, pp. 1–22. ISBN: 978-0-12-817891-7
14. Li X, Zhu J, Wei B, Guo Y (2019) Biomass-Derived carbon materials for high-performance supercapacitors. ACS Appl Energy Mater 2(8):5693–5714. https://doi.org/10.1021/acsaem.9b00584
15. Yang T, Lai C, Teng C, Jeng RJ (2020) Biomass-Derived carbon materials: preparation and applications in supercapacitors. Catalysts 10(6):616. https://doi.org/10.3390/catal10060616
16. Liu R, Wu D, Feng P, Kou T, Wang G (2018) Recent advances in the synthesis and application of biomass-derived carbons for supercapacitors. J Mater Chem A 6(36):17333–17354. https://doi.org/10.1039/C8TA05988B
17. Jia Y, Chen L, Zhou X (2019) Biomass-derived carbon materials: recent advances in synthesis and application in supercapacitors. Front Chem 7:280. https://doi.org/10.3389/fchem.2019.00280
18. Zhang Y et al (2021) Waste-derived graphene for supercapacitors: a review. J Mater Sci Technol 77:86–103

19. Zhang Y et al (2020) Waste-to-energy conversion with graphene-based materials: a review. Chem Eng J 388:124209
20. Chen L et al (2020) Nitrogen-doped hierarchical porous carbon derived from waste tea residue for high-performance supercapacitors. J Alloy Compd 832:154922
21. Liu Q et al (2018) Three-dimensional hierarchical porous carbon derived from waste pomelo peel for high-performance supercapacitors. J Power Sources 402:23–32
22. Wang H et al (2019) Sustainable synthesis of nitrogen-doped hierarchical porous carbon from waste pomelo peel for high-performance supercapacitors. J Mater Sci 54(18):11738–11751
23. Simon P, Gogotsi Y (2008) Materials for electrochemical capacitors. Nat Mater 7(11):845–854
24. Zhang LL, Zhao XS (2009) Carbon-based materials as supercapacitor electrodes. Chem Soc Rev 38(9):2520–2531
25. Winter M, Brodd RJ (2004) What are batteries, fuel cells, and supercapacitors? Chem Rev 104(10):4245–4270
26. Burke A (2000) Ultracapacitors: why, how, and where is the technology. J Power Sources 91(1):37–50
27. Miller JR, Simon P (2008) Electrochemical capacitors for energy management. Science 321(5889):651–652
28. Gao Y et al (2018) Waste-derived hierarchical porous carbon for high-performance supercapacitors. ACS Appl Mater Interfaces 10(12):10216–10224
29. Li W et al (2017) Highly flexible and sustainable supercapacitors based on nitrogen-doped waste biomass-derived activated carbon. Carbon 117:427–436
30. Pour GB, Ashourifar H, Aval LF, Solaymani S (2023) CNTs-supercapacitors: a review of electrode nanocomposites based on CNTs, graphene, metals, and polymers. Symmetry 15:1179. https://doi.org/10.3390/sym15061179
31. Anand A, Unnikrishnan B, Mao J-Y, Lin H-J, Huang C-C (2018) Graphene-based nanofiltration membranes for improving salt rejection, water flux and antifouling—a review. Desalination 429:119–133
32. Wen Y, Mei N, Xin R, Yafen T, Ning L, Yuefeng S, Xiaoling Z (2015) Graphene in supercapacitor applications. Curr Opin Colloid Interface Sci 20(5–6):416–428. https://doi.org/10.1016/j.cocis.2015.10.009

Index

N. Talreja et al. (eds.), *Waste-Derived Carbon Nanostructures*, Nanostructure Science and Technology, https://doi.org/10.1007/978-3-031-75247-6

The manufacturer's authorised representative in the EU is Springer Nature Customer Service Centre GmbH, Europaplatz 3, 69115 Heidelberg, Germany. If you have any concerns regarding our products, please contact ProductSafety@springernature.com

Printed and bound by CPI Group (UK) Ltd, Croydon, CR0 4YY
01/10/2026
02225958-0010